Computer–Mediated Relationships and Trust:
Managerial and Organizational Effects

Linda L. Brennan
Mercer University, USA

Victoria E. Johnson
Georgia Gwinnet College, USA

INFORMATION SCIENCE REFERENCE

Hershey · New York

Acquisitions Editor:	Kristin Klinger
Development Editor:	Kristin Roth
Senior Managing Editor:	Jennifer Neidig
Managing Editor:	Sara Reed
Copy Editor:	Jeannie Porter and April Schmidt
Typesetter:	Cindy Consonery
Cover Design:	Lisa Tosheff
Printed at:	Yurchak Printing Inc.

Published in the United States of America by
Information Science Reference (an imprint of IGI Global)
701 E. Chocolate Avenue, Suite 200
Hershey PA 17033
Tel: 717-533-8845
Fax: 717-533-8661
E-mail: cust@igi-pub.com
Web site: http://www.igi-global.com/reference

and in the United Kingdom by
Information Science Reference (an imprint of IGI Global)
3 Henrietta Street
Covent Garden
London WC2E 8LU
Tel: 44 20 7240 0856
Fax: 44 20 7379 0609
Web site: http://www.eurospanonline.com

Library of Congress Cataloging-in-Publication Data

Computer-mediated relationships and trust : managerial and organizational effects / Linda L. Brennan and Victoria E. Johnson, Editors.

p. cm.

Summary: "This book examines trust in a third dimension. It considers how building trust is different for managers developing "virtual" relationships. Questions answered include: To what extent can we inform the way: remote workers are managed; electronic commerce is used to sell products and services to unseen consumers; IT is relied on to interface with organizations, virtual or otherwise?"--Provided by publisher.

Includes bibliographical references and index.

ISBN 978-1-59904-495-8 (hardcover) -- ISBN 978-1-59904-497-2 (ebook)

1. Virtual reality in management. 2. Trust. 3. Virtual work teams--Psychological aspects. 4. Electronic commerce--Psychological aspects. 5. Telematics--Social aspects. I. Brennan, Linda L. II. Johnson, Victoria (Victoria Elizabeth)

HD30.2122.C66 2008

658.4'022--dc22

2007007292

British Cataloguing in Publication Data
A Cataloguing in Publication record for this book is available from the British Library.

All work contributed to this book set is original material. The views expressed in this book are those of the authors, but not necessarily of the publisher.

Table of Contents

Detailed Table of Contents

Chapter I
Trust in Computer-Mediated Communications: Implications for Individuals
and Organizations / *Susan K. Lippert* 1

Chapter I explores the concepts of trust as they relate to computer-aided communications. The author defines trust for interpersonal and for inanimate technology, identifies common factors affecting quality of meaning in communication, and identifies ways businesses can foster and enhance trust independent of the medium of communication.

Chapter II
Trust Types and Information Technology in the Process of Business
Cooperation / *Alfonso Miguel Márquez-García, and Sebastián Bruque-Cámara* 14

In this chapter, the authors describe several different types of trust. These are examined in the context of business cooperation to suggest what types of trust are needed at each point in the process of cooperation and whether IT has an enhancing effect.

Chapter III
Virtual Teams: The Role of Leadership in Trust Management / *Nabila Jawadi,*
Mohamed Daassi, Michel Kalika, and Marc Favier 34

This chapter analyzes the impact of leadership on trust management and development in virtual teams. E-leaders have to adopt new roles and to build strategies to manage task achievement, individual team members' needs, and team cohesion. At the same time, these e-leaders may face problems related to distortion of the communication processes, member diversity, technology problems, and time pressures which can inhibit trust development

Chapter IV

Chapter IV provides another perspective of virtual teams, emphasizing the need for trust in team-building. The author examines and discusses best practices in face-to-face teams, with suggestions as to how these may be used effectively in a virtual environment.

Chapter V

This chapter explores the role media effects and familiarity play in the development of trust in computer-mediated environments. As team members interact with one another via technology, each team member assesses information and makes assessments about the trustworthiness of their teammates. This research uses media synchronicity theory and the concept of interpersonal familiarity to examine virtual team interactions and the formation of trust.

Chapter VI

Through partnering and outsourcing, organizations are exposed to managing simultaneous organizational trust and distrust. This chapter describes how technology can facilitate these computer-mediated relationships by leading to what the authors describe as "virtual assurance." It is also suggested that the presence of virtual assurance will ultimately provide a competitive advantage to firms in making contractual agreements, tracking progress, imposing penalties, and shielding organizations from potential harm.

Chapter VII

Chapter VII examines the concept of trust as it relates to the relationship between managers and remote workers within teleworking organizations. The author discusses the bases of trust and the different levels of trust that can support teleworking relationships. The chapter concludes with a conceptual model that illustrates levels and conditions of trust that can exist in a remote working arrangement.

Chapter VIII presents an in-depth discussion of different kinds of trust, as well as factors influencing trust. Included in the chapter is a discussion of virtual multicultural teams and their inherent challenges. The authors use a case study on experiences of international business students during virtual team projects. They use these individual reflections as a tool to illustrate the development of trust in multicultural teams.

Chapter IX discusses the literature on trust and trust brokers. The authors suggest that the recognition and use of trust brokers can ameliorate the "paradox of trust" in distributed work teams. Focusing on trust building in groups, they argue that trust brokers can establish trust more quickly and contribute to the sustainability of the team. A case study illustrates the feasibility of their conclusions.

Chapter X discusses e-negotiation and identifies types and subtypes of virtual relationships and the challenges inherent in building trust relationships, which are established and maintained online. The author focuses on trust-building in negotiation processes, which are conducted via any text-based channel allowing for both contextual and relational communication.

Chapter XI identifies the major antecedents of a consumer's trust in electronic commerce and develops a reference model summarizing those antecedents. The author presents a research study that identifies several determinants of consumer trust in a Web site (i.e., privacy, protection, quality, reputation, reliability), indicates the likelihood of purchasing from a Web site, and provides data regarding the direct effects of the determinants on trust and purchase intentions.

Chapter XII examines the importance of trust in business-to-consumer e-commerce. The author explores the issue of trust in the development and implementation of e-commerce and focuses on the context and role of users and consumers in transactions. The author contends that trust is more than a technical consideration and emphasizes the non-technical components such as community, identity, and experiences and their relevance to e-commerce.

Chapter XIII addresses the saliency of below-the-Web technologies for trust-based marketing messages. The authors present a research study on the transmission of a trust-based marketing message, which indicates that below-the-Web technologies allow creation of below-the-Web communities that can be appropriate tools for globally disseminating trust-building messages.

Chapter XIV discusses paradoxes that have arisen in response to society's dependence on the Internet and the new business models that attempt to resolve the contradictions. The authors present a Trust Model illustrating a series of hurdles that must be overcome in order for an individual or an organization to exhibit trust, that is, risk-taking behavior. The chapter addresses the formation of new information providers, "infomediaries," which can inform risk-taking behavior in computer- mediated relationships.

Knowledge management is increasingly reliant on information systems to identify, collect, and disperse information and knowledge. Moreover, such systems are stretching across the borders of the firm to include collaborators and their knowledge assets in e-networks. This scenario has important implications for trust between the organization and individuals who contribute to and/or use knowledge management systems. Organization-to-organization trust issues are also apparent as valuable proprietary information and knowledge are shared across the borders of firms.

Chapter XVI focuses on the dynamics of trust and distrust through presenting a qualitative field study that illustrates practices of communication between parties engaged in collaboration on IT projects in Poland. The analysis, centered on the process of cooperation in project work, provides an interesting insight into the role of trust in cooperation and offers a reflective account of actual practice of cooperation in a distrustful environment.

This chapter explores the issues and challenges faced in establishing trust among individuals and teams participating in offshore outsourcing of software development projects. The chapter discusses the special characteristics of offshore software outsourcing relationships which make the establishment of trust a challenge and offers suggestions for overcoming these hurdles.

Preface

In 1859, Charles Dickens opened *A Tale of Two Cities*, his classic book set during the French Revolution, with the statement, "It was the best of times, it was the worst of times … it was the age of wisdom, it was the age of foolishness" (Dickens, 1960). That statement succinctly described the upheaval and uncertainty of France during the Revolution. However, it also accurately depicted the conditions and challenges of London during the Second Industrial Revolution. To be certain, it may be argued that almost every era can be characterized in such terms, and the post-modern era is no exception.

TWIN CHALLENGES

Although no exception, the modern world must nevertheless contend with two major developments, globalization and its handmaiden, technology, which have blurred permanently the boundaries between business and the community. These two developments have not only influenced and changed society at particular places and points in time, but they have also changed the social landscape and the face of commerce with breathtaking speed.

In years past, social and organizational interaction was conducted sometimes with little more than a handshake. When shaking hands, two parties could look each other in the eye and presume the honesty and trustworthiness of the other. Additionally, within this personal connection, one could also assess the other's body language, tone of voice, level of attention, mood, and so forth. Moreover, businesses and individuals were more comfortable dealing with those whom they knew or with those who had had prior relationships with friends or business associates. The old workplace consisted of a gathering of employees in one or more centralized locations and of clients and vendors with whom one could interact on a personal and a regular basis. In this context, an individual could assess another's behavior and elements of character by observing the person in social and organizational settings. That evaluation could then lead to the propensity to trust, or not trust, other individuals.

While it is acknowledged that business today is rarely conducted on the strength of a handshake, we live and work in a complex and uncertain global environment where organizations and their employees now must navigate an intricate organizational reality (Johnson & Brennan, 2000). Decision-making frameworks have changed, necessitating new business paradigms and new social contracts. Thus, the old buffers of time and place have disappeared, transforming not only relationships, but also the entire business arena (Post, 2000).

THE CONTEXT OF TRUST

Today's manager has three interrelated tasks: (1) to prevent and resolve conflicts of interest between and among stakeholders; (2) to co-ordinate the efforts of internal and external groups in achieving firm goals; and (3) to manage resources in order to attain sustained competitive advantage in the marketplace (Johnson-Cramer, Berman, & Post, 2003, p. 154). Achievement of each of these tasks depends on the establishment and maintenance of solid, cohesive relationships. Human interaction, human judgment, and human purpose breathe life into the heart of business (Tung, 2001). Moreover, successful relationships depend on the elements of authenticity, character, and integrity, in other words on trust. Although technology and social systems are inextricably bound in the new economy, it is the social context of information that will determine which tools will work and which will not (Agle, Mitchell, & Sonnenfeld, 2000). Time and space may be transcended, but trust as the key element of social capital remains the lifeblood of successful relationships.

Trustworthiness is an assessment of the ability, benevolence, and integrity of the one to be trusted (Mayer & Schoorman, 1995). That being the case, within the global and digitized environment of today's environment, how is trust established and maintained? How can individuals and groups coalesce in an atmosphere of trust to accomplish organizational goals? In other words, within a virtual context, how can we trust that people are who they say they are and that they will do what they said they would do? And, can we have high touch and high tech relationships simultaneously?

Rapid advances in technology always outpace a society's ability to align, to control, and to manipulate technological capabilities for the common good. Being freed from the bounds of physical presence, interaction in computer-mediated relationships offers ample opportunities for opportunistic behavior and dishonest claims (Walther & Bunz, 1996). Nevertheless, trust is important in exchange relations because it is a key element of social capital and is related to firm performance, employee satisfaction, competitive advantage, and other economic outcomes (Mayer, Davis, & Schoorman, 1995).

The role of trust in social interaction cannot be overstated and the concept of trust has been examined in countless studies in a myriad of disciplines. However, very few studies regarding computer-mediated trust relationships and the concomitant issues and challenges have been conducted. Furthermore, because national cultures influence individual and organizational trust development processes (Doney, Cannon, & Mullen, 1998; Hofstede, 1994), and because of the all-encompassing reach of the digital highway, it is important to broaden the inquiry. Therefore, this book not only examines these relationships within numerous contexts, but also does so from multi-cultural perspectives. Scholars from the United States, United Kingdom, Australia, Austria, France, Israel, Italy, Norway, Poland, and Spain contributed chapters to this project. Their contributions include such topics as trusting remote workers, building team trust in virtual communities, e-negotiation and trust, antecedents of consumer trust, and the role of leadership in virtual teams, among others.

Margaret Wheatley (1999) argues that current and future conditions call for fluid, permeable "self-designing forms which intermingle and flow. Thus, technological innovation involves harnessing human imagination to create new approaches to the needs, problems and concerns of the modern global world (Post, 2000). We believe that the topics explored in this volume contribute to the important but illusive quest to assure that the best of times and the time of wisdom are accurate descriptors of society in the 21st century.

ORGANIZATION OF THE BOOK

Each chapter provides a thorough review of the relevant research on trust and offers analysis and insight into the differences for organizations and management practice in a virtual environment. The authors also take up the challenge of considering future directions with thought-provoking commentary. The book is organized around the type of computer-mediated relationships, whether among individuals, between individuals and organizations, or between organizations. Written in a style intended to be approachable to practitioners, the work should also be of value to academicians as a solid grounding in the extant literature on trust.

REFERENCES

Agle, B.R., Mitchell, R.K., & Sonnenfeld, J. (2000). *A report on stakeholder attributes and salience, corporate performance, and CEO values* (Research in stakeholder theory, 1997-1998: The Sloan Foundation Minigrant Project). Toronto: Clarkson Centre for Business Ethics.

Dickens, C. (1960). *A tale of two cities.* London: Penguin Books, Ltd.

Doney, P.M., Cannon, J.P., & Mullen, M.R. (1998). Understanding the influence of national culture on the development of trust. *Academy of Management Journal, 3*(3), 601-620.

Hofstede, G. (1980). Motivation, leadership, and organization: Do American theories apply abroad? *Organizational Dynamics, 9*(1), 42-63.

Johnson-Cramer, M., Berman, S., & Post, J. (2003). Re-Examining the concept of stakeholder management. In *Unfolding stakeholder thinking.* Greenleaf, UK: Greenleaf Publishing Limited.

Johnson, V., & Brennan, L. (2002). Examining the impact of technology in social responsibility practices. In *Re-Imagining business ethics: Meaningful solutions for a global economy*, 4. Oxford: Elsevier Science Ltd.

Mayer, R.C., Davis, J.H., & Schoorman, F.D. (1995). An integrative model of organizational trust. *Academy of Management Review, 20*(3), 709-734.

Post, J. (2000). Moving from geographic to virtual communities: Global corporate citizenship in a dot. com world. *Business and Society Review, 105*, 27-46.

Walther, J.B., & Bunz, U. (2005). The rules of virtual groups: Trust, liking and performance in computer-mediated communication. *Journal of Communication, 55*(4), 828-846.

Wheatley, M. (1999). *Leadership and the new science: Discovering order in a chaotic world* (2nd ed.). San Francisco: Brett-Koehler Publishers, Inc.

Acknowledgment

The editors would like to thank all of the authors of chapters included in this volume for their insightful and excellent contributions. Each of these authors also served as referees, and we greatly appreciate their comprehensive, constructive, and collegial participation. We are also indebted to Dr. Spero Peppas at Georgia Gwinnett College for his diligent support and assistance. The camaraderie, dedication, trustworthiness, and output of these individuals illustrate the extraordinary possibilities and achievements of successful computer-mediated relationships.

We also thank the staff at IGI Global for the opportunity to edit another book. A special thank you goes to Mr. Ross Miller for his patience during the project and to Ms. Jessica Thompson for her commitment to getting it done.

Finally, Linda thanks her son Jonathan for the great backrubs while at the computer and her mother Barbara for her sharp proofreading skills in the review process.

Victoria thanks Larry for his love and support and dedicates this book to Ariana, Sophia, and Pierce, the next generation of scholars.

Chapter I
Trust in Computer–Mediated Communications:
Implications for Individuals and Organizations

Susan K. Lippert
Drexel University, USA

ABSTRACT

This chapter explores the concepts of trust as they relate to computer-aided communications. The author defines trust for interpersonal and for inanimate technology, identifies common factors affecting quality of meaning in communication, and identifies ways businesses can foster and enhance trust independent of the medium of communication.

INTRODUCTION

The purpose of this chapter is to explore trust as it relates to computer-aided communications, commonly electronic-mail. There are four objectives for this chapter: (a) to comprehensively define trust as it appears in the scholarly literature, both for human interpersonal trust and for inanimate technology trust; (b) to explore the essence of communications, identifying the common factors that affect the quality of meaning in communicative interactions; (c) to compare and contrast trust in different forms of communication media;

and (d) to offer some thoughts on what can be done in the business world to foster and enhance trust, independent of the communication medium chosen.

TRUST

Trust is a contextual phenomenon commonly applied to casual conversation without conscious knowledge of what the construct means or how it manifests in daily interactions. This suggests that trust is a latent variable in the communications process since many people are not conscious that what they say and how they say it can affect their trust relationships with others. Scholars often debate how to operationalize trust and, consequently, how to measure whether an individual displays and demonstrates trust. While the definitions of trust vary, there is agreement that its meaning is situationally or contextually based (McKnight & Chervany, 1996; McKnight, Cummings & Chervany, 1998), which further suggests that how trust is operationalized is partly a function of the object upon which trust is being placed.

Rotter (1971, p. 443) suggested that "the entire fabric of our day-to-day living, of our social order, rests on trust—[from] buying gasoline, paying taxes, going to the dentist, flying to a convention—almost all of our decisions involve trusting someone [or something] else." Other scholars have noted that trust is central to all social transactions (Dasgupta, 1988), reduces transaction effort (Bromiley & Cummings, 1995), and is classified as an important component of social systems (Arrow, 1974). Trust has been cited as a vital form of social capital within social systems (Coleman, 1990; Fukuyama, 1995), since "without trust ... everyday social life ... is simply not possible" (Good, 1988, p. 32).

Much of the trust literature, particularly in the area of organizational theory and management, has focused on interpersonal trust where the object of trust is another individual. This form of trust, interpersonal trust, is most commonly defined using the research of Mayer, Davis, and Schoorman (1995, p. 712), who suggest that interpersonal trust is:

... the willingness of a party to be vulnerable to the actions of another party based on the expectation that the other will perform a particular action important to the trustor, irrespective of the ability to monitor or control that other party.

Other definitions frame trust as "a psychological state comprising the intention to accept vulnerability based upon positive expectations of the intentions or behavior of another" (Rousseau, Sitkin, Burt, & Camerer, 1998, p. 395). The trust that one individual places in another may fluctuate during the course of a relationship based on a variety of external stimuli. That is, trust is not a naïve faith that a party takes for granted, based on an interaction that occurred in the distant past (McEvily, Perrone, & Zaheer, 2003). Instead, individuals examine new information about those with whom they interact and decide if they should increase or decrease their trust in that individual (McEvily et al., 2003). Using Mayer et al.'s (1995) interpersonal trust definition, we can state that the evaluation of another's trustworthiness is a function of three antecedents: ability, benevolence, and integrity. The key difference between trustworthiness and trust is that trustworthiness is a *perceived characteristic of the trustee*, while trust is a *psychological state of the trustor* (Saparito & Lippert, 2006). Trust becomes relevant when individuals develop dependencies on, and vulnerabilities to, the actions and decisions of others (McEvily et al., 2003).

Levin, Whitener, and Cross (2006) found differences in an individual's willingness to trust another when they classified relationships into new, intermediate, and older relationships. In their research, they also found that in newer relationships, the basis for trust was gender parity, perhaps as a function of communication or personal style, that behavioral expectations that result from moderate social interaction affected intermediate relationships, and that a personal knowledge of shared perspectives (values, beliefs, perceptions, and environments) was linked to older relationships. This suggests that how individuals behave and reinforce trust, by communicating, may serve as a moderator for trusting relationships. Zahra (2003, 2005) suggested that familial and kinship ties encourage trust as a function of the depth of the relationship. In a recent study, Zahra, Yavuv, and Ucbasaran (2006) suggested that managers can build trust relationships with individuals and groups through solicitation of ideas, problems, and questions. This communication strategy has the capacity to enhance trust relationships in an organizational setting.

In an empirical investigation of 18 software vendor companies based in India, Oza, Hall, Rainer, and Grey (2006) found that communication was a key component necessary to maintain trust in established outsourcing relationships. In fact, Oza et al. (2006, p. 352) reported that one of their interviewees stated that "there should be enough communication between [the] vendor and client to prosper trust in the relationship." Henttonen and Blomqvist (2005) proposed that communication behavior that includes timely responses, open communication, and providing useful feedback were considered to build trust in virtual teams. Lewicki and Bunker (1996) along with Mayer et al. (1995) assert that trust-building is an experiential process that occurs over time as individuals engage in continual and repeated conversations. Social communication such as exchanging greetings, interests, and other personal information is important to the evolution of trust in both virtual and traditional teams (Järvenpaa & Leidner, 1998). Henttonen and Blomqvist contend that in the early stages of a relationship, both the content (social vs. work-related) and context (face-to-face vs. intranet, or e-mails) should be mixed rather than being treated separately because the combination of communication mechanisms appears to have a strong tie and connection to emotional commitment and trust.

Pepper and Larson (2006) discuss differences in the use of information communication technologies (ICTs) and face-to-face communication in geographically dispersed organizations that were recently acquired. They identified four concerns regarding the use of ICTs in organizational settings as shown in Table 1.

Each concern demonstrates that the use of technology to facilitate communications may indeed result in unexpected outcomes such as a paucity of interpersonal trust. What one can learn from these concerns is that managed use of ICTs is important if trust between employees is to be maintained and nurtured. Trust may form in relationships in which interpersonal contact ranges from extensive to nonexistent (Wilson, Straus, & McEvily, 2006).

The above studies suggest that what is said, how it is said, and the media used to facilitate communication between two individuals in an organizational setting will serve as moderators to the trust relationship. Therefore, conscious and thoughtful reflection on these elements, prior to and during the communication process, should offer organizational members the opportunity to facilitate trust in their interactions.

While trust between individuals has received the most attention in the scholarly literature, other forms of trust exist that also require investigation. Research has shown that the object of trust determines the type of trust relationship under consideration (Giffin, 1967). Trust may develop between individuals (Johnson-George & Swap, 1982) in both professional and personal settings. Additionally, trust can also develop between individuals and organizations (Zaheer, McEvily, & Perrone, 1998); between organizations (Gulati, 1995); between individuals and social institutions (Barber, 1983); between individuals and technol-

Table 1. Four communication technology concerns compiled from Pepper and Larson (2006)

Concern	Description
1	Improper ICT use exacerbates commitment problems
2	Cultural differences influence technology use
3	Over-reliance on ICTs can lead to a lack of trust
4	The timing of ICT choices influences employee perceptions

Table 2. Examples of theorists within specific disciplines investigating trust

Discipline	Studies
Management	Ferrin, Dirks & Shah, 2006; Roy & Dugal, 1998; Kramer & Tyler, 1996
Economics	Bertrand, Duflo & Mullainathan, 2004; Dasgupta, 1988
Information Systems	Bekkering & Shim, 2006; Lippert, 2001, 2007; Lippert & Davis, 2006; Lippert & Forman, 2006; Lippert & Swiercz, 2005; Muir, 1994; Gefen, Karahanna & Straub, 2003; Oza et al., 2006
Psychology	Overwalle & Heylighen, 2006; Rotter, 1967, 1971
Sociology	Sitkin & George, 2005; Welch, Rivera, Conway, Yonkoski, Lupton & Giancola, 2005; Zucker, 1986
Political Science	Barber, 1983; Leach & Sabatier, 2005; Letki & Evans, 2005
Anthropology	Blum, 1995; Carrithers et al., 2005
Organizational Theory	McKnight et al., 1998; Lewicki & Bunker, 1996; McAllister, 1995; Mayer et al., 1995; Wilson et al., 2006
Communication	Coppola, Hiltz & Rotter, 2004; Hubbell & Chory-Assad, 2005; Walther & Bunz, 2005

ogy (Lippert, 2001, 2007, Lippert & Davis, 2006; Lippert & Forman, 2006; Lippert & Swiercz, 2005); as a general characteristic of different societies (Fukuyama, 1995); and as a personal trait (Rotter, 1971). The principle differentiator between the forms of trust is the object of trust, which Giffin (1967), asserts can be a person, place, event, or object.

Trust has been investigated in many social sciences including management, economics, information systems, psychology, sociology, political science, anthropology, organizational theory, and communication. Although not all inclusive, Table 2 offers a view of disciplines in which the trust phenomenon is studied.

Technology Trust: An Alternative Object of Trust

Technology trust is defined as:

the extent to which an individual is willing to be vulnerable to the information technology (IT) based on expectations of technology predictability, technology reliability and technology utility and influenced by the individual's predilection to trust technology. (Lippert, 2001, p. 9)

Trust assessments are often based upon a single interaction with a technology and then reinforced or diminished each time an information system is used (Denning, 1993). If the technology is operational when needed, a positive assessment of system performance is recorded. Users may consider frequent or inconvenient downtimes as negative experiences with the system. Past experiences with the technology, both positive and negative, influence an individual's assessment of that system as a whole.

This suggests that the use of technology to facilitate communication may offer some additional insights into what causes individuals to develop or lose trust in others based on exchanges undertaken via ICTs. An investigation into trust in technology has the potential to offer further insight into whether a user's trust of a specific technology is likely to impact use of that system. This suggests that dependence upon technology for the completion of daily tasks makes individuals *vulnerable* to the technology. Employees depend upon technology to process financial transactions, maintain corporate Web sites, schedule meetings with clients, and communicate with geographically dispersed coworkers. However, computers are fallible, experience downtimes, and sometimes fail to function consistently from day to day. Depending upon the task, the lack of functional technology can, at a minimum, temporarily disrupt employee performance, or, at worst, put a halt to any productivity for an indeterminate length of time. These potential problems may have an impact on employees' willingness to use the technology to communicate information resulting in possible trust issues.

THEORETICAL BACKGROUND

Theory Overview

Proposed by Daft and Lengel (1984, 1986), *media richness theory (MRT)*, also known as *information richness theory*, establishes a scale to rank communication media based on the quality, breath, and depth of information in order to explain managerial media selection behaviors. MRT research is a foundation for later media selection theories and a direct base for modern communication theory. MRT explains how organizations contend with uncertainty and equivocality, or ambiguity of interpretation within organizational communication.

Classification Criteria of Media Richness Theory

Media richness theory asserts that individuals use media selection to reduce ambiguity in communication. Richness, in this context, is the ability of the communication channel to not only transfer data but also, more importantly, to impart meaning. Richness is achieved by carrying equivocal information, which modifies the communication participants' understanding. Understanding occurs when different conceptual frames of reference converge or ambiguous issues are resolved. If a particular medium provides new understanding or carries equivocal information effectively, the medium is deemed rich; otherwise, it is considered a lean medium.

Organizational tasks differ in terms of ambiguity (Daft & Macintosh, 1981) and communication reduces task ambiguity, thereby providing alternative solutions to a given problem (Guinan & Faraj, 1998). Likewise, communication media vary in their ability to reduce ambiguity in communication. According to MRT, ambiguity is reduced through a blending of four criteria based on the communication medium's ability to: (a) facilitate feedback; (b) convey meaning through multiple cues; (c) use a variety of language; and (d) present personalized messages. Medium richness results from this blending of criteria used to rank media along a richness continuum. "Rich" media reduce high levels of ambiguity whereas "lean" media are sufficient for tasks of low ambiguity.

Communication Media and Information Richness

The classification criteria in Table 3 operationalize qualities used to assess media richness.

Table 3. Media richness classification criteria compiled from Daft and Lengel (1984, 1986)

Criteria	Description
Feedback	Instant vs. delayed feedback—offers an opportunity to ask questions and make corrections
Multiple cues	The capacity to convey meaning through multiple cues including visual cues, audio cues, body language, tone of voice, facial expression, words, numbers, and graphic symbols
Language variety	The capability to customize the message by using different words to increase understanding
Personal focus	The extent to which a person can convey personal or impersonal feelings in the communication

Table 4. Communication media and information richness compiled from Daft and Lengel (1984)

Communication Media	Information Richness
Face-to-face	Highest
Telephone	High
Electronic mail *	Moderate
Written, Personal (letters, memos)	Moderate
Written, Formal (bulletins, documents)	Low
Numeric, Formal (computer output)	Lowest
* Added by Daft et al. (1987)	

Communication media with more classification features rank higher on the richness scale then media with fewer richness features. The level of understanding between the communication participants determines the richness of the communication. Daft and Lengel (1984, 1986) used these criteria to rank five communication media on a richness continuum (Table 4). Media, in richness identity, rank from high to low: face-to-face, telephone, personal written text (letters, memos), formal written text (documents, bulletins), and formal numeric text (computer output). Although not included in early studies, electronic mail was later ranked between telephonic communications and written personal documents in the context of richness (Daft, Lengel, & Trevino, 1987).

Face-to-face communication ranked highest in information richness for two reasons. First, face-to-face communication provides for multiple cues (verbal and nonverbal) and second, offers instant feedback between communication participants. In face-to-face communication, a high degree of message personalization is supported.

The telephone offers fast feedback, and audio communication is available via the telephone. Visual and nonverbal communications including hand, facial, or body gestures are non-existent with this medium. However, nonverbal cues derived from voice inflection, tone, and speaking style still are active in telephonic communications. Message personalization is available, although not to the degree of face-to-face communication. Communication is dependent upon language content and audio cues for understanding rather than written word or behavioral cues.

Personal written letters or memoranda offer limited visual cues. Feedback is slow and dependent upon receiver response time. Message personalization is available although less than telephone and face-to-face communication. Meaning must arise from the written word without the benefit of the ancillary nonverbal cues available in richer media. Audio cues are absent and richness is conveyed only through words on the page and the structure of the written word.

Formal written documents such as bulletins are less rich, since feedback is very slow, audio cues and nonverbal communications are absent, and visual cues are limited. Richness is conveyed through written words from which meaning must arise. The nature of this type of communication is often viewed as impersonal.

Formal numeric documents are considered lean. Computer generated reports provide text and numbers which offer limited visual cues. Neither audio nor nonverbal cues are available to enhance understanding. Feedback is very slow and communication is seen as impersonal.

COMMUNICATIONS

A Definition of Communication and Medium

McCroskey and Richmond (1999, p. 7) offer a definition of communication as "the process by which one person stimulates meaning in the mind(s) of another person (or persons) through verbal and nonverbal messages." Verbal messages consist of language where language is "a set of symbols or codes used to represent certain ideas or meanings" (McCroskey & Richmond, p. 7). Nonverbal messages are any messages other than verbal such as tone of voice, vocal pitch, body position, eye movements, hand gestures, and facial expressions. The authors also emphasize the importance of meaning rather than the physical exchange of messages (McCroskey & Richmond, 1999, p. 7). For purposes of this discussion, their definition is appended to include the notion of an appropriately rich communication channel. As such, the definition of communication becomes:

... the process by which one person stimulates meaning in the mind(s) of another person(s) through verbal and nonverbal messages via an appropriate communication channel.

This definition recognizes the role that the communication channel plays in human exchanges. The channel can be technologically based, such as with electronic-mail, or the channel can be face-to-face contact. This definition encompasses both the presence of technology and the human element. Noise is a mediating factor in the communication channel since it has the capacity to interrupt the message resulting in distorted meaning. For example, in electronic messaging, slow speed might impede the message and meaning that is intended to be communicated.

A Model of Communication

There are various models and diverse definitions of communication. Borden (1971) introduced eight different schemas including a behavioral communication and telecommunications schema. Later models of communication incorporated a "medium" or communication channel. George and Jones' (1996) model of communication is used to explain the process of communication,

since it allows for the potential influence of the communication medium on the process.

The George and Jones (1996) model includes a communication medium in both the initial transmission and the feedback loop. The *message* represents the information the sender wants to share with another individual, group, or organization. The message consists of the sender's thoughts, either conscious and/or unconscious, and can be any form of data or information. Through the *encoding process*, the sender expresses thoughts in symbols or language, through either written or oral form to the receiver.

The Media of Communication

The medium is the mechanism through which the communication is transmitted. The presence of a computer-mediated medium such as electronic-mail creates an artificial communication barrier between the sender and the receiver. This potential message distortion may be either a technological, behavioral, or interpretational error. If the data being exchanged between the sender and receiver is unclear for any reason in the transmission process, a technological error occurs. A behavioral error exists when the actions of the sender and/or receiver are distorted. The interpretational error occurs if the receiver misinterprets the intended message. This medium, regardless of the type, sets up the possibility of additional noise during transmission.

The degree of understanding shared between what the sender intended and what the receiver decoded measures communication effectiveness. Frequently, miscommunication results from noise distortions between the sender's intent and the receiver's understanding of the message. Noise may be positive or negative and may include a history of previous communications with the sender, the receiver's history in decoding communications of this type, the receiver's current mental state, the method of communication used to convey the message, or the receiver's health. Both the sender and the receiver construct or deconstruct the message as influenced by noise.

A challenge occurs in ensuring that the message sent and message decoded are congruent. The congruity problems might include: (a) *coding and decoding problems* (Moorhead & Griffin, 1995); (b) *lack of common experience*—a lack of common experience occurs when there is no shared language between the sender and the receiver. A shared language experience arises from the use of mutually understood symbols; (c) *semantics* – semantic problems occur when people attribute different meanings to the same words or language forms; (d) *jargon* – jargon is specialized or technical language that is specific to a field or profession. Jargon is usually a hybrid form of the standard language where words hold special and atypical meaning to the members of the group; and (5) *medium problems* – medium problems arise from selecting an inappropriate communication medium for the message being communicated.

Another issue in the communication process is the assumption that the receiver accurately decodes the message (Moorhead & Griffin, 1995) and nothing interferes with the message translation. This problem set might include: (a) *selective attention*—the problem of selective attention occurs when the receiver focuses on only selected parts of a message; (b) *value judgments* – value judgments occur when a valiance is place on the message received. If the message corresponds to the receiver's personal beliefs, the receiver may accept the message without reservation; otherwise, the receiver may disregard the message in its entirely; (c) *lack of source credibility* – if the sender is viewed as incompetent, unknowledgeable or lacking in credibility, the receiver may partially or completely disregard the message. Likewise, if the source is an expert in the field, the receiver may accept the message without question. Vital information may be discounted if

the receiver questions the sender's credibility; and (d) *overload* – communication overload occurs when an individual receives a greater amount of information than they can reasonably process. Organizations are overwhelmed with information being disseminated, such as computer printouts, electronic mail, and voice mail messages.

The medium of communication becomes a complex variable when assessing the quality of a message. In the modern information age, the use of technological systems to transmit and decode messages places an additional burden on the individuals involved in the process. These burdens include the recording, organizing, and interpreting the message.

THE IMPORTANCE OF EFFECTIVE COMMUNICATION TO FACILITATE TRUST

In the communication process, trust becomes an ever increasingly more important phenomenon. Problems such as selective attention, value judgments, lack of credibility, and information overload become manageable elements in the development of trust in an online environment. Trust is enhanced when the receiver is encouraged to attenuate the entire message. Recognition of value judgments as untested evaluations contributes to communication errors and can negatively affect trust. Credibility is closely aligned with trust in interpersonal exchanges in which the receiver views the sender as lacking in source knowledge or competency. Trust becomes an analog to credibility. Information overload becomes extraneous noise that affects the communication process by having the receiver attend to too much information. The intended outcome of the communication process is the development of shared meaning which enhances the predictability and reliability of the information exchange.

Enhancing Trust in Communications

Trust is a concept that permits the formation of perceptions based on the degree of predictability, reliability, and utility an individual exhibits toward a communication medium and the degree of faith in the integrity, honesty and benevolence of others. Trust is an underlying precept for all social interactions within both personal and professional relationships. Perceptions of trust and assessments of trustworthiness are oftentimes unconscious and frequently misplaced when individuals use erroneous data to evaluate an exchange with another. The potential noise that can occur, as a function of using a computer-mediated communication mechanism, can generate additional distortions to an otherwise clear exchange of ideas and intentions. In the process of seeking clarity, individuals must guard against making judgmental trust errors, since adjustments in trust perceptions, as a function of these errors, may result in degradations of trust rather than a refinement of trust accuracy. Interactions that are trust dependent serve as a continuous learning device to upgrade or downgrade the trustworthiness dimensions.

One way to minimize these errors is to select a communication medium based on its richness that is matched to the content of the intended exchange. More ambiguous data should be conveyed through a richer communication medium to help support the other's interpretations of the original intentions. Conscious selection of media to convey rather than using an ICT because it is convenient is a simple process to assist with noise reduction.

Communications is one of the foundations of social interaction. Humans have developed sophisticated systems to convey thought, feeling, and meaning. In the technology age, the evolution of technological systems permits the development and use of mechanical and electronic devices to aid in the speed, clarity, and meaning that is conveyed between two individuals. Communication technologies are subject to their own trust evaluation through the assessments of predictability, reliability, and utility offered by the intended user. Hence, trust in technological communication systems is both important and profound since it may offer further explanations for the presence of noise, the potentials for miscommunication, and the unintentional affects on trust. Each use of a system generates an experience that serves as the basis for an individual's trust judgments about that technology.

The challenge in the modern information age is to be able to trust the conveyed meaning shared with another and to accurately recognize meaning from others, in order to improve the efficiency and effectiveness of message dissemination. Technology provides an added challenge to trust by introducing an alternate dimension, which links our perceptions and judgments about the predictability, reliability, and utility to a communications system. An objective to improve both interpersonal and technology trust through conscious recognition of the factors that affect the evaluation of trust are oftentimes difficult to operationalize. However, provided below are a few recommendations for managers and their employees to facilitate accurate communication that may lead to enhanced trust:

1. Include trust as a conscious consideration in all forms of communication rather than something that is considered when a problem in communication occurs;
2. Recognize that trust perceptions of people and technology are oftentimes distorted and in error;
3. Consistently work to say what we mean and then mean what we say;
4. Conduct continual assessments of the qualities of the communications and change the medium, the message, or the delivery to optimize transfer of meaning;

5. Clarify as many variables as possible when a technological system is used to convey a message;
6. Select media based on richness to reduce ambiguity in communications; and,
7. Take full advantage of modern information technology to aid in the communication process by using technology where feasible and desired.

FUTURE TRENDS

Sometimes the popular press suggests that we have reached the limits of our capability to effectively communicate, whether by direct human interaction or through new and more complex technology. The future, however, holds promise for the enhancement of communication styles, structures, patterns, and practices, through greater understanding of the many factors that affect the communication process. We can expect to see improvement in the quality of workplace communication, through the use of more advanced and sophisticated technologies to help the message get through accurately, effectively, and efficiently. At the same time, the workplace recognizes that because communication is the mechanism in which business is conducted, better understanding, improved skills, and more sophisticated technology will all contribute to increase capacity. The breath and depth of communications will increase and the technological system used to assist in the operationalization of information sharing will become more accurate, user friendly, predictable and reliable, and more useful, leading to an ever-increasing trust which aids human interaction. As we become more conscious of the affect of trust on communications, we need to continue the development of our knowledge, skills, and attitudes toward improved practices both within and outside the work environment.

REFERENCES

Arrow, K. (1974). *The limits of organization.* New York: Norton.

Barber, B. (1983). *The logic and limits of trust.* New Brunswick, NJ: Rutgers University Press.

Bekkering, E., & Shim, J. P. (2006). Trust in videoconferencing. *Communications of the ACM, 49*(7), 103-107.

Bertrand, M., Duflo, E., & Mullainathan, S. (2004). How much should we trust differences-in-differences estimates? *The Quarterly Journal of Economics, 119*(1), 249-275.

Blum, S. D. (2005). Five approaches to explaining "truth" and "deception" in human communication. *Journal of Anthropological Research, 61*(3), 289-315.

Borden, G. A. (1971). *An introduction to human communication theory.* Dubuque, IA: W.C. Brown Company Publishers.

Bromiley, P., & Cummings, L. L. (1995). Transactions costs in organizations with trust. In R. J. Lewicki, R. J. Bies, & B. H. Sheppard (Eds.), *Research on negotiation in organizations,* 5 (pp. 219-247). Greenwich, CT: JAI Press.

Coleman, J. S. (1990). *Foundations of social theory.* Cambridge, MA: Harvard University Press.

Coppola, N. W., Hiltz, S. R., & Rotter, N. G. (2004). Building trust in virtual teams. *IEEE Transactions on Professional Communication, 47*(2), 95-104.

Daft, R. L., & Lengel, R. H. (1984). Information richness: A new approach to managerial behavior and organization design. In L. L. Cummings & B. M. Staw (Eds.), *Research in organizational behavior,* 6 (pp. 191-223). Greenwich, CT: JAI Press.

Daft, R. L., & Lengel, R. H. (1986). Organizational information requirements, media richness and structural design. *Management Science, 32*(5), 554-571.

Daft, R. L., Lengel, R. H., & Trevino, L. K. (1987). Message equivocality, media selection, and manager performance: Implications for information systems. *MIS Quarterly, 11*(3), 355-366.

Daft, R. L., & Macintosh, N .B. (1981). A tentative exploration into the amount and equivocality of information processing in organizational work units. *Administrative Science Quarterly, 26*(2), 207-224.

Dasgupta, P. (1988). Trust as a commodity. In D. Gambetta (Ed.), *Trust: Making and breaking cooperative relations* (pp. 49-72). New York: Basil Blackwell Ltd.

Denning, D. E. (1993). A new paradigm for trusted systems. In *Proceedings of the 1993 Association for Computing Management SIGSAC on New Security Paradigms Workshop* (pp. 36-41).

Ferrin, D. L., Dirks, K. T., & Shah, P. P. (2006). Direct and indirect effects of third-party relationships on interpersonal trust. *Journal of Applied Psychology, 91*(4), 870-883.

Fukuyama, F. (1995). *Trust: The social virtues and the creation of prosperity.* New York: Free Press.

Gefen, D., Karahanna, E., & Straub, D. W. (2003). Inexperience and experience with online stores: The importance of TAM and trust. *IEEE Transactions on Engineering Management, 50*(3), 307-321.

George, J. M., & Jones, G. R. (1996). *Understanding and managing organizational behavior.* Reading, MA: Addison-Wesley Publishing Company.

Giffin, K. (1967). The contribution of studies of source credibility to a theory of interpersonal trust in the communication process. *Psychological Bulletin, 68*(2), 104-120.

Good, D. (1988). Individuals, interpersonal relations, and trust. In D. Gambetta (Ed.), *Trust: making and breaking cooperative relations* (pp. 31-48). New York: Basil Blackwell Ltd.

Gulati, R. (1995). Does familiarity breed trust? The implications of repeated ties for contractual choice in alliances. *Academy of Management Journal, 38*(1), 85-112.

Guinan, P. J., & Faraj, S. (1998). Reducing work related uncertainty: The role of communication and control in software development. In *Proceedings of the 31ˢᵗ Annual Hawaii International Conference on Systems Sciences* (pp. 73-82). Retrieved February 21, 2007, from http://ieeexplore.ieee.org/iel5/5217/14260/00654761.pdf

Henttonen, K., & Blomqvist, K. (2005). Managing distance in a global virtual team: The evolution of trust through technology-mediated relational communication. *Strategic Change, 14*(2), 107-119.

Hubbell, A. P., & Chory-Assad, R. M. (2005). Motivating factors: Perceptions of justice and their relationship with managerial and organizational trust. *Communication Studies, 56*(1), 47-70.

Järvenpaa, S. L., & Leidner, D. E. (1998). Communication and trust in global virtual teams. *Journal of Computer-Mediated Communication, 3*(4). Retrieved February 21, 2007, from http://www.ascusc.org/jcmc/ vol3/issue4/jarvenpaa.html

Johnson-George, C., & Swap, W. C. (1982). Measurement of specific interpersonal trust: Construction and validation of a scale to access trust in a specific other. *Journal of Personality and Social Psychology, 43*(6), 1306-1317.

Kramer, R. M., & Tyler, T. R. (1996). *Trust in organizations: Frontiers of theory and research.* Thousand Oaks, CA: Sage Publications.

Leach, W. D., & Sabatier, P. A. (2005). To trust an adversary: integrating rational and psychological models of collaborative policymaking. *The American Political Science Review, 99*(4), 491-503.

Letki, N., & Evans, G. (2005). Endogenizing social trust: Democratization in east-central Europe. *British Journal of Political Science, 35,* 515-529.

Levin, D. Z., Whitener, E. M., & Cross, R. (2006). Perceived trustworthiness of knowledge sources: The moderating impact of relationship length. *Journal of Applied Psychology, 91*(5), 1163-1171.

Lewicki, R. J., & Bunker, B. B. (1996). Developing and maintaining trust in work relationships. In R. Kramer & T. R. Tyler (Eds.), *Trust in organizations: Frontiers of theory and research* (pp. 114-139). Thousand Oaks, CA: Sage Publications.

Lippert, S. K. (2001). *An exploratory study into the relevance of trust in the context of information systems technology.* Doctoral dissertation, The George Washington University, Washington, DC.

Lippert, S. K. (2007). Investigating post-adoption utilization: An examination into the role of inter-organizational and technology trust. *IEEE Transactions on Engineering Management.*

Lippert, S. K., & Davis, M. (2006). Synthesizing trust and planned change initiatives to enhance information technology adoption behavior. *Journal of Information Science, 32*(5), 434-448.

Lippert, S. K., & Forman, H. (2006). A supply chain study of technology trust and antecedents to technology internalization consequences. *International Journal of Physical Distribution and Logistics Management, 36*(4), 271-288.

Lippert, S. K. & Swiercz, P. M. (2005). Human resource information systems (HRIS) and technology trust. *Journal of Information Science, 31*(5), 340-353.

Mayer, R. C., Davis, J. H., & Schoorman, F. D. (1995). An integrative model of organizational trust. *Academy of Management Review, 20*(3), 709-734.

McAllister, D. J. (1995). Affect- and cognition-based trust as foundations for interpersonal cooperation in organizations. *Academy of Management Journal, 38*(1), 24-59.

McCroskey, J. C., & Richmond, V. P. (1996). *Fundamentals of human communication: An interpersonal perspective.* Prospect Heights, IL: Waveland Press, Inc.

McEvily, B., Perrone, V., & Zaheer, A. (2003). Trust as an organizing principle. *Organization Science, 14*(1), 91-103.

McKnight, D. H., & Chervany, N. L. (1996). *The meanings of trust* (MISRC Working Paper Series No. 96-04). University of Minnesota. Retrieved February 21, 2007 from http://www.misrc.umn.edu/wpaper/wp96-04.htm

McKnight, D. H., Cummings, L. L., & Chervany, N. L. (1998). Initial trust formation in new organizational relationship. *Academy of Management Review, 23*(3), 473-490.

Moorhead, G., & Griffin, R. W. (1995). *Organizational behavior: Managing people and organizations* (4th ed.). Boston: Houghton Mifflin Company.

Muir, B. M. (1994). Trust in automation: Part I. Theoretical issues in the study of trust and human intervention in automated systems. *Ergonomics, 37*(11), 1905-1922.

Overwalle, F. V., & Heylighen, F. (2006). Talking nets: A multiagent connectionist approach to communication and trust between individuals. *Psychological Review, 113*(3), 606-627.

Oza, N. V., Hall, T., Rainer, A., & Grey, S. (2006). Trust in software outsourcing relationships: An empirical investigation of Indian software companies. *Information and Software Technology, 48*(5), 345-354.

Pepper, G. L., & Larson, G. S. (2006). Overcoming information communication technology problems in a post-acquisition organization. *Organizational Dynamics, 35*(2), 160-169.

Rotter, J. B. (1967). A new scale for the measurement of interpersonal trust. *Journal of Personality, 35*(4), 651-665.

Rotter, J. B. (1971). Generalized expectancies for interpersonal trust. *American Psychologist, 26*(5), 443-452.

Rousseau, D. M., Sitkin, S. B., Burt, R. S., & Camerer, C. (1998). Not so different after all: A cross-discipline view of trust. *Academy of Management Review, 23*(3), 393-404.

Roy, M. H., & Dugal, S. S. (1998). Developing trust: The importance of cognitive flexibility and co-operative contexts. *Management Decision, 36*(9), 561-567.

Saparito, P. A., & Lippert, S. K. (2006). *A typology for building trust in interpersonal relationships within an organizational setting* (Working Paper Series). Drexel University.

Sitkin, S. B., & George, E. (2005). Managerial trust-building through the use of legitimating formal and informal control mechanisms. *International Sociology, 20*(3), 307-338.

Walther, J. B., & Bunz, U. (2005). The rules of virtual groups: Trust, liking, and performance in computer-mediated communication. *Journal of Communication, 55*(4), 828-846.

Welch, M. R., Rivera, R. E. N., Conway, B. P., Yonkoski, J., Lupton, P. M., & Giancola, R. (2005). Determinants and consequences of social trust. *Sociological Inquiry, 75*(4), 453-473.

Wilson, J. M., Straus, S. G., & McEvily, B. (2006). All in due time: The development of trust in computer-mediated and face-to-face teams. *Organizational Behavior and Human Decision Processes, 99*(1), 16-33.

Zaheer, A., McEvily, B., & Perrone, V. (1998). Does trust matter? Exploring the effects of interorganizational and interpersonal trust on performance. *Organization Science, 9*(2), 141-159.

Zahra, S. (2003). International expansion of US manufacturing family business: The effect of ownership and involvement. *Journal of Business Venturing, 18*(4), 495-511.

Zahra, S. (2005). Entrepreneurial risk taking in family firms. *Family Business Review, 18*(1), 23-40.

Zahra, S. A., Yavuz, R. I., & Ucbasaran, D. (2006). How much do you trust me? The dark side of relational trust in new business creation in established companies. *Entrepreneurship Theory and Practice, 30*(4), 541-555.

Zucker, L. G. (1986). Production of trust: Institutional sources of economic structure, 1840-1920. In B. M. Staw & L. L. Cummings (Eds.), *Research in organizational behavior, 8* (pp. 53-111). Greenwich, CT: JAI.

Chapter II
Trust Types and Information Technology in the Process of Business Cooperation

Alfonso Miguel Márquez-García
University of Jaén, Spain

Sebastián Bruque-Cámara
University of Jaén, Spain

ABSTRACT

This chapter deals with how IT influences the different levels and types of trust that arise in business cooperation relationships. Trust is specially important for cooperation and communication is essential along its different stages. So, the role of IT is key for trust development, because IT offers greater possibilities of access to more and better information, and increases the chances of interaction between the agents who use these technologies. We analyze how IT can affect trust from different perspectives and typologies along the process of business cooperation which begins with the initial decision to cooperate, and follows with the selection of potential partners, the negotiation and structure definition, management, evaluation and relationship evolution. At each stage we comment some possibilities of promoting different trust types through IT to improve cooperation performance.

INTRODUCTION

Information technology (IT) has become one of the motors of the economic and organizational change in the last decades. The weight of computer, robotic, and telecommunications technologies in the developed countries during the last four decades has reached figures higher than 7% of the GDP. Also, it has been esteemed that in the last years of the 20th century, IT has contributed to a 33% of the growth of the Western economies (Gual & Ricart, 2001).

Changes in technologies and market structures have shifted competition among organizations to a global level. This has resulted in the need for new organizational structures. Traditional organizational structures may not be adequate for the new business trends. In the information era, a responsive IT infrastructure is crucial to the flexibility and constantly changing needs of a business organization. This turbulent business environment is forcing organizations to re-evaluate totally their processes and structures, indicating an increasing need for networking and cooperative arrangements (DOMINO, 2005).

Although as a whole the impact of IT has been extensively studied, at organizational, group, and individual levels, there still are many questions that arise or are related to the form of adopting, managing, and renewing IT in organizations. Among them is how IT influences the different levels and types of trust that arise in a business cooperation relationship. According to Lane (1998), "more knowledge-intensive products and a more information-based mode of production, necessitating more sharing of often sensitive information, have made trust a highly desirable property" (p. 1), and it is specially important for cooperation relationships.

Research has shown that throughout the phases of the cooperation process (Child, 1998; Parkhe, 1998; Smith, Carrol, & Ashford, 1995), communication is essential. This is true for the initial decision to cooperate, the selection of potential partners, the negotiation and structure definition, management, evaluation, and relationship evolution. Communication channels should be kept open along the whole cooperation's life and along all the organizational levels involved in the cooperation agreement. In this context, organizations can use IT to increase the possibilities of access to more and better information, and to increase the chances of interaction between the agents who use these technologies. Such applications may reduce the cost of communications, increase accessibility of information, reduce response times, and facilitate a more agile and dynamic cooperative

arrangement. Altogether, IT use allows firms to manage in a globalized context, whereas blurring geographic barriers and facilitating the connection and collaboration with other agents.

It is important to consider, however, that trust is a key factor in the process of technology adoption and implementation (Premkumar, Ramamurthy, & Crum, 1997; Soliman & Janz, 2004), as well as in some processes of collaborative relationships (Child, 2001; Contractor & Lorange, 1988; Kramer & Tyler, 1996; Lane, 1998; Márquez & Casani, 2001; Nielsen, 2004). It is the aim of this chapter to examine the influence of IT on different dimensions of trust in the context of the phases of business cooperation.

This chapter is organized as follows: In the incoming section, we study the IT role on trust perspectives, trust types, and trust levels. In the subsequent section, we describe the effect of IT and trust in the process of business cooperation. Finally, in the final two sections, we provide future trends and conclusions.

IT ROLE ON TRUST

To analyze the IT role on trust we have distinguished among trust perspectives, trust types, and trust levels. This threefold point of view allows us to better explaining the effects that IT mediated interaction and information-sharing have on the separate trust dimensions. We will first address the IT effects on trust according different perspectives, second on trust types, and third on trust levels.

IT and Trust Perspectives

Trust can be considered from an interpersonal, institutional, or systemic perspective, although any theoretical approach must consider it globally as a multidimensional social reality (Lewis & Weigert, 1985) to be able to join these perspectives.

IT and Interpersonal Trust. Two elements are necessary so that interpersonal trust exists: risk and some information on the individual which one will trust on or the specific situation in which trust will appear. According to Luhmann (1979), personal trust is based on familiarity and taking things for granted. The greater communication that IT makes possible favors the access to more and better information on the other part of the relationship. Also, the greater interaction possibilities that IT allows, eliminating the geographic restrictions, increasing the frequency of communication, thus reducing response time. In this way it is possible to contribute to the rising of trust personal relationships that serve as a basis to establish more solid relations among the firms for which each individual works.

Some theoretical reasoning (Bolton, 1991) suggests that face-to-face communication allows parties to understand and empathize with one another, which affects the utility each places on the other's outcome and contributes to the building of trust. Valley, Moag, and Bazerman (1998) interpret their experimental results as providing evidence for the contribution of face-to-face communication to trust-building by increasing the incentive for truth-telling. In their experiments, verbal exchanges emphasized the interpersonal aspect of communication in ways that other communication mechanisms did not. The theoretical reasoning and experimental evidence suggest that the act of information exchange through two-way communication media is more likely to be the source of an information-trust correlation than is information exchange through one-way communication media. So the arising of trust through electronic means depends on the degree in which users accept the use of electronic equipment in their work (Chen & Dhillon, 2003; Keat & Mohan, 2004; Kim & Prabhakar, 2004).

In order to promote interpersonal trust through IT, some companies have begun to facilitate interactions between online consumers and customer service representatives (CSRs) using computer-generated text-to-speech (TTS), voice and 3-dimensional humanoid avatars to embody CSRs. The results (Qiu & Benbasat, 2005) demonstrated that the presence of TTS voice significantly increases consumers' cognitive and emotional trust toward the CSR. These findings offer practitioners guidelines to improve the interface design of real-time human-to-human communications for e-commerce Web sites.

IT and Institutional Trust. According to organization theory, it is considered that interorganizational relationships go beyond the agents who create or break them (Barney & Hansen, 1994). Trust is institutionalized in the mechanisms of decision-making, and it is perpetuated by means of control systems that reward reliable behavior. Trust types will be affected by the institutional environment in which they operate. In her study of trust production and destruction in the United States, Zucker (1986) discusses a number of institutions that facilitate trust production.

Online reputation mechanisms are emerging as a promising alternative to more established mechanisms for promoting trust and cooperative behavior, such as legally enforceable contracts. As information technology dramatically reduces the cost of accumulating, processing, and disseminating feedback, it is plausible to ask whether such mechanisms can provide an economically more efficient solution to a wide range of moral hazard settings where societies currently rely on the threat of litigation in order to induce cooperation. Comparing online reputation to legal enforcement as institutional mechanisms in terms of their ability to induce cooperative behavior, we find that although both mechanisms result in losses relative to the maximum possible social surplus, under certain conditions online reputation outperforms litigation in terms of maximizing the total surplus, and thus the resulting social welfare (Bakos & Dellarocas, 2003).

New virtual business models provide interesting settings to build online reputation and institutional trust. In contrast to consumer electronic marketplaces, the raters in B2B communities are skilled and connected, necessitating a reputation mechanism to account for the relationship between the user and the rater. To solve this problem, TrustBuilder, a prototype rating tool, incorporates a methodology to calculate a weighted rating aggregating ratings from different sources. It also uses validated scales for measuring a source's (rater's) credibility. Finally, the weights of a rater's ratings depend on user preferences instead of rater behavior, which decreases the amount of data required to calibrate the model. The experiment by Ekstrom, Bjornsson, and Nass (2005) showed that the use of a credibility-weighted tool led to increased user confidence as well as more varied evaluations. So, we get evidence that incorporating source credibility theory in a rating tool adds value in the process of evaluating service providers by increasing the decision maker's confidence in the accuracy of the information.

IT and System Trust. Although to trust always supposes to face the problem of insufficient information on the trusted object, for Luhmann (1979) the knowledge acquisition on the structural properties that one shares with others surpasses the need for information and provides supports to construct trust. IT provides information on these structural properties to fortify the trust in the system, at the same time representing an additional element of the structure that is shared with others. Among others, problems of security and privacy that affect to the transactions in the electronic world represent a challenge for the construction of trust in the new virtual system.

Harrison and Falvey (2001) point out the democratizing effects of new technologies in four major areas: interpersonal, organizational, government-political, and community networking contexts. The Internet and globalization are invoked to signify sweeping social, cultural, and institutional change (Khiabany, 2003). Currently, we can see that some IT-based tools have become so frequent in our daily life that we trust in sending e-mails instead phoning, we book flights

and hotels rooms through the Internet, and we fulfill legal duties in e-government initiatives through IT-based devices, such as computers, mobile telephones, and so forth. Also teleworking is a more habitual practice, the home-based Internet businesses become frequent and online communities expanding everywhere to share information and experiences (Edley, Hylmö, & Newsom, 2004).

IT and Trust Typologies

In the literature, we find a varied typology attending to the bases on which trust is founded and the targets where trust is placed. Focusing on the contents of expectations, Parsons (1969) distinguishes between trust in the integrity of, and trust in the competence of the trustee. Barber (1983) identifies three expectations as the basis for trust: expectations of the persistence and fulfillment of the natural and moral social order; expectation of "technically competent role performance" from those we interact with in social relationships and systems, and expectations that partners in interaction will "carry out their fiduciary obligations and responsibilities, that is, their duties in certain situations to place others' interests before their own" (p. 9). The trust typology more used when referring to business cooperation relationships is that of Sako (1992, 1998), who differentiates between contractual trust (will the other party carry out its contractual agreements?), competence trust (is the other party capable of doing what it says it will do?) and goodwill trust (will the other party make an open-ended commitment to take initiatives for mutual benefit while refraining from unfair advantage taking?).

IT and Contractual Trust. The role of IT in the configuration of the contractual trust must be to facilitate both parts knowing the terms of the agreement, especially others' obligations. For example, it is usual that office computer systems as well as intranets serve to spread partners' obligations among the members of the organization with greater capacity to judge the reliable will of the company with which it is desired to cooperate. Thus, it is possible to increase the trust in that partners which will fulfill the contractual agreements. IT may foster information spreading during the stages of the contractual relationship, improving information previous to signing, easing the conditions under which negotiation is developed and providing new forms of guaranteeing contractual terms (Águila, Bruque, & Padilla, 2002; Amit & Zott, 2001).

IT and Competence Trust. More interesting is the relation that can appear between the use of IT and competence trust. Current inter-organizational networks as well as the Internet can serve as a basis to know with greater guarantees each potential partner trajectory, its organizational structure, its economic and financial achievements, and its production structure. This information could be useful to know the real partners' competencies. In this sense, an important advance has taken place as a consequence of the progressive implantation of the denominated inter-organizational information systems, by means of which the internal flows of information are linked with the ones of the commercial partners, suppliers, clients, or technology providers. Therefore, it is necessary that the partners involved in the cooperation agreement allow a transparent access to their own information systems as well as to the ones that are created *ad hoc* when cooperation starts.

IT and Goodwill Trust. This trust type has greater subjective connotations due to the not strictly rational links that arise between cooperation partners. The role of IT in these links has been the subject of controversy, because it is not yet clear if relationships mediated by electronic means facilitate or inhibit the positive responses related to the goodwill or the commitment among people or organizations. In any case, the effect of IT on partners' goodwill trust can depend to a great extent on the own psychological characteristics of the people who lead the cooperation relationship. In this sense, it is interesting to in-

dicate that the intensive use of IT can be seen as a positive attribute, able to harmonize partners' visions and, therefore, to inspire the goodwill in the relationship, but also as a negative attribute, promoting distrust among agents who prefer face-to-face relationships and the use of means that allow direct verbal and nonverbal interactions (body-language, touching, etc.). The age, common values sharing (Cazier, Shao, & St. Louis, 2006), the individual social environment (Bruque, Moyano, & Eisenberg, 2006), the time of previous contact with the technology or personality traits tending to technological introversion can explain the presence or absence of goodwill as an element of trust in cooperation relationships (Bruque-Cámara, 2002; Kielser, 1987) in which electronic means are used.

Considering elements from the interpersonal trust literature, Faulkner (1999) distinguishes among calculative trust, predictive trust, and friendship trust in business alliances. Calculative trust: one partner calculates that the other can help it and trusts the other in the hope that matters will work out well. An element of calculation may be present in most trusting behavior (Zucker, 1986) and the calculations weigh the cost and benefits to either the trustor or the trustee. Predictive trust: one partner comes to believe that the other will behave as it says it will, since it has been as good as its word in the past. Friendship trust: here the partners get to like each other as people, and trust takes on a more personal aspect. Successful alliances do not need friendship trust to be successful but if it exists alliances are likely to be more robust and flexible when problems arise.

IT and Calculative Trust. According to Williamson (1993), calculative trust, defined as a valuation of the expected benefits and costs from cooperation, is "a contradiction in terms" (p. 463), and the term *trust* should have to be restricted solely to the personal scope, in which control or supervision does not exist. Nevertheless, for Sako and Helper (1998), inter-organizational trust, as it happens in the scope of personal relationships, usually is associated to a periodic and intense mutual observation (Sabel, 1993), although this not necessarily implies that all firms are always calculating the benefits and costs of each action that carry out with respect to the rest of firms. IT can play a notable role in the configuration of calculative trust, since they can increase the volume of available information. Thus, cooperation partners can increase their analysis capacity on the future behavior of the other partners in the cooperation relationship by means of the information available through electronic means, external databases, inter-organizational information systems, and inter-personal communication devices.

IT and Predictive Trust. Current information systems, both intra-organizational and inter-organizational, also allow to increase the available information and the interaction among partners, so it facilitates the arising of predictive trust. As an example, tracking systems of the partners' past economic and financial behavior can serve as a valid precedent in the conformation of a solid predictive trust or, on the contrary, in the final rejection of the possibility of cooperating.

IT and Friendship Trust. The same what it happened with Sako's goodwill trust, we find greater difficulties to explain IT role on friendship trust. The function that IT can potentially develop can be limited by the own character of the friendship trust, focused in the informal relationship among the individuals that participate in the cooperation relationship. Again, organizational psychologists and sociologists do not agree to conclude that if the electronic intermediation favors or makes difficult the maintenance and development of friendship bonds that underlie friendship trust nor if information technology usage is favored by a high orientation towards information technology within the individual's friendship network (Bruque et al., 2006). Research results suggest that the construction of friendship relationships through electronic means seems to be influenced by diverse contingencies (participants' personality, age, type of technology, previous image of the individuals with respect to the technological change, etc.). Moreover, diverse studies affirm that technological settings, under certain circumstances, can give rise to phenomena like the "*burnout syndrome*" and techno-stress, circumstances that do not help to establish a solid friendship bond (Salanova & Schaufeli, 2000).

In order to illustrate the effect of IT on the different trust types, we show the role that FON's Weblog (http://blog.fon.com.en/) plays as trust enhancer. FON tries to create a world-wide Wi-Fi network made up of users who share their bandwidth.

FON's blog encourages IT-based contractual trust since there we can find real-time details about the firm's activities and its commitments with clients, partners, and the community. In this blog, we can also check FON's team which is made up of outstanding IT professionals, with wide international experience. Doing so, FON's blog stimulates competence trust. It also contributes to create goodwill trust since from there the founder makes community a party to the project, emphasizing that FON is based in collaboration and has a social interest.

Also, the founder posts in the blog the favorable and unfavorable opinions about the project allowing that trust can have calculative basis, weighing the risks and possibilities of the project for investors, suppliers, clients, and so forth. In this blog we can find links and information related to the founder's and team's enterprise trajectory, so that it becomes possible to trust with a predictive base considering the previous experiences. Finally, FON's blog also serves as an IT-based tool to develop friendship trust, since the founder posts himself everyday and replies the doubts, critics, and suggestions of readers and users, generating an image of proximity, even more when sharing not only professional but also personal information.

IT AND TRUST LEVELS

Trust level goes from complete distrust to complete trust. According to Granovetter (1985), Sako (1992), and Barney and Hansen (1994), a high level of trust removes the need for any contractual and monitoring devices because personal obligation and/or value-consensus are seen to ensure against opportunism. A high level of trust between exchange partners is said to incline them towards expanding the amount of knowledge they make available to each other (Child, 1998; Sako, 1998). As Lane (1998) points out, when trust exists information exchanged among partner may be more accurate, comprehensive, and timely (Chiles & McMakin, 1996)

According to Burt and Knez (1996), trust is increasing the frequency of interaction among individuals embedded in a broader social network; therefore, IT can increase trust levels by allowing more frequent communication and favoring the creation of Barney and Hansen's (1994) strong-form trust. But interactions with third-parties (indirect connections) reinforce previously held beliefs about whether or not the other party will cooperate in future interactions, thus affecting trust intensity, not direction. Thus, some types of information flows reinforce distrust as well as trust.

Tomer and Ken (2005) use a similar approach when designing a "trust meter". They investigate the use of robots to perform complex tasks by transmitting information between them. A trust meter can be established where the robots trust their values less as time passes and trust them more after interacting with another robot, especially if the other robot's values are close to ours. Another example of trust levels use in IT environments is the trust meter of YOUPowered Orby Privacy Plus (http://www.w3.org/P3P/orbypic.jpg). It was mainly a browser toolbar created to support the user during Web browsing. Its features included cookies and password manager, a one-click form fill, and a security manager. Orby also included a trust meter that analyzes a site's P3P policy and evaluates it on a number of factors and computes a rating. Users can click on the trust meter to see the various factors that went into a particular rating. Besides the meter metaphor, green and red colors were used to visualize the level of privacy to the user.

Altogether, IT can increase or diminish trust due to their increasing effect of the available information and the interaction among partners. At first, it is expected that greater information availability affects more types of trust with rational basis (calculative and contractual) and the greater interaction affects the construction of trust with emotional basis (goodwill and friendship). Both competence trust and predictive trust can be affected by the information exchange and the interaction that IT facilitates (Figure 1).

Figure 1. IT and trust types

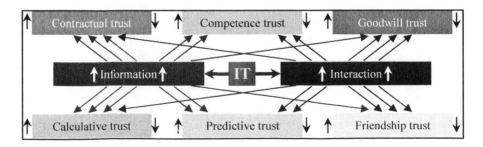

THE ROLE OF TRUST AND IT IN THE PROCESS OF BUSINESS COOPERATION

The analysis of the influence of each trust type on cooperation becomes rich if we consider a dynamic perspective in the analysis of the cooperative process. According to Lane (1998), many theorists hold that the nature of trust will vary with the stage of a relationship reached. Lewicki and Bunker (1996) propose a model of "the stage-wise evolution of trust" (p. 124), in which trust develops gradually as the parties move from one stage to another. Cooperation relationships can develop over time and this development may be associated with the deepening of trust based on an evolution of its foundations (Child, 1998). The available information also evolves throughout the relationship life cycle, so IT can influence the way a trust relationship among partners in a cooperation relationship is created and maintained (Márquez-García, Fuentes-Lombardo, & Bruque-Cámara, 2006; Rodríguez, 2003).

From the outset, certain types and levels of trust exist when creating the relationship and it must be sufficient so that the agreement could be negotiated without restrictions. Later, this climate should have to be maintained and reinforced so that the relationship could become successful and enduring (Barney & Hansen, 1994; Mohr & Spekman, 1994; Zaheer & Venkatraman, 1995).

The decision to start a cooperation relation implies a joint and coordinated bet to reach an objective more accessible collectively. In this phase, trust can have a contractual foundation if partners believe that everybody will behave according to that which is expected considering the commitments they assume when establishing the relationship. Also, trust can come from the competences, resources, abilities, and so forth, that each partner can contribute to the cooperation relationship. Also, a more exhaustive knowledge of the partners can cause that they trust on each

other's goodwill to do their best during the cooperation process.

According to Faulkner's (1999) typology, trust in the initial decision to cooperate can be calculative if trust is the result of an exhaustive valuation of its possible advantages and disadvantages; predictive, if trust is a result of previous successful relationships; and friendship, in case that personal relations of mutual affection exist among the potential partners in the agreement. Lewicki and Bunker (1996) argue that trust first develops on the basis of calculation. This is the stage at which people are prepared to take some risk in entering into dependence on others because they are aware of some institutional safeguards or deterrents against reneging. For some relationships trust may remain of this kind and at this level.

In this phase IT can play a key role to obtain information about the cooperation advantages and disadvantages. Also, it is possible that the access to information on other experiences of success and failure serves to have a more real perspective of the possibilities and difficulties that imply to establish a cooperation relationship.

An element that facilitates or inhibits the interactions is distance and the decision to coordinate with another organizations is easier if the involved organizations are physically close (Schermerhorn, 1975) since proximity promotes familiarity (Hall, 1996). IT allows eliminating the physical distance by the virtual proximity, favoring the interchange. So, the intensive use of IT may lead to cooperation among firms even when they are distant. The interconnection possibilities IT offers cause that it is more likely to establish collaboration relationships or to receive a cooperation proposal from institutions that are not in the same geographical area. Thus, managers should directly promote systematic implementation of IT-based tools in order to broaden their range of possible partnerships.

In the phase of partners' selection, it is essential to find the best candidate that adjust to the relationship needs, complementing weaknesses and harnessing strengths. It is difficult to distinguish among partners who really are trustworthy from those that only affirm to be trustworthy (Arrow, 1974; Barney & Hansen, 1994; Williamson, 1985). Thus, the attention is centered in the valuation of the level of commitments fulfillment that are arranged to assume (contractual trust), the evaluation of the competences and abilities that potential partners say to have (competence trust), and their will to collaborate openly to benefit the relationship beyond their initial commitments (goodwill trust).

Trust also can have a calculative origin if it comes from an exhaustive knowledge and the corresponding valuation of advantages and disadvantages of trusting each one; predictive, if the fact to trust or not in them comes from previous relationships; and friendship, when the origin of the trust in partners has a foundation in the friendship relation that can exist among the firms' representatives who can become partners of the cooperation relationship.

In the stage of partner selection, IT also can provide a suitable vehicle in looking for new partners. Among the electronic business models, vertical virtual communities may constitute a worthy source of potential partners. Vertical virtual communities usually include search engines that may be useful for partner finding and selection. The information and the interaction possibilities that IT allow facilitate a better knowledge of the potential partners. Child (1998) stated that "however, information about prospective partners will be limited, especially that relating to their internal cultures, competences, and values. This means that judgments will have to be made 'on the basis of the partners' reputations'" (p. 250).

In the negotiation stage firms show their mutual expectations with the aim of favoring their attainment through cooperation. If the possible partners do not trust that the motivations others declare fit the real ones and suspect that hidden agendas exist, the relationship will be more difficult to create and to maintain, in spite of using control mechanisms. In case of trust lack or distrust among partners, the negotiation stage can be so complex and delicate that it can fatally disrupt the cooperation process.

If the cooperation process continues, negotiation will equip the relationship with a structure, more or less formal according to the partners' needs and desires. Trust level will be one of the determining factors of the cooperation structure. According to Hirsch (1978), the more the contracts are detailed, the less can be expected outside of them. In this way, a written detailed agreement seems incompatible with trust, although Barber (1983) insists on its complementary character. Thus, he indicates that within organizations in which trust among members is always important, there are other alternatives and complements to trust

In the negotiation stage trust can have a contractual foundation in case that partners trust the commitment that each one assumes in the negotiation and in the agreement formalization. Also competence trust represents each partner's trust to do what they say can do in the relationship. If there is a deeper and personal knowledge among partners it is possible that trust has a goodwill basis, so partners believe each one will contribute its real reasons to cooperate and will make available its resources and competences to favor the cooperation success, beyond the minimums to which they have committed formally.

In this stage trust partner will be calculative when it arises as a result of the valuation of declared and possible hidden expectations throughout the relationship negotiation and structuring. The greater it is the commitment than each one assumes (e.g., investing in specific assets) a greater trust is more likely. According to Child (1998), the agreement to cooperate is an act of trust based primarily upon calculation. If some favorable information or own experience of previous successful negotiations with these partners exist, it is possible that trust can be described as predictive, and even as friendship if personal relations exist among the members of the firms that decide to cooperate and are carrying out this negotiation stage.

Within this negotiation phase IT can favor the interconnection of potential partners allowing a more flexible and frequent communication among them to settle dynamically any aspect of the relationship. The new cooperation structure should be equipped with a parallel IT infrastructure. During the negotiation process partners should consider the costs related to the creation and maintenance of the new common information system, considering that these costs could be higher for firms without suitable information systems to interconnect. They also have to take into account the time needed to start the system. The inter-connection among partners should be made in two levels. The first level deals with management of communication by means of standard communication devices such as e-mail, video-conference, forums, and so forth. The second level is the operative one. This deals with the elements in the value chain that should become transparent to other partners. Transparent elements in the value chain should be connected (at a transactional level or at scorecard level) with the partners' operational information systems. Other key issues that have to be addressed during negotiation are related to what information the partners are willing to share and the different access levels (open vs. limited access) for each type of partner. Virtual private networks (VPN) can be established through IP and must fulfill certain security requirements, such as tunneling (security protocols), encryption, information integrity (integrity packet), firewalls, and user and system authentication.

In the daily cooperation management, trust will facilitate information exchange and better decision-making. However, the impact of trust on cooperation management will vary according to the foundation on which it is sustained.

In this way, if trust is contractual, trust will arise as a result of the established regulations in the cooperation agreement, in which partners includes dissuasive and punitive mechanisms to prevent possible opportunistic behaviors. If partners trust their experience, knowledge, resources, and so forth, to collaborate of valuable form to the relationship, the trust foundation would their competence. When the conviction exists about partners really want to collaborate to reach coordinately a common objective, and about they will contribute their better effort to get a successful relationship, trust can be described as goodwill trust.

If partners' trust arises as a result of the calculation of advantages and disadvantages of trusting and the possibilities that this trust does not be betrayed by the existence of safeguard mechanisms, we can describe trust as calculative. If trust is based on information or previous experiences of successful cooperation management with these partners, the foundation of trust can be predictive, and the trust can be friendly if it is the result of personal relationships among the members of the companies that are in charge of the relationship management.

In this relationship management stage, information systems created *ad hoc* to cooperate should be started to make possible interaction and information exchange among partners. It is very likely that during the first steps of cooperation the *ad hoc* information system have performance, inconsistence or integrity problems. If the problems identified are relevant, it may appear a pessimistic feeling about the effectiveness of the information system. The change of transactions from the physical world to the electronic world makes Castelfranchi and Tan (2001) ask how electronic transactions can mimic the trust-building elements of physical-world transactions because this change produces a temporal and spatial separation that increases fears of opportunism, security and privacy. Besides, there is a concern about the reliability of the underlying technology and related infrastructure. This negative, pessimistic phase usually disappears when initial adjustment problems among the partners' systems are sorted out. It may be useful to interconnect the transactional systems as well as data bases involved in the cooperation agreement. The interconnection may be achieved based on a cooperative extranet that eases partners' access to shared systems. If cooperation is formalized through a new organization and the partners are geographically distant, the creation of a liaison committee would be useful.

To evaluate a cooperation relationship, expectations about its outcomes and the real ones obtained are compared. The satisfaction level with the current results will influence the trust perception. Since no partner has a total and exact knowledge of the real advantages and disadvantages that it supposes to participate in the cooperation relationship, a lack of trust among partners, and especially distrust, can take them to overvalue the benefits that others receive and to minimize the costs and disadvantages which they incur.

In this stage trust can be based in the commitment they accepted when involving in the relationship and the belief that they will make a suitable evaluation of the cooperation outcomes as a result of the agreement structure (contractual trust). In the evaluation of the relationship also it is balanced if partners are doing what they said and its competence to suitably evaluate the results they are obtaining from cooperation. Goodwill trust implies the conviction that the evaluation of the own and other partners' results in the cooperation relationship, explicit and implicit, direct and indirect, will be made loyally and balanced, trying as far as possible not to overvalue partners' benefits and to hide the own ones as a justification to renegotiate the conditions of collaboration and for increasing the advantages that cooperation reports.

If trust comes from an exhaustive and rational valuation of all the results partners obtain from the cooperation, this trust has a calculative foundation. When the evaluation of the partners' behavior in current or previous relationships has given favor-

able results trust can have a predictive base. When a personal relationship exists among partners or among the representatives in charge of the cooperation evaluation, it is possible that trust has a friendship base.

In this phase information is essential to contribute to generate trust, since if players are less able to monitor the action of others, there will be less cooperation, and hence increased opportunistic behavior, because this makes it more difficult to ascertain whether or not cheating has occurred. The cooperative extranet that is the basis for cooperation management may be also useful to create a joint balanced scorecard adapted to particular features of cooperation. Thus, firms could have a global view about each partner's contributions, tangible and intangible, and results, present and future, achieved through the cooperation relationship.

As a result of the relationship management and evaluation partners will decide on their evolution. It is possible even that although the relationship has not been successful, the greater knowledge of some partners whom now can be trusted more or less than before serves to restructure the relationship, making it evolves. The expectation of mutually advantageous future transactions can end in a reliable behavior. This expectation can have contractual, competence, goodwill, calculative, predictive, or friendship foundations.

Trust in the fulfillment of future contractual obligations can help to maintain the cooperation relationship. Since partners' contributions to the relationship do not have to be simultaneous, it is likely that cooperation stays because they trust the other partners will maintain their collaboration in the terms they agreed and committed. Competence trust also helps to explain the evolution of the cooperation because it makes more likely that the relationship stays in the future when cooperating with other firms that have valuable resources and capabilities to contribute to the relationship. Nevertheless, it is not only competence trust which can drive to maintain the relationship, but fundamentally trust in partners' willingness to cooperate. Thus, goodwill trust represents the conviction that partners will do what could be necessary in benefit of the relationship, beyond their commitments, when understanding them in ample sense not restricted to the possible detailed agreement specifications. Therefore, trust in the evolution of the relationship can mean to trust in partners will do what they must because they have committed to do it; partners will do what they say can do, because they have competence to do it; and partners will want to do more than they say can do, because its cooperative will is real and not only formal.

From a calculative perspective, without more or less subjective ethical valuations, and following the theory of the rational man who looks for optimizing own benefits, the decision to trust in partners can be a suitable rational option if we consider it in the long term. If the decision to trust is within a context of arm's length contract, it is more likely that the reliable behavior could be defrauded by the partner's opportunism. Nevertheless, it is different when this decision is tackled from a process perspective. This way, an opportunistic behavior that prevails over possible future collaborations will not be rational, since partners lose the possibility of benefiting from the advances that each firm could have achieved, in addition to the reputation loss of the opportunistic partner (Ariño, Abramov, Skorobogatykh, Rykounina, & Vilá, 1997; Gulati, 1995), which constitutes an intangible asset that can be difficult to recover because it is based on perceptions. Thus, trust in the evolution of the relationship will have a calculative base when taking into account these aspects.

Also trust can have a predictive foundation in this stage. Thus, if the behavior of partners throughout the previous cooperation experience has been honest, the decision to trust them in the future could have a predictive base. When a friendship relationship exists among partners, trust in the cooperation evolution can be sustained on this base, beyond calculation and prediction. According to Child (1998), as relationships de-velop over time with successful results, "there is a natural tendency for those concerned to identify increasingly with another's interests as well as for emotional ties to grow. In this way, 'bonding' can form between partners" (p. 252).

Also, IT may improve the information used in the evolution stage. In addition, IT may also reduce the time needed to evaluate the relationship. Thanks to IT, firms involved in the cooperation agreement are in better conditions to decide about the future of cooperation. If cooperation is highly related to technology, managers should not be dis-appointed by negative results of IT implementation in the short term. If the change related to IT has been relevant, it is likely not to obtain positive results for a period of 6 months to 2 years after the implementation. If firms do not evaluate the cooperation results in the short-term, it is likely that cooperation evolves to fulfill each partner's expectations in the long-term. According to Sako (1998), easy exchange of information makes exchange partners more open to each other and thus inclines them to explore new opportunities of collaboration.

Summing up, we can say that throughout the cooperation process IT can help partners with obtaining information as well as promoting interaction among them. IT can lead to a trust increase and the maintenance of the cooperation, as to a smaller trust and the relationship break-down, based on the content of the information, the result of the interactions and the cooperation stage. This way, if the greater information and interaction that IT allows provides incentives can be stimulated trust among partners, and could evolve from system trust to interpersonal trust as the relationship become stronger and more ma-ture, using throughout the process the necessary mechanisms of institutional trust. This effect is illustrated in the top part of Figure 2.

Although the initial conditions in cooperation relationships can be varied, when there are no previous relationships among potential partners it is more likely that trust will be based on contractual and calculative basis, as much as on partners' competence evaluation and predictions about their possible behavior (Figure 2, top part).

As the relationship evolves, calculation and legal mechanisms can give way to a greater competence trust and improve predictive trust. A possible result from frequent interaction among partners and collaborative decision-making is the development of interpersonal trust, based on goodwill and friendship.

When the information that is acquired along the process and the interactions does not stimulate trust (Figure 2, bottom part), we can witness a trust reduction throughout the cooperation process. Interpersonal trust is more difficult to appear and the institutional trust based in contracts and rational calculations is not enough to maintain the relationship.

Graphically (Figure 2), we draw both situations, although the relevance of each trust type in each phase can adopt other forms based on the level of previous knowledge regarding the relationship, the disposition to cooperate, and so forth.

FUTURE TRENDS

Analyzing IT evolution in the next years, it is possible that diverse elements appear affecting significantly the consolidation of trust relationships among organizations. After several decades in which the main firms' concern has been to equip their processes with basic computer infrastructures, in the last years we are witnessing a greater interest by efficiency in IT adoption and implementation, a greater investment rationalization and a search of integration as a means to improve the electronic resources distributed at the different departments in the organization. Although this rationalization of the technological

Figure 2. IT, trust, and the cooperation process: Summary

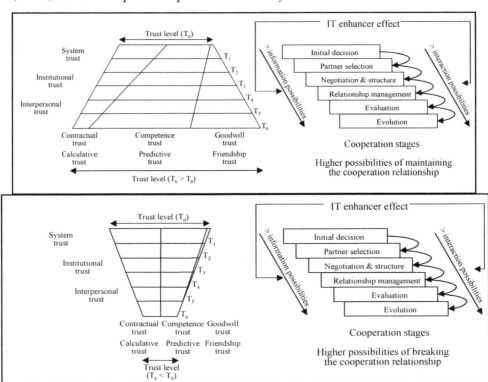

investment and integration are movements that have had a strictly technological origin, they will end up influencing clearly the way to manage and to organize firms. In addition, they are exerting a greater influence in the definition of the relationships that appear among members that share tasks in the organization, as well as between formal and informal work groups.

Changing the analysis from an internal point of view towards an inter-organizational one, the mentioned phenomena of rationalization and integration will have a series of interesting repercussions on trust. These implications can be seen strengthened if we consider, at the same time, the modifications that the labor mass (work force) of the developed countries is going through, as changing its social behavior. The greater integration and resources virtualization derived from the generalization of Grid technologies (Foster & Kesselman, 1999) and wireless technologies will surely allow a greater workers' autonomy. They also will increase the proportion of the working day that people remain isolated in interaction, practically exclusive, with electronic means of communication. The effect of physical isolation at a local level but of high connectivity at a global level will bring up new opportunities for some aspects of business management, as, for example, the increase of contacts among potential partners from different geographic areas, circumstance that can give rise later to the formation of strategic alliances. Also, if companies develop an alliance strategy (Gomes-Casseres, 1997) instead of strategic alliances, it seems to be more likely the possibility of making the most of technological resources that can favor the establishment of successful cooperation relationships.

However, the great question to solve is the level of consolidation that reaches the interpersonal relationships that, after all, sustain the inter-organizational relationships. One of the parameters that will measure that consolidation level of the interpersonal relationships will be the level and type of trust able to be developed under these technological settings. According to current

international sociological trends (Castells, 1998), the greater degree of technological formation, the globalized culture and the convergence of uses, customs, and social and individual values can be indicators of that there are ways to maintain trust relationships based on stable and lasting electronic bonds among organizations. Nevertheless, we understand that these trust relationships will be favored if the digital interaction goes with traditional face-to-face interactions, based on personal interchange of information, values, experiences, and points of view. This circumstance would make certain trust types with a high subjective component arise (as goodwill trust, friendship trust, etc.).

Another very interesting issue that may be addressed in the future regards how new information technology infrastructure may affect deep psychological phenomena that eventually drive to new ways of trust (and distrust) construction. One of these phenomena has to do with the arising of new personality traits among the youths that rely almost exclusively on electronic media to maintain interpersonal interactions. In extreme cases, a very technology-orientated personality may provoke the lack of some inter-personal skills that may lead to a disruption in the way individuals construct different types of trust. Whatever the psychological process involved in this way of life, it could be worthy to analyze if this extreme orientation towards IT in the personal, non-working life may have side effects on the way individuals interact to achieve work-related and firm-related purposes under different trust levels and types.

CONCLUSION

IT can strengthen or reduce any type of trust between partners in a business cooperation. The impact depends on the nature of the information exchanged and interaction experienced by the partners, whether favorable or unfavorable.

IT affects the interpersonal trust scope and also IT influences the impersonal one (institutional and systemic), because IT provides a source of trustworthiness (institutional) when allowing a greater information access and diffusion. At the same time, IT generalization is generating a social trust (systemic) in the technological tools, so that what is made through computer or electronic means, usually is not questioned and it is assumed optimally done.

IT tends to have more impact on trust types based on a rational view of cooperation, in which the detailed analysis of the information referring to the cooperation constitutes an essential element. We can say, therefore, that IT can be appropriate means to clear the doubts that can affect the contractual trust or competence trust. In the same way, IT can be constituted as effective means to gather enough information to sustain a suitable calculative and predictive trust. The existence of online reputation mechanisms can improve contractual trust as being easier the diffusion of opportunistic behaviors. At the same time, competence trust can also be favored by IT when having more and better information on the partners, their past behavior and the resources and capacities that they have and which they can contribute to the relationship.

From a dynamic point of view IT can exert a significant role in practically all the cooperation stages. However, this influence will depend to a great extent on the trust type more relevant for the relationship development as well as of the contingent circumstances that affect the relationship. Considering the mentioned conditions, IT can be specially valuable at the previous moments to the cooperation beginning, specially when the initial decision to cooperate is taken and the more suitable partners are chosen, using systems of business intelligence and starting exchanging information directly (Child, 1998). Greater knowledge about partners not necessarily generates trust, but it also can make appear distrust. Previous interactions or information obtained from third parties cause that trust and distrust have more important rational bases, and even emotional in case of existing personal relationships. During the cooperation process, IT can become effective means of information exchange and interaction, stimulating trust development.

The increase of the available information through IT can be crucial in the determination of the positive and negative aspects in evaluating the cooperation relationship. Here, the strategic information systems can be especially useful, currently included in ERP systems. In general, the electronic environments allow continuous information *refreshment*, so it is easier to readjust the parameters that affect trust types and level.

This chapter has some implications that may prove useful for practitioners aiming to foster trust before, during, and after IT implementation. Regarding trust perspectives, managers should know that success of IT adoption and implementation may be mediated each one of trust perspectives (interpersonal, institutional, and system). IT implementation success and trust perspectives are two interrelated processes that may benefit one from each other. First, managers ruling an IT implementation process can design this implementation process in order to promote interpersonal relationships and interpersonal trust. Some technical and managerial tools may be effective in promoting interpersonal trust, such as complementary mechanisms that facilitates more complete interpersonal interactions (instant messaging, text to speech voice interactions, multi-media interactions). Also, managers may strengthen interpersonal trust providing a complementary agenda for IT users in which these users have the opportunity to have face-to-face interactions. Second, managers may shape new IT systems in such a way as to promote the trust the user has on the institution that holds this new IT system and also on the new IT system itself. As we have pointed out in section two, there are some best-practices that may prove useful to improve institutional trust through an intensive use of IT. Among them, inter-rating systems that use inter-agent evaluations coming from different and credible sources (inter-rating systems usually include evaluations of sources' credibility) may help new users to build a stronger institutional trust when they approach a new service provided by *a priori* unknown institution. Third, managers should be able to take advantage of efforts originating in systemic institutions such as public administrations that daily promote IT use as an intrinsic positive value in modern societies. One efficient way to foster systemic trust consists of prioritizing those IT previously legitimated by public (systemic) initiatives, aligning new IT strategies to overall systemic trends.

There are also some managerial implications with regards trust typologies. Overall, managers should give priority to interaction-intensive IT vs. information-intensive IT depending on the type of trust they want to promote. For instance, information-intensive IT, such as comprehensive data-bases gathering detailed information of partners' background during a cooperation process may prove useful to foster highly rational types of trust such as contractual or calculative trust. Otherwise, interaction-intensive IT, such as multimedia meeting management systems, collaborative blogs or corporate wikis may prove useful to settle and strengthen more emotional types of trust such as goodwill or friendship trust. These types of interaction-based types of trust may flourish in new collaborative virtual organizations that have been recently named as "enterprise 2.0" (McAfee, 2006).

In any further analysis of the influence of IT on trust and the cooperation process should consider the level of partners' IT implementation. Literature differences between the following phases: office automation, information, interaction, transaction and digitalization. It is also necessary to consider the experience level and familiarity with these technological tools that facilitate interaction and information access. The use of certain technological platforms is a tool that favors collaboration or limits it when not having access to them or knowledge to use them. A significant difference between the partners can represent an obstacle for the relationship and for trust development. Also, a variety of different IT technologies that can be used in a cooperation relationship should be considered, as well as the different cooperation relationship types (complementary, competitive and symbiotic; equity and non equity; domestic and international; dyadic and multiple; etc.). So, Lewicki and Bunker (1996) believe that trust based on shared values and identification may be less common specially in business transactions where some difference of interest is usually inherent in the relationship.

Diverse future research lines arise from the previous discussion as well as the exposed limitations. Among them, it would be necessary to analyze, from an empirical point of view, the relative influence of IT on trust and control, considering the crossed effects that can arise among these three variables. Second, it could be very interesting to establish an IT taxonomy based on the influence they exert on trust and control, considering the type of cooperation. Third, research can be carried out to analyze the circumstances under which IT, or a certain group of IT, could behave as trust inhibitors.

REFERENCES

Águila, A. R., Bruque, S., & Padilla, A. (2002). Global information technology management and organizational analysis: Research issues. *Journal of Global Information Technology Management, 5*(4), 18-38.

Amit, R., & Zott, C. (2001). Value creation in e-business. *Strategic Management Journal, 22*(6), 493-520.

Ariño, A., Abramov, M., Skorobogatykh, I., Rykounina, I., & Vilá, J. (1997). Partner selection and trust building in Western European-Russian joint ventures: To Western perspective. *International Studies of Management and Organization, 27*(1), 19-37.

Arrow, K. J. (1974). *The limits of organization.* New York: W. W. Norton & Co.

Bakos, Y., & Dellarocas, G. N. (2003). Cooperation without enforcement? To comparative analysis of litigation and online reputation ace quality assurance mechanisms (MIT Sloan Working Paper No. 4295-03). Retrieved February 21, 2007, from http://ssrn.com/abstract=393041

Barber, B. (1983). *The logic and limits of trust.* New Brunswick, NJ: Rutgers University Press.

Barney, J. B., & Hansen, M. H. (1994). Trustworthiness as a source of competitive advantage. *Strategic Management Journal, 15* (Winter Special Issue), 175-190.

Bolton, G. (1991). A comparative model of bargaining: Theory and evidence. *American Economic Review, 81,* 1096-1135.

Bruque-Cámara, S. (2002). *The paradox of IT productivity. The case of the pharmaceutical distribution industry.* Jaén: University of Jaén.

Bruque, S., Moyano, J., & Eisenberg, J. (2006). The effects of social networks on worker's adaptation to a major technological change. *Academy of Management Meeting Proceedings,* Atlanta, GA, electronic version.

Burt, R. S., & Knez, M. (1996). Trust and third-party gossip. In R. M. Kramer & T. R. Tyler (Eds.), *Trust in organizations. Frontiers of theory and research* (pp. 68-89). Thousand Oaks, London, New Delhi: Sage Publications.

Castelfranchi, C., & Tan, Y. (Eds.). (2001). *Trust in virtual societies.* Amsterdam: Kluwer Academic Publishers.

Castells, M. (1998). *The information era: Economy, society and culture.* Madrid: Alianza.

Cazier, J. A., Shao, B. B. M., & St. Louis, R. D. (2006). E-business differentiation through value-based trust. *Information and Management, 43*(6), 718-727.

Chen, S. C., & Dhillon, G. S. (2003). Meeting dimensions of to consumer trust in e-commerce. *Information Technology and Management, 4*(2-3), 303-315.

Child, J. (1998). Trust and international strategic alliances. In C. Lane & R. Bachmann (Eds.), *Trust within and between organizations. Conceptual issues and empirical applications* (pp. 241-272). Oxford: Oxford University Press.

Child, J. (2001). Trust – The fundamental bond in global collaboration. *Organizational Dynamics, 29*(4), 274-288.

Chiles, T. H., & McMackin, J. F. (1996). Integrating variable risk preferences, trust and transaction cost economics. *Academy of Management Review, 21*(1), 73-99.

Contractor, F. J., & Lorange, P. (Eds.). (1988). *Cooperative strategies in international business.* New York: Lexington Books.

DOMINO. (2005). Network information infrastructures management in the construction industry: Emergence and impact on work and management arrangements. Retrieved February 22, 2007, from http://www.ist-domino.net/dynamic/public.php?action=outline4

Edley, P. P., Hylmö, A., & Newsom, V. A. (2004). Alternative organizing communities: Collectivist organizing, telework, home-based Internet businesses, and online communities. *Communication Yearbook, 28*(1), 87-125.

Ekstrom, M. A., Bjornsson, H. C., & Nass, C. I. (2005). A reputation mechanism for business-to-business electronic commerce that accounts for rather credibility. *Journal of Organizational Computing and Electronic Commerce, 15*(1), 1-18.

Faulkner, D. (1999, November 29). Trust and control in strategic alliances. *Financial Times.* Retrieved February 21, 2007, from http://www.sbs.ox.ac.uk/sbs/newco6jl.html

Foster, I., & Kesselman, C. (1999). *The grid: Blueprint for a new infrastructure.* New York: Morgan Kaufmann.

Gomes-Casseres, B. (1997). *The alliance revolution. The new shape of business rivalry.* Cambridge, MA: Harvard University Press.

Granovetter, M. (1985). Economic action and social structure: A theory of embeddedness. *American Journal of Sociology, 91*(3), 481-510.

Gual, J., & Ricart, J. E. (2001). *Enterprise strategies in telecommunications and Internet.* Barcelona: Retevisión Foundation.

Gulati, R. (1995). Does familiarity breed trust? The implications of repeated ties for contractual choice in alliances. *Academy of Management Journal, 38*(1), 85-112.

Hall, R. H. (1996). *Organizations, structures, processes, and outcomes.* New York: Prentice Hall.

Harrison, T. M., & Falvey, L. (2001). Democracy and new communication technologies. *Communication Yearbook, 25*(1), 1-43.

Hirsch, F. (1978). *Social limits to growth.* Cambridge, MA: Harvard University Press.

Keat, T. K., & Mohan, A. (2004, September). For integration of TAM based electronic commerce models trust. *The Journal of the American Academy of Business,* 404-410.

Khiabany, G. (2003). Globalization and the Internet: Myths and realities. *Trends in Communication, 11*(2), 137-153.

Kielser, S. (1987). The hidden messages of the computer science networks. *Harvard Deusto Business Review, Trim. 1,* 69-78.

Kim, K. K., & Prabhakar, B. (2004). Initial trust and the adoption of B2C e-commerce: The marries of Internet Banking. *Databases advances in information systems, 35*(2), 50-65.

Kramer, R. M., & Tyler, T. R. (1996). *Trust in organizations. Frontiers of theory and research.* Thousand Oaks, CA: Sage Publications.

Lane, C. (1998). Introduction: Theories and issues in the study of trust. In C. Lane & R. Bachmann (Eds.), *Trust within and between organizations. Conceptual issues and empirical applications* (pp. 1-30). Oxford: Oxford University Press.

Lewicki, R. J., & Bunker, B. B. (1996). Developing and maintaining trust in work relationships. In R. M. Kramer & T. R. Tyler (Eds.), *Trust in organizations. Frontiers of theory and research* (pp. 114-139). Thousand Oaks, London, New Delhi: Sage Publications.

Lewis, J. D., & Weigert, A. (1985). Trust as a social reality. *Social Forces, 6*(4), 967-985.

Luhmann, N. (1979). *Trust and power.* Chichester, UK: Wiley.

Márquez, A. M, & Casani, F. (2001). *Trust: The newest trend in co-operation relationships study.* Paper presented at the Foundational Conference of the European Academy of Management (EURAM).

Márquez-García, A. M., Fuentes-Lombardo, G., & Bruque-Cámara, S. (2006). The role of IT in family firm internationalization through strategic alliances. In S. Martínez-Fierro, J. A. Medina-Garrido, & J. Ruiz-Navarro (Eds.), *Utilizing Information Technology in developing strategic alliances among organizations* (pp. 170-202). Hershey, PA: Idea Group Publishing.

McAfee, A. P. (2006). Enterprise 2.0: The dawn of emergent collaboration. *Sloan Management Review, 47*(3), 21-31.

Mohr, J., & Spekman, R. (1994). Characteristics of partnership success: Partnership attributes, communication behavior, and conflict resolution techniques. *Strategic Management Journal, 15*(2), 135-152.

Nielsen, B. B. (2004). The role of trust in collaborative relationships: A multi-dimensional approach. *M@n@gement, 7*(3), 239-256.

Parkhe, A. (1998). Building trust in international alliances. *Journal of World Business, 33*(4), 417-437.

Parsons, T. (1969). Research with human subjects and the professional complex. In P. A. Freund (Ed.), *Experimentation with human subjects* (pp. 116-151). New York: George Braziller.

Premkumar, G., Ramamurthy, K., & Crum, M. (1997). Implementation of electronic dates interchange. *Journal of Management Information Systems, 11*(2), 157-186.

Qiu, L. & Benbasat, I. (2005). Online consumer trust and live help interfaces: The effects of text-to-speech voice and three-dimensional avatars. *International Journal of Human-Computer Interaction, 19*(1), 75-94.

Rodriguez, F. (2003). Influence of the new technologies on the organizational behavior. In F. Gil, & C. Alcover (Eds.), *Introduction to the psychology of the organizations* (pp. 179-228). Madrid: Publishing alliance.

Sabel, C. (1993). Studied trust: Building new forms of cooperation in a volatile economy. *Human Relations, 46*(9), 1133-1170.

Sako, M. (1992). *Prices, quality and trust, inter-firm relations in Britain & Japan.* Cambridge: Cambridge University Press.

Sako, M. (1998). Does trust improve business performance? In L. Christel & R. Bachmann (Eds.), *Trust within and between organizations. Conceptual issues and empirical applications* (pp. 88-117). Oxford: Oxford University Press.

Sako, M., & Helper, S. (1998). Determinants of trust in supplier relations: Evidence from the automotive industry in Japan and the United States. *Journal of Economic Behavior & Organization, 34*(3), 387-417.

Salanova, M., & Schaufeli, W. B. (2000). Exposure to information technologies and its relation to burnout. *Behavior and Information Technology, 19*(5), 385-392.

Schermerhorn, J. R. (1975). Determinants of interorganizational cooperation. *Academy of Management Journal, 18*(4), 846-856.

Smith, K. G., Carroll S. J., & Ashford, S. J. (1995). Intra and interorganizational cooperation: Toward a research agenda. *Academy of Management Journal, 38*(1), 7-23.

Soliman, K. S., & Janz, B. D. (2004). An exploratory study to identify the critical factors affecting the decision to establish Internet-based interorganizational information systems. *Information & Management, 41*(6), 697-706.

Tomer, A., & Ken, F. (2005). *Improvement of mapping by odometry.* Retrieved February 21, 2007, from Computer Science Department, Intelligent Systems Laboratory, Israel Institute of Technology: http://www.cs.technion.ac.il/Labs/Isl/MARS/FProjects/previous/ Mapping/rep/odo_report.pdf

Valley, K. L., Moag, J., & Bazerman, M. H. (1998). A matter of trust: Effects of communication on the efficiency and distribution of outcomes. *Journal of Economic Behavior in Organizations, 34*(2), 211-238.

Williamson, O. E. (1985). *The economic institution of capitalism*. New York: Free Press.

Williamson, O. E. (1993). Calculativeness, trust and economic organization. *Journal of Law and Economics, 36*(1), 453-486.

Zaheer, A., & Venkatraman, N. (1995). Relational governance as an inter-organizational strategy: An empirical test of the role of trust in economic exchange. *Strategic Management Journal, 16*(5), 373-392.

Zucker, L. G. (1986). Production of trust: Institutional sources of economic structure 1840-1920. In B. M. Staw, & L. L. Cummings (Eds.), *Research in organizational behavior* (pp. 53-111). Greenwich, CT: JAI Press.

Chapter III
Virtual Teams:
The Role of Leadership
in Trust Management

Nabila Jawadi
Paris Dauphine University, France

Mohamed Daassi
University of Grenoble, France

Michel Kalika
Paris Dauphine University, France

Marc Favier
University of Grenoble, France

ABSTRACT

This chapter analyses how to build and maintain trust in virtual teams. Based on literature on trust and leadership in virtual teams, the purpose of this chapter is to identify e-leaders' roles and behaviours related to trust management and development. The authors argue that trust is qualified as swift in virtual context and that it relies heavily on leaders' contribution to be established and maintained. E-leaders have to adopt new roles and build strategies to manage task achievement, individual team members' needs, and team cohesion to face problems related to distortion of the communication processes, member diversity, technology problems, and time pressure that inhibit trust management.

INTRODUCTION

Trust in virtual environments becomes an increasingly important and accepted topic in both computer-supported cooperative work (CSCW) and in e-business research. In *virtual collaboration*, trust is identified as a key factor for successful interactions and is associated with cooperative behaviours, coordination, and high performance of *virtual teams* (Jarvenpaa, Knoll, & Leidner, 1998; Jarvenpaa & Leidner, 1999; Kanawattanachaï & Yoo, 2002). However, the specific characteristics of the virtual context inhibit its establishment and development. This derives from virtual team members' reliance on computer-mediated-communication (CMC) that eliminates the face-to-face interactions, physical proximity, verbal cues, and facial expressions that contribute to interpersonal relationship development (Bell & Kozlowski, 2002; Dubé & Paré, 2002; Handy, 1995; Townsend, DeMarie, & Hendrickson, 1998). This is why most studies consider the virtual context to be a barrier to trust building and attempt to face this problem by identifying factors facilitating trust building in virtual teams.

Current literature on the topic shows that leadership plays an important role in fostering trusting relationships between remote members. Many studies have revealed that effective leaders develop high levels of trust, which in turn results in high performance in teams (Jarvenpaa et al., 1998; Kayworth & Leidner, 2001, 2002). Yet, less is known about how e-leaders build and reinforce trust in virtual teams, as well as the mechanisms helping them to do so. Previous studies state strategies and determinants for establishing trust without specifying leaders' contributions, despite their important role in dealing with challenges facing virtual teams.

In addition, e-leaders do not take into account the characteristics of different virtual teams' configurations, instead considering them as a single type. On the other hand, they neglect the specific form of trust that develops in virtual teams, which is swift and *ex ante* (Iacono & Weisband, 1997; Jarvenpaa et al., 1998).

Our purpose is to identify e-leaders' roles and behaviours related to trust management and development based on the current literature on both trust and leadership in virtual teams. To serve this purpose, this chapter will begin by clarifying the concept of virtual teams through identifying their specificities and their implications on various forms and dynamics of trust. The second section will analyse characteristics of trust in the virtual context. It will be primarily a question of swift trust, a specific form of trust that develops in temporary systems (Meyerson, Weick, & Kramer, 1996). We then shall explain how e-leaders implement mechanisms and strategies for building and maintaining trust in their teams through their functions, roles, and behaviours. A body of relevant managerial practices for trust management will be presented simultaneously with these developments. The conclusion will sum up our findings and present some limits and potential future extensions.

BACKGROUND

Towards a Better Understanding of Virtual Teams

A critical literature review on virtual teams reveals noteworthy limits concerning their definition and the identification of their characteristics and specificities (Bell & Kozlowsky, 2002; Jarvenpaa & Leidner, 1999; Larsen & McInerney, 2002; Lipnack & Stamps, 1997; Lurey & Raisinghani, 2001; Montoya-Weiss, Massey & Song, 2001; Townsend et al., 1998). These limits result from the confusion existing between virtual teams and other virtual work forms, as well as from considering virtual teams monolithically, different from traditional teams yet with similar characteristics.

In addition, virtual teams are different from virtual groups, virtual communities, virtual organisation, and telecommuting (Dubé & Paré, 2002). All of these work arrangements nevertheless share a common element, the computer mediated communication allowed by information and communication technologies (ICT). These forms of work differ in their objectives, nature of relationship, work organisation, and size. To better apprehend these differences, we refer to Katzenbach and Smith's (1993) definition of a team as "a small number of people with complementary skills who are committed to a common purpose, set of performance goals, and approach for which they hold themselves mutually accountable" (p. 112). This definition emphasises four dimensions in defining a team, namely: the number of individuals, interdependence, common goals, and shared responsibilities. Given these elements, we can define virtual teams as a group of more than two interdependent individuals, separated in space and time, using ICT to achieve a common short-term or long-term goal (Townsend et al., 1998). With regard to this definition, we can first distinguish virtual teams from traditional teams and, second, from other virtual arrangements. Indeed, the two distinctive dimensions of virtual teams are their substantial use of ICT and the distance between team members (Massey, Montoya-Weiss, & Hung, 2003; Maznevski & Chudoba, 2000; Montoya-Weiss et al., 2001). These two characteristics justify the virtuality of teams.

Moreover, we can distinguish virtual teams from (a) virtual groups, which do not have interdependent members, (b) virtual communities, whose members do not share responsibilities nor common goals, (c) virtual organisations, which are bigger in size and can be composed of many virtual teams, and (d) telecommuting, which involves only one individual and thus eliminates the collective dimension of virtual teams (Dubé & Paré, 2002).

Considering virtual teams as a single, uni-dimensional category impoverishes the concept and reduces the significance of study results adopting this approach. Indeed, initial first research on

the topic considered that the short lifecycle, the temporal and spatial dispersion of members, and their reliance on information and communication technologies to accomplish collective tasks were the main characteristic of virtual teams (Poltrock & Englebeck, 1999; Townsend et al., 1998; Zigurs, 2003). Townsend et al. defined virtual teams as "groups of geographically and/or organizationally dispersed co-workers that are assembled using a combination of telecommunications and information technologies to accomplish an organizational task" (p. 18). However, these researches did not take into account what creates distinctions between various virtual teams, and they overlooked the existence of several possible configurations.

To fill this gap, recent studies have adopted a multidimensional approach to describe and analyse issues related to virtual teams and their management. They assume that they may be constituted according to multiple modalities and generate many types of virtual teams. For this purpose, some studies use taxonomies to enrich their approach (Bell & Kozlowski, 2002; Cascio & Shurygailo, 2003; Dubé & Paré, 2002; Jarvenpaa et al., 1998). They assume that the variability of virtual teams' characteristics can generate several configurations. Each configuration has its own characteristics which determine both the way the team works and the nature of its organisational mechanisms.

We find that the two most relevant taxonomies are that of Bell and Kozlowski (2002) and Dubé and Paré (2002). Indeed, they introduce and explain several types of virtual teams contrary to other typologies, which concentrate on only one type characterised by short lifespan, dispersion, and high interdependence between members and reliance on CMC (Casio & Shurygailo, 2003; Jarvenpaa et al., 1998). In the first typology, the authors establish four criteria to define different types of virtual teams: lifespan, time-distribution, a team's organisational, functional and cultural boundaries, and member roles. They also introduce the complexity of the task as a key variable in the nature of the team as it influences and shapes all the other criteria (Bell & Kozlowski, 2002). In the second one, Dubé and Paré (2002) identify the characteristics common to any virtual team, as well as those that can help determine different types of virtual teams. In the latter category, the authors introduce more criteria than the aforementioned: size, geographical scattering, duration of the task, shared prior experience, role of the members, nature of their relations, interdependence of activities, and cultural diversity.

Both typologies admit that the team's nature varies along a continuum determined by the variability of their characteristics. A team can then be defined by any combination of identified properties. Two extreme cases are the archetype of the virtual team (short lifespan, members' geographical scattering, and intensive information and communication technologies use) and relatively permanent virtual teams (stable framework, unique role of the members, and real-time communication).

Despite the considerable contribution of these typologies, dimensions that they use do not really help to distinguish different type of virtual teams and may be common to many configurations of virtual teams such as cultural diversity or members' roles. That's the reason why we propose another typology, one that results from the comparison the two previous ones. In our typology, we retain the following characteristics: geographical distribution, lifespan, interdependence, and previous shared work experience. Among these characteristics, we tried to put together those that are common to the two previous typologies. Our choice was also guided by the role played by each of these variables in determining the design of the virtual team and by their impact on the nature and development of trust (Jarvenpaa & Leidner, 1999; Kanawattanachai & Yoo, 2002; Meyerson et al., 1996).

Like Bell and Kozlowski (2002), we acknowledge from the outset that these criteria vary in

a continuum taking different value (from high to low or from short to long) and resulting in different configurations that can consist of any combination of these criteria. Three representative configurations can be identified: traditional, hybrid, and pure virtual teams. Figure 1 presents possible configurations of a virtual team with the three representative cases.

First we find pure virtual teams whose members are geographically scattered, who do not know one another and never meet face to face. In this type of team, there is a strong interdependence and a short lifespan. Given these characteristics, it is this type of team which throws actual management models into question and calls for managerial innovation. We can find such a type in some R&D teams that are assembled by a company for a specific objective and that dissolve after accomplishing the task.

Second, we find traditional virtual teams whose characteristics are comparable to those of traditional teams. Members of such a team are dispersed but have already worked together in the past, have a long time period to achieve the task they were entrusted with, and rely heavily on information and communication technologies. The usual management practices are still valid

Figure 1. A proposed typology of virtual teams

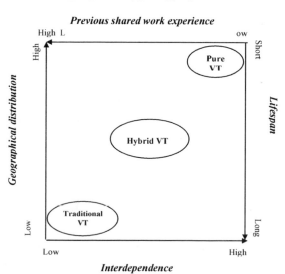

and organisational mechanisms see their usual development modes unchanged. These teams are usually used by multinational firms that have scattered sites and that adopt teamwork.

Third, mixed or hybrid virtual teams share the characteristics of the two previous categories. In such a configuration, members can combine two modes of communication (face-to-face or CMC), they can share previous work experiences, and they have a moderate deadline for the achievement of their task. Some research shows that in practice, it is the latter type that prevails most of the time.

These cases are not the only ones, as we can find other types that may, for example, have a short lifespan, and be composed of dependent members who know each other but who are geographically dispersed.

The existence of several types of virtual teams has a considerable impact on the analyses related to the way they work and on the organisational processes that are developed. Indeed, a prevailing feature in the study of virtual teams is the notion that they form only one type diverging from traditional teams considering the existence of several types of virtual teams and the variability of their characteristics leads us to accept that their functioning and their organisational mechanisms can diverge according to their characteristics. This observation affects trust, its nature, and its dynamics. In the following section, we analyse to what extent trust is influenced by virtual teams' characteristics.

Trust in Virtual Teams

Trust is a key element to building successful interactions and to overcoming selfish interests. It plays an important role in the construction and stability of interpersonal relationships. Trust represents a means for coping with complexity and uncertainty in contexts where high levels of interdependence and interaction between different actors exist. It helps create a climate of coopera-

tion and understanding both on the individual and collective levels. It also encourages citizenship behaviours and improves the quality of decisions made (Kanawattanachaï & Yoo, 2000; Mayer & Davis, 1995; McAllister, 1995; Ring & Van de Ven, 1992, 1994).

Trust requires that specific conditions be met in order for it to appear and develop, such as physical proximity, mutual information exchange (Handy, 1995), time, a shared social context, common values, and similar cultures (Meyerson et al., 1996). Yet if we refer to the specificities of the virtual teams' contexts, these conditions are not always met. Trust in virtual teams has indeed been regarded as paradoxical so far (Wilson, Straus, & McEvily, 2006). More precisely, Wilson et al. (2006) found "one of the fundamental factors that are believed to be important in determining the success or failure of virtual teams is trust. ... This is because trust functions like the glue that holds and links virtual teams together" (p. 188). In addition, the absence of physical proximity, a shared social context, and the limited lifespan of virtual teams themselves hinder the development of trust (Handy, 1995; Hummels & Roosendaal, 2001; Townsend et al., 1998). Underlying these results is the assumption that virtual teams are single entities. Yet, according to the typologies previously presented, these results cannot be applied to all types of virtual teams. This is why we suppose that the variability of virtual teams' characteristics entails the variation of the nature of trust and its development mechanisms.

Therefore, the conditions required for trust building (time, cultural similarities, physical proximity, and face-to-face interactions) are fulfilled in teams belonging to the second category (traditional virtual teams). As a result, it is difficult to confirm the paradoxical dimension of trust in these teams. However, it seems that trust takes a traditional form in these teams as all the factors required to its development through time can be found. Concerning the two other types (pure and hybrid virtual teams), a recent trend in research in the field has introduced a particular form of trust. "Swift trust" has been developed by Meyerson et al. (1996) to explain the behaviours of temporary systems' members. The analogy drawn between virtual teams and temporary systems finds its source in the similarities between the two notions. Temporary systems are indeed represented, according to Meyerson et al., by "groups of individuals with various skills working together to achieve a complex task during a short period" (p. 168). Members of a temporary system do not know each other, they have never worked together in the past, and do not plan to do so in the future; there is a strong interdependence between them and the lifespan is limited to the achievement of the task. Unlike virtual teams, members of a temporary system communicate face-to-face.

These similarities have led some researchers to consider that trust develops swiftly in pure and hybrid virtual teams (Jarvenpaa et al., 1998; Kanawattanachaï & Yoo, 2002; Piccoli & Ives, 2003). Swift trust can be defined, according to Meyerson et al. (1996) as:

... to trust and to be trustworthy within the limits of a temporary system means that people have to wade in on trust rather than wait while experience gradually shows who can be trusted and with what. Trust must be conferred presumptively or ex ante. (p. 177)

Swift trust in virtual teams has not been sufficiently explored. Research projects in this area are limited and lack empirical bolstering and applicability. Yet, an analysis of the few studies that have been thus far carried out enables us to identify the characteristics of swift trust and the factors influencing its development. According to its definition, swift trust is based on an assumption that occurs right from the start. Indeed, in virtual teams, members do not have enough information about each other. Moreover, they do not have time to collect the information that would help them assess other members' behaviour. For this reason,

they suppose that other members are trustworthy in order to limit the uncertain and risky dimension inherent to the opposite hypothesis. They then discover later if their presumptions were right or wrong (Meyerson et al., 1996).

Concerning its development, Jarvenpaa et al. (1998) have shown that swift trust has the same determinants as traditional trust: ability, benevolence, and integrity. Nevertheless, these determinants do not follow the same evolution in both cases. In the virtual context, ability and integrity are more important than benevolence during the first stages of a team's creation. These variables may intensify or weaken as the work progresses, depending on the information collected by members.

On the other hand, Kanawattanachaï and Yoo (2002) identified two dimensions of swift trust: cognitive and affective. They appear to agree with Meyerson et al. (1996) when they characterise trust as a "depersonalised form of action" to express the predominance of the cognitive dimension over the affective one. It indeed develops under the effect of actions and achievements rather than feelings. Meyerson et al. found that "feelings, engagement and exchanges are less important than actions and the achievement of the task" (p. 180). In addition to these factors, swift trust operates through social and psychological mechanisms and cognitive processes related to categorisation and the creation of stereotypes imported from similar working situations and applied to the virtual context (Jarvenpaa & Leidner, 1999; Meyerson et al., 1996; Zigurs, 2003).

In this regard, swift trust is more task-related than socially-oriented (Cascio & Shurygailo, 2003; Iacono & Weisband, 1997; Johnson, Suriya, Yoon, Berrett, & La Fleur, 2002; Yoo & Alavi, 2004). Its development relies more on actions related to activity planning and achievement, deadline respect, and task distribution than on social exchanges and interactions. Jarvenpaa et al. (1998) identified strategies and behaviours facilitating trust reinforcement and performance enhancement in virtual teams. They include: the style of actions, the focus of dialogue, team spirit, tasks' goal clarity, role division and specificity, time management, patterns of interaction, and the nature of feedback. Although leaders' contributions to establish these strategies were not analysed in this study, it would seem that leaders play an important role and have made considerable contributions in trust building. Based on the results of the aforementioned study and others analysing leadership roles, we attempt in the following section to provide some insight into how leaders manage trust in the virtual context.

E-Leadership and Trust Management in Virtual Teams

New parameters for virtual contexts introduce considerable changes in leadership. E-leaders have to deal with challenging issues generated by a distortion of the communication processes, member diversity, technology problems, and, in some cases, time pressure. According to Kayworth and Leidner (2001, 2002), "given these challenges with communication, technology, logistics, and culture … virtual environments may be more complex than their traditional counterparts." Leadership nevertheless remains a social influence process for producing a change in attitudes, feelings, thinking, behaviour, and/or performance with individuals, groups, and/or organisations (Avolio & Kahaï, 2003; Avolio, Kahai, & Dodge, 2001). These elements are mediated by information and communication technology, as they are the main means of interaction in the virtual context. E-leadership styles and functions have to change and integrate virtuality in order to be effective and to ensure virtual teams' success and performance.

For e-leaders, trust becomes particularly significant and is highly emphasised for relationship building, especially given that control mechanisms existing elsewhere lose their importance and become inoperative. Trust becomes an essential factor for ensuring cohesion, cooperation, and citizenship behaviours among team members. This is precisely why it is important to establish the conditions required for its proper its development.

As discussed above, trust in virtual teams takes a particular form where it develops swiftly and where factors influencing its maintenance are different from those of traditional trust-building scenarios. E-leaders have to take them into consideration when formulating strategies for building and reinforcing it.

Consistent with behavioural complexity theory on leadership in traditional organisational settings, leaders have to develop a portfolio of complementary and at the same time paradoxical roles and behaviours to manage their subordinates and to be effective. These roles are: innovator, broker, producer, director, coordinator, monitor, facilitator, and mentor (Denison, Hooijberg, & Quinn, 1995). Yoo and Alavi (2004) validate behavioural complexity theory in virtual contexts in their study of emergent leaders, and noted that e-leaders differ from other team members by their roles of initiator, integrator, and scheduler. For example, e-leaders were the first members that sent e-mails to constitute the teams. They also plan work, encourage other members to respect deadlines, intervene to resolve problems, and so forth.

The results of the study reveal that e-leaders are most concerned with three main fields of team management: task achievement, individual team members' needs, and team cohesion. All of these factors either directly or indirectly influence trust management. The remainder of this section explains how e-leaders build and maintain trust via these factors.

First, as swift trust is more related to *doing* than *feeling*, it is influenced by interactions focused on activity planning and goal achievement. Hence, e-leaders have to schedule work, set deadlines, and control workflows in order to respect them. They also have to establish coordination mechanisms facilitating information sharing and work exchanges between team members, according to Paul, Sheetharaman, Samarah and Mykytyn (2004), as "the issue of coordination becomes more complex when virtual teams interact asynchronously" (p. 317).

This implies that e-leaders must pay particular attention to technology problems. Indeed, accessibility and the effective use of information and communication technologies are the conditions required for ensuring proper work achievement, as they are the exclusive means of communication in the absence of face-to-face interactions. In addition, e-leaders have to consider differences in time zones when preparing activities and organising virtual meetings. Attendance of all team members or a majority of them is essential to resolving work problems, to explaining and clarifying task schedules, and to guaranteeing that there are no "free riders." In that way, e-leaders may be able to uphold a dynamic, positive, and optimistic team spirit that reinforces trust level (Jarvenpaa et al., 1998).

The second important intervention domain of e-leaders is that of individual team members' needs. This factor becomes increasingly important given the characteristics of virtual contexts. Indeed, virtual team members can feel isolated, as "most of the interaction in virtual teams occurs not in physical places, but in electronic spaces" (Sarker & Sahay, 2004, p. 4). They do not see or know each other because of physical separation and electronic communication. This may create a feeling of isolation, inhibit their commitment to the team and their goals, and damage potential collective social context, which naturally builds and fosters trust relations. This is precisely the reason why e-leaders' contributions are focused on managing and satisfying team members' needs.

For this purpose, e-leaders must encourage social information exchanges between members to allow them to better know each other and to assess their behaviours. E-leaders have two main ways to do this. On one hand, they can organise one face-to-face meeting at the least at the onset of the project. This meeting aims to introduce team members to each other and allows for social face time with colleagues. This also facilitates future electronic interactions and provides more visibility and vividness among team members (Zigurs, 2003). On the other hand, when holding face-to-face meetings is impossible, e-leaders may use team building exercises before the effective work begins which "should not only reveal information about the members, but also create a team identity, which is an important facilitator of trust in a collective context" (Jarvenpaa et al., 1998, p. 3). Team-building exercises can also be useful when e-leaders face time pressure in short lifecycle teams. In this context, leaders have to act rapidly to establish effective communication patterns and trusting relationships.

In addition to this, e-leaders can enhance member participation and commitment to the team by setting clear task goals and specifying role divisions and contributions for each member. Combined with immediate and substantial feedback, the previous actions help reduce the uncertainty prevailing in virtual context related to a lack of information about team members, and can increase beliefs that other team-mates will not take advantage of others' vulnerability.

E-team trust follows from the beliefs and expectations that members have of each other, that each member will live up to agreed upon commitment, that each member is acting with good intentions on behalf of the group, and that each will work hard on behalf of the group (Zaccaro & Bader, 2003, p. 382).

The three behaviours described above were identified by Jarvenpaa et al. (1998) as those which reinforce strategies focused on increasing trust levels.

The third field requiring leaders' intervention is team cohesion management, which is specifically important given its direct influence on trust. Team cohesion management in virtual teams includes cultural diversity management and conflict resolution.

On one hand, as virtual teams spread out over organisational, functional, and professional boundaries, they may be constituted by members from different cultures that speak different languages and that have different perceptions and referents. E-leaders have to deal with this diversity and find a common ground of understanding. They should establish a set of collective and accepted norms to guide task behaviours. They also have to intervene at the appropriate moment to resolve misunderstandings related to language barriers or conflicting perceptions. To avoid these problems, e-leaders can organise training sessions on cultural differences and language instruction at the pre-project level according to the resources available.

On the other hand, team cohesion management requires developing suitable strategies for conflict resolution. This implies the proper identification of conflict sources which can originate in the spatial and temporal dispersion of team members or from organisational and cultural diversity (Shin, 2005). To avoid the damaging effects of conflicts, e-leaders have to develop collaborative conflict management styles as identified by Paul et al. (2004) as the most effective strategy for conflict resolution in virtual teams especially heterogeneous ones. This strategy allows e-leaders and members to overcome problems that may hinder task progression and inhibit team performance. It enhances team cohesion, as it contributes to creating a favorable climate for collaboration and cooperation directly related to trust.

The previous analyses reveal the importance of leaders' contributions in trust management in virtual teams as mentioned in the current literature on the subject. It appears that trust building in CMC requires leaders to adopt a wide variety of roles and behaviours necessary to accomplishing their job functions. In this regard, virtual teams do not differ from their traditional counterparts. However, some roles are more highly emphasised in the virtual context such as director, coordinator, and facilitator (Yoo & Alavi, 2004). E-leaders have also to pay particular attention to team dynamics and individuals concerning member satisfaction, task achievement, and team cohesion (Lurey & Raisinghani, 2001). These roles and functions can be summarised in Figure 2.

As cognitive dimension is more prevalent in swift trust, e-leaders have to develop specific roles and functions ensuring its management. These roles and functions are related to setting clear goals and tasks, developing coordination mechanisms that take into consideration geographical and temporal dispersion of members, encouraging social information exchange between members, and overcome cultural barriers such language or

Figure 2. Leaders' contribution in trust management within virtual teams

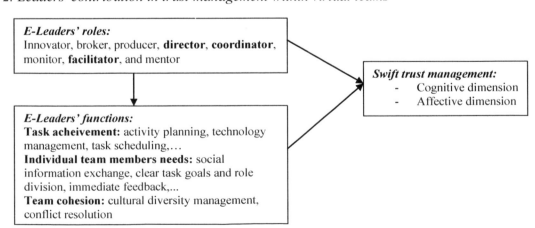

different referent. Affective dimension of swift trust may become as important as cognitive one in late stages of the project, that's the reason why e-leaders have to provide required conditions to its management. They have to encourage team cohesion and to build and reinforce team identity and common shared norms. Their roles of facilitator and innovator help them to do that. Collaborative conflict resolution strategies also contribute to interpersonal relation development including trust.

To achieve these objectives, e-leaders need to develop new skills specific to the virtual context in general and to trust management in particular. According to Cascio and Shurygailo (2003), these skills are related to virtual collaboration, virtual socialisation, and virtual communication. Specialised and targeted training sessions for e-leaders are hence essential prior to a team's constitution.

CONCLUSION

The purpose of this chapter was to identify leaders' contributions to trust management in computer-mediated-communication through an examination of virtual teams. We attempted to draw up an integrative framework based on the current literature on both trust and leadership in virtual teams in order to develop a body of relevant managerial actions.

While there is currently a lack of research on the topic, it is growing in fields such as information systems or organisational behaviour. For this reason, some limits must be addressed and may be further developed in future studies. First, it would be interesting to further clarify the two concepts of swift trust and e-leadership and better identify their distinctive characteristics in order to give more parsimony and robustness to research studying them. This implies a profound analysis of changes on team-based work introduced by virtuality with its two dimensions of intensive ICT use and distance.

Second, our developments were based mainly on findings of experimental studies conducted with university students. It would be interesting to test them in a real organisational setting and to check if these findings remain valid. Third, additional variables not analysed here should be added in future extensions of leaders' contributions to trust management. We can mention, for example, influences in the style of interaction (Poltrock & Engelbeck, 2002; Potter & Balthazard, 2002), training programs (Beranek, 2000), or psychological mechanisms (Lee-Kelly, 2006). We can also consider the question in terms of the decline of trust in virtual teams in order to investigate its causes and e-leaders' role in managing it (Piccoli & Ives, 2003).

REFERENCES

Avolio, B.J., Kahaï, S.S., & Dodge, G.E. (2001). E-leadership: Implications for theory, research, and practice. *Leadership Quarterly, 11*(4), 615-668.

Avolio, B.J., & Kahaï, S.S. (2003). Adding the "E" to e-leadership: How it may impact your leadership. *Organizational Dynamics, 31*(4), 325-338.

Bell, B., & Kozlowski, S.W. (2002). A typology of virtual teams, implications for effective leadership. *Group & Organization Management, 27*(1), 14-49.

Cascio, W.F., & Shurygailo, S. (2003). E-leadership and virtual teams. *Organizational Dynamics, 31*(4), 362-376.

Denison, D.R., Hooijberg, R., & Quinn. R.E. (1995). Paradox and performance: Toward a theory of behavioural complexity in managerial leadership. *Organization Science, 6*(5), 524-540.

Dubé, L., & Paré, G. (2002). The multi-faceted nature of virtual teams. *Cahier du GreSI, N° 02-11*, 1-33.

Handy, C. (1995). Trust and virtual organization. *Harvard Business Review, 73*(3), 40-50.

Hummels, H., & Roosendaal, H.E. (2001). Trust in scientific publishing. *Journal of Business Ethics, 34*(2), 87-100.

Iacono, C.S., & Weisband, S. (1997). Developing trust in virtual teams. In *Proceedings of the 30th Annual Hawaii International Conference on System Sciences*.

Jarvenpaa, S.L., Knoll, K., & Leidner, D.E. (1998). Is there any body out there? Antecedents of trust in global virtual teams. *Journal of Management Information Systems, 14*(4), 29-64.

Jarvenpaa, S.L., & Leidner, D.E. (1999). Communication and trust in global virtual teams. *Organization Science, 10*(6), 791-815.

Johnson, S.D., Suriya, C., Yoon, S.W., Berrett, J.V., & La Fleur, J. (2002). Team development and group processes of virtual learning teams. *Computers & Education, 39*, 379-393.

Kanawattanachaï, P., & Yoo, Y. (2002). Dynamic nature of trust in virtual teams. *Strategic Information System, 11*, 187-213.

Katzenback, J.R., & Smith, D.K. (1993). The discipline of teams. *Harvard Business Review,* 111-120.

Kayworth, T., & Leidner, D. (2001, 2002). Leadership effectiveness in global virtual teams. *Journal of Management Information Systems, 18*(3), 7-40.

Larsen, K.R.T., & McInerney, C.R. (2002). Preparing to work in virtual organization. *Information & Management, 39*(6), 445-456.

Lee-Kelly, L. (2006). Locus of control and attitudes to working in virtual teams. *International Journal of Project Management*.

Lipnack, J., & Stamps, J. (1997) *Virtual teams: Reaching across space, time, and organizations with technology.* New York: John Wiley & Sons.

Lurey, J.S., & Raisinghani, M.S. (2001). An empirical study of best practices in virtual teams. *Information & Management, 18*, 523-544.

Massey, A.P., Montoya-Weiss, M.M., & Hung, Y.T. (2003). Because time matters: Temporal coordination in global virtual project teams. *Journal of Management Information Systems, 19*(4), 129-155

Mayer, R.C., & Davis, J.H. (1995). An integrative model of organizational trust. *Academy of Management Journal, 20*(3), 709-734.

Maznevski, M.L., & Chudoba, K.M. (2000). Bridging space over time: Global virtual team dynamics and effectiveness. *Organization Science, 11*(5), 473-492.

McAllister, D.J. (1996). Affect and cognition based trust as foundations for international cooperation in organizations. *Academy of Management Journal, 38*(3), 24-59.

Meyerson, D., Weick, K.E., & Kramer, M.R. (1996). Swift trust and temporary groups. In M.R. Kramer & T.R. Tyler (Eds.), *Trust in organizations, Frontiers of theory and research* (pp. 166-195). Thousand Oaks, CA: Sage Publications.

Montoya-Weiss, M.M., Massey, A.P., & Song, M. (2001). Getting it together: Temporal coordination and conflict resolution in global virtual teams. *Academy of Management Journal, 44*(6), 1251-1262.

Paul, S., Sheetharaman, P., Samarah, I., & Mykytyn, P.P. (2004). Impact of heterogeneity and collaborative conflict management style on the performance of synchronous global virtual teams. *Information & Management, 41*, 303-321.

Piccoli, G., & Ives, B. (2003). Trust and the unintended effects of behaviour control in virtual teams. *MIS Quarterly, 27*(3), 365-395.

Poltrock, S.E., & Engelbeck, G. (1999). Requirement for virtual collocation environment. *Information and Software Technology, 41*, 331-339.

Potter, R.E., & Balthazard, P.A. (2002). Understanding human interaction and performance in the virtual teams. *Journal of Information Technology Theory and Application, 4*(1), 1-23.

Ring, P.S., & Van de Ven, A.H. (1992). Structuring cooperative relationships between organizations. *Strategic Management Journal, 13*(7), 483-498.

Ring, P.S., & Van de Ven, A.H. (1994). Developmental processes of cooperative interorganizational relationships. *The Academy of Management Review, 19*(1), 90-118.

Sarker, S., & Sahay, S. (2004). Implication of space and time for distributed work: An interpretive study of US-Norwegian systems development teams. *European Journal of Information Systems, 12*, 3-20.

Shin, Y. (2005). Conflict resolution in virtual teams. *Organizational Dynamics, 34*(4), 331-345.

Townsend, A.M., DeMarie, S.M., & Hendrickson, A.R. (1998). Virtual teams: Technology and the workplace of the future. *Academy of Management Executive, 12*(3), 17-29.

Wilson, J.M., Straus, S.G., & McEvily, B. (2006). All due in time: The development of trust in computer-mediated and face-to-face teams. *Organizational Behaviour and Human Decision Processes, 99*, 16-33.

Yoo, Y., & Alavi, M. (2004). Emergent leadership in virtual teams: What do emergent leaders do? *Information and Organization, 14*, 27-58.

Zaccaro, S.J. & Brader, P. (2003). E-leadership and the challenges of leading e-teams: Minimizing the bad and maximizing the good. *Organizational Dynamics, 31*(4), 377-387.

Zigurs, I. (2003). Leadership in virtual teams: Oxymoron or opportunity. *Organizational Dynamics, 31*(4), 339-351.

Chapter IV
An Examination of Team Trust in Virtual Environments

Martha C. Yopp
University of Idaho, USA

ABSTRACT

The purpose of this chapter is to examine ways to help build trust in virtual environments. More business and decision-making is being accomplished using virtual teams. These people seldom meet face-to-face, but they work together toward common goals. A crucial factor in determining the success or failure of virtual teams is trust. Successful techniques for promoting and building an atmosphere of trust within virtual teams and maintaining that trust are a primary focus of the chapter. Learning organizations are discussed as a vehicle for promoting attentive listening, sharing information, mutual scholarship, and meeting expectations through innovation and networking.

INTRODUCTION

More and more business and decision-making is accomplished using virtual teams. Talented, experienced people from around the state, the nation, and the world are working together to solve common problems and engage in creative and innovative collaboration. Although it is unlikely these people have met face-to-face, they are working together toward common goals. Virtual teams provide advantages over face-to-face teams by bridging time and space, therefore reducing the need for physical travel. One crucial factor

in determining the success or failure of virtual teams is trust. Kanawattanachai and Yoo (2005) suggest that trust promotes open communication, encourages cooperation, facilitates higher quality decision-making, supports risk-taking, fosters commitment, and increases satisfaction during the decision-making processes. Higher levels of trust are associated with advanced performance.

As a population working towards socioeconomic stability, a strong social fabric, collaboration, and cooperation, we depend on trust to navigate everyday challenges. Internal, emotional, and psychological comfort is derived through relational trust, mutual respect, openness, and by valuing and appreciating other people.

The purpose of the chapter is to examine ways to help build trust in virtual environments. The objective is to examine and discuss best practices in face-to-face teams and to suggest how these may be used effectively in a virtual environment.

BACKGROUND

For conducting business in computer-mediated environments, information and communication technology permits geographically disconnected employees to become collaboratively interconnected (Hossain & Wigland, 2004, p. 1). Online "webinars" are utilized for continuing education and professional recertification, and although virtual learning is increasingly gaining acceptance, apprehension still exists about quality and effectiveness as well as the development of trust.

Advocate researchers and educators studying virtual workplaces reported deficiencies in online relational trust but, compared to the advantages, consider most negative arguments to be resolvable challenges. Establishing, building, and maintaining trust are not considered overshadowing impediments to winning teamwork or trusting relations with online virtual peers.

Walther and Bunz (2005) suggest "analyses of computer-mediated communication groups achieved more positive levels on several dimensions of interpersonal communication than did face-to-face groups." Walther's controlled study analysis aptly shows that electronic communication can promote surprisingly positive communication among online peers. Whereas some research on the interpersonal effects of computer-mediated communication has deemed it "impersonal, task-oriented, and hostile," Walther and Bunz's ongoing research reveals the existence of warm associations with progressively positive interpersonal "adjustments over time" (p. 186). Researchers also suggest that computer-medi-

ated communication media are better suited for long-term interaction rather than short-term meetings.

Walther and Bunz (2005) continue to support computer-mediated communication's demonstrated likelihood for long-term organizational trust, warmth, attentiveness, concern, and other interpersonal dimensions that positively affect working relationships and organizational outcomes. Furthermore, professional and online learners participating in various virtually networked communication activities frequently report appreciation for flexible scheduling, which, in turn, better accommodates the multidimensional aspects of real life. Additionally, opportunities for professional development and higher education are offered online for those who could not participate otherwise.

Trust can be divided into at least two areas. Cognitive trust includes competence, reliability, integrity, and professionalism. Affective trust involves caring, emotional connections, and bonding among team members (Kanawattanchai & Yoo, 2005). Teams with high levels of trust frequently engage in continuous and recurrent communications while focusing on relevant issues and adequately socializing during the project's early stages. Successful teams exchange background and personal information and, albeit virtually, are truly interested in developing relationships with their collective team members.

The Advantages of Virtual Teams

Virtual workplaces are not environments for dominance or control but trust and empowerment—organized zones for creative endeavors, networking, forming relationships, experimentation, education, personal development, and communal encouragement.

Blackwell (2006), author of *Working with Virtual Teams*, reveals that working with virtual teams involves rethinking the nature of work as well as the development of a thorough understanding of team dynamics. Blackwell defines a team as "a group of people with common interests and at least one common goal" (p. 1). To be successful, there must be a collaborative spirit and a strong commitment to share work and responsibilities. According to Blackwell, "virtual teams exhibit all the confusion and organizational predicaments of face-to-face teams, plus they bring their own set of concerns" (p. 1). Two benefits of virtual teams include reduced cost in travel and conference fees, and work can continue around the clock if you have people in different time zones.

To help virtual teams become successful and positive, it is important to identify challenges and obstacles in addition to establishing common ground, setting priorities, and developing trust as the means for effective communication. Trust is very important to team effectiveness, yet it may be difficult to establish in virtual environments. Nevertheless, it is not impossible. Blackwell (2006) strongly recommends that organizations "consider every opportunity for kick-off and in-person meetings" (p. 2), to establish the social capital and interaction so vital for future team success. He also suggests that virtual teams never be more than 20 people and that they never address more than two anticipated or expected outcomes at one time. In summary, Blackwell believes virtual teams are great for companies and individuals and that virtual teaming is definitely worth the effort.

Essential elements of trust include feelings of inclusion and belonging in addition to overall acceptance and receiving respect from others. When expectations are clear and protocols are defined, contributory members can confidently acknowledge personal experiences and apply their expertise to tackling the underlying tasks.

Getting along well with people is a powerful tool for proactively bringing about positive interaction but particularly in virtual learning environments. Distances physically separating diverse learners and instructors make trustworthy cohesive teamwork imperative to a quality virtual experience.

Peter Andrews, writing for IBM, identifies three elements of trust, which are necessary for people to work as a team. The three elements he discusses are value, commitment, and thoroughness. Andrews states that virtual teams have many benefits in bringing people with unique talents together, providing international perspectives, reducing travel dollars, and time. He believes that unless trust is established teams run the risk of becoming dysfunctional (Andrews, 2006).

Developing Models of Trust

In early childhood, we develop relationships that have direct bearing on our ability to trust and engage in group bonding. By examining common traits of adults who trust, we are better prepared to identify successful candidates for participation in virtual learning environments.

Studies reveal secure adults possess high personal esteem, are comfortable developing trust and closeness, and function better in relationships with "secure models of self and others." Contextual attachment forming and trust is difficult for lower-esteem, insecure adults with inherent tendencies toward anger, resentment, suspicion, defensiveness, and/or destructiveness daring conflict (Berson, Dan, & Yammarino, 2006, p. 169).

Low-esteem behaviors, which can occur among well-educated, knowledgeable people, can sabotage satisfactory relations at every level and create dysfunctional groups. In semi-engaged virtual media education at a distance, it is challenging to observe what an encouraging peer, mentor, or instructor can do to mitigate negative tendencies. Some positive and proactive responses to conflict include (a) talking, (b) clarifying information, and (c) finding common ground. To determine which works best, members can agree to disagree, look for other options, create a third alternative, negotiate, compromise, and/or follow two paths. Outcomes are more positive if virtual teams are comprised of individuals with relatively high levels of confidence and self-esteem. People must desire to make this work and be equally willing to absorb a certain level of risk and uncertainty. Ideally, team hierarchies are relatively flat. Designated leaders must set agendas, develop project criteria, facilitate virtual meetings, assign tasks, and delegate responsibilities. A removal or rotation mechanism should be in place for any team member who firmly believes the process cannot and will not work. It is vital that the facilitator, as well as team members, become voices of persuasion not voices of protest. A proactive and positive approach is clearly more effective over the long-term.

Commitment implies realistic expectations. Relevant questions include: What is needed, by whom, and when? Moreover, are milestones in place, meetings scheduled, and can the work be done in the allotted time? Thoroughness entails checking and double-checking everything. Listening for concerns and questions and checking on the progress of team members is continual. In the virtual working world, the most common response to serious problems is silence.

Once trust is established on a virtual team, interaction is generally smooth. As teammates cultivate positive attitudes and become productive, team members will grow in capabilities and confidence.

Techniques for Creating Atmospheres of Trust

Virtual trust is only borne through good working relations and communication. The following conditions tend to promote trust:

- Focus on assignments and tasks
- Do not allow blaming, criticizing, or finger pointing

- Advance team missions with constructive behaviors
- Reflect on the significance of trust
- Enrich the atmosphere with loyalty and trust
- Establish clear conduct, communication, participation, and privacy policies
- Do not allow members to procrastinate, make excuses, or delay action
- Value your peers in the same manner in which you desire to be valued
- Be courteous and responsible

For instigating critical and positive thought, virtual team members should draft a learning agreement. The expectations should be clear, but, at the same time, flexible and emerging.

Value to the team should be evidenced by members sharing an updated resume, Web page, or vitae which clearly validates their value to the team by highlighting experience, qualifications, and past successes. During the first group meeting, members should be introduced to each other, describe their jobs and what they want to accomplish, in addition to discussing member roles, responsibilities, and expectations.

A community learning agreement, or a plan of action, is useful for all team projects including virtual teams. Although virtual formats promote flexibility, the agreement should be written. Some elements of the agreement helping to promote harmony and success include the empowerment of all members, making it a mutually good experience for everyone, encouraging people to have some fun, play fair, and strive for excellence and effectiveness. Enjoyment is encouraged along with creativity and humor. Within the spirit of collaboration and cooperation, each team member should be empowered to make decisions, experiment, take risks, and try new things. Honor the uniqueness of each member and assign each person areas of responsibility where they can excel. Strive for authenticity, focus, and respect.

The intent is to release the human potential in all members and encourage them to speak up and to share their ideas without fear of ridicule or rejection. Encourage all members to listen for understanding, to avoid being judgmental, and to seek consensus, enjoyment, and results which exceed expectations.

Participants are encouraged to practice shared leadership and co-responsibility. They must honor their agreements. Everyone is expected to do their best work while working collaboratively and persisting until a successful outcome is achieved.

In virtual environments, some leaders express concern that, because peers and facilitators do not have regularly scheduled, direct one-on-one interaction, communication is negatively impacted and trust is compromised. This is not necessarily true. Research is beginning to reveal that analyses of computer-mediated communication groups achieved more positive levels on several dimensions of interpersonal communication than did face-to-face groups (Walther, 2005). Walther's analysis and controlled study shows that electronic communication can promote some surprising, positive relational communication among people, whereas some research on the interpersonal effects of computer-mediated communication has found it to be impersonal, task-oriented, and hostile. Walther's work shows warm personal relations and the gradual adjustments in interpersonal relations over time. Walther's research indicates that computer-mediated communication may be better suited to long-term interaction rather than short-term meetings. Computer-mediated communication (CMC) may be a more satisfying medium when it is used for task forces or teams which have a longer-term association. Walther (1995) continues to support CMC and reports that it has the potential to convey organizational trust, warmth, attentiveness, concern, and other interpersonal dimensions positively affecting work relationships and organizational outcomes.

From another perspective, professionals and learners participating in virtual networks frequently report that they appreciate the flexibility, which, in turn, better accommodates the multi-dimensional aspects of real life.

The Magnitude of Trust

Let's examine what we know about trust, convivial teamwork, and computer-mediated study. How do you define trust? What skills do you bring to a remote team? Are your values in concert (Andrews, 2006) with your virtually-networked colleagues? Studies suggest that team trust must be developed immediately "within parallel, project, action, and networked teams" (Team building, p. 3a). Additionally, three factors are shown to help build trust: competent performance, integrity, and, concern for the welfare of others (Team building, p. 3b).

A virtual team requires that someone is appointed as the director or leader. The leader is a facilitator, not a dictator. Successful practices and processes emphasize that all team members are equal and valuable and the team is only as strong as its individual members.

The environment is not one of dominance and control. The atmosphere promotes trust and empowerment, networking and forming relationships, and encourages creativity and experimentation. Team members are encouraged to clarify their personal values, to develop shared values, and to envision the future by imagining exciting and enabling possibilities.

The group is asked to develop a common vision and to search for opportunities by seeking innovative ways to change, grow, and improve. Participants experiment and take risks by recognizing small wins and learning from mistakes. Collaboration is fostered by promoting cooperative goals and building trust, strengthening others by sharing power, recognizing contributions by showing appreciation for individual excellence, and celebrating victories and creating a spirit of community.

A feeling and belief is developed that members are all in this together and they will help each other and have fun and success along the way.

Those engaged in the development of trust and cooperation must ask themselves if they are listening to all voices, valuing all visions, and expressing genuine interest in the team's diversities? Advanced interpersonal skills help to maintain workgroup trust while valuing differences and areas of expertise. The following example of establishing, building, and maintaining trust is offered for analysis, discussion, and imagined application in a virtual environment:

Kevin was the first to go through the formal introduction process. As instructed by the program facilitators, he shared introductory information: his name, what he did, something of which he was proud, and something about himself that was not on his resume. He responded to the last prompt by saying, "This may be a bit controversial, but think of a cookie that has only one chocolate chip in it. That's me. I'm usually the only chocolate chip in the cookie. You can make of that what you want." (Livers & Caver, 2003, p. 1)

Now, consider the following:

- With no visual image or verbal interface, how would you perceive Kevin's introduction?
- If working in a virtual environment, how would you respond to Kevin's comment?
- Was this introduction conducive to building trust?
- Would you trust Kevin?
- Was Kevin playing a race card? Or, was he merely making a statement about who he is?

Kevin later admitted that, through his presentation, he was purposely pushing his colleagues to acknowledge his differences. After getting to know and trust him, one colleague described Kevin as a great guy.

Establishing Trust

Domestically and globally, electronic communication is thriving. By covering wide geographic areas, virtual learning is becoming an efficient cost effective medium for mass delivery of information and curriculum. In computer-mediated environments team trust is constructed from and supported by virtual etiquette including commitment, timely responses, respect, inclusion, camaraderie, open communication, credibility, and trustworthiness, to name a few.

Timely responses imply quick turn around time to all postings and discussion. Respect and inclusion means that every team member will be valued and appreciated. Camaraderie facilitates trust and quality work sessions enhanced by effective communication strategies that promote credibility and trustworthiness.

Building Trust

In typical learning environments, trust, a primary tenet for successful teamwork, evolves and grows like a good friendship. Collective adaptations of trust building theory include familiarity-based trust and relational-based trust.

Familiarity-based trust grows as interaction increases, comfort levels rise, and team member behaviors and performance levels become predictable. Relational-based trust results as discussions progress and teammate camaraderie evolves into closeness and shared feelings. Bonding occurs.

Maintaining Trust

Above all, in virtual learning environments, adherence to group rules, statements of purpose, respect, privacy, and confidentiality are central to establishing and maintaining trust. Virtual colleagues should never discuss teammates or reveal personal stories or identifiable anecdotes, nor should anyone outside the workgroup be allowed to participate in team projects for ulterior motives.

The Learning Organization

To achieve success, virtual teams should develop into learning organizations. For ideas and strategies, we will examine the work of Peter Senge. In his book, *The Fifth Discipline: The Art and Practice of the Learning Organization,* Senge (1990) identifies elements and processes that enhance learning organizations including:

- Personal mastery
- Shared vision
- Systems thinking
- Mental models
- Team learning

Personal mastery is the continual clarifying and deepening of one's personal vision and responsibility about obvious and perceived creative tensions motivating us to change. Virtual teams are more successful if individual members personally commit to the team and are willing to become active members sharing intrinsic responsibilities necessary to achieve successful outcomes.

Personal mastery reaches beyond competence and skills. It involves frequently clarifying what is important to us and continually learning how to see current realities more clearly, according to Senge (1990, p. 141). People with high levels of personal mastery operate with a sense of purpose. For them, visions are a calling vs. good ideas. They function best in continual learning modes but never completely "arrive" at the optimum (Senge, 1990, p. 142). Senge suggests that as individuals practice the discipline of personal mastery, they begin to (a) integrate reason with intuition, (b) continuously see and acknowledge more of their connectedness to the world, (c) exhibit increased compassion, and (d) commit to the whole (Senge, 1990, p. 167). Team members are comfortable

becoming part of something bigger and more significant than self-interests.

Shared vision is the discovery of commonly held organizational views for the future by cultivating genuine commitment rather than merely compliance. It is very important that virtual teams develop an exact shared vision about what they need to accomplish and how. A shared vision is not an idea, rather it is a powerful force in people's hearts. It may be inspired by an idea but it is the answer to the question, "What do we want to create?" (p. 206). When people truly share a vision, they are bound together by a common aspiration. Shared visions derive power from mutual caring and commitments. Because increasing clarity, enthusiasm, communication, and commitment develops, visions grow and spread. Enthusiasm builds and people get excited.

Systems thinking include complex participatory interrelationships and assigned tasks. Virtual teams must assign responsibility and tasks to individual members which then must be brought back and integrated into the total group effort. A systems approach is important to the division of labor followed by successful integration of the parts into a meaningful whole. Systems thinking is a conceptual framework, a body of knowledge and tools that have been developed to make the big picture clearer and to help develop strategies for effective change over time. People tend to blame outside circumstances for problems. Someone or something else is to blame. Systems thinking show us there is no outside. The cause of the problem is part of a single system. The cure lies in your "relationship with your enemy" (p. 67).

Systems thinking is a discipline for seeing a whole: a framework for seeing interrelationships rather than things. It is a discipline for seeing the structures that exist in complex situations. By seeing the whole we learn better how to create the future.

Mental models determine how we make sense of the world and how we take action. Mental models affect what we do because they affect what we see. Mental models are internal images of how the world works and provide us with images that limit us to familiar ways of thinking and acting. Individuals and teams need to examine deeply held assumptions, generalizations, and images that influence how we view the world and respond to various situations, and the evolution of our ever emerging vision for the future. The members of virtual teams can benefit from Senge's (1990) work on examining deeply held assumptions and taking a close look at generalizations and images that influence one's view of the world and predictable responses to various situations. Healthy organizations are those that can bring people together, face-to-face and/or virtually, to develop the best possible mental models for facing any situation at hand.

Team learning is the dialogue and continuous learning in which members move out of merely self-interest into collaboration that embraces the common bond and organizational alignment towards reaching goals. It is the process of aligning and developing the capacity of a team to create the results its members truly desires. It builds on personal mastery and on the discipline of developing shared vision. Mastering team learning has never been more important. Team learning is an essential element of virtual teams if they are going to bond together and produce good work. Team learning involves mastering the practices of dialogue and discussion along with active listening to one another and the suspension of one's individual views. Team members must see each other as colleagues in a mutual quest for deeper insight and clarity. Team learning must be practiced.

Along the same lines, Kouzes and Posner's (2002) book, *The Leadership Challenge*, identifies five practices of exemplary leadership, which provide a model for virtual team building and collaboration. These include:

- Challenging the process
- Inspiring a shared vision
- Enabling others to act
- Modeling the way
- Encouraging the heart

Challenging the process involves looking for opportunities to shake things up and change the status quo. Change agents investigate and look for innovative ways to improve organizations and, for these professionals, experimentation and risk taking are common operational behaviors. Inevitably, mistakes are made but are also considered learning opportunities. The same approaches to innovation and change exist within virtual teams.

Inspiring a shared vision implies the creation of transformational strategies, unique and ideal images of what the organization can become. Through magnetism and/or quiet persuasion, team leaders enlist others to share visions and identify with exciting future possibilities. When working with virtual teams, sharing an optimistic vision about the future is desirable, as well.

Enabling others to act builds spirited teams. Each team member has meaningful responsibilities, which are accomplished in an atmosphere of mutual respect, trust, and human dignity.

Team members strengthen one another in ways that make each person feel capable, powerful, and contributory to the tasks.

Modeling implies that team leaders create standards of excellence and set an example for others to follow. Team outcomes are more successful if small achievable goals are set so team members enjoy small wins as they work toward larger objectives. When working with virtual teams, creating opportunities for success and celebration is important to maintaining motivation and momentum.

Encouraging the heart means that all team members should feel important, appreciated, and acknowledged by recognizing and celebrating contributions. By sharing in all rewards, teammates experience a sense of accomplishment, and are proud to be a group member. In both virtual and traditional environments, helping peers to feel good about themselves and sharing a sense of mutual pride are important elements of team success (Kouzes & Posner, 2003).

CONCLUSION

When using virtual environments, it is important to promote effectiveness as well as efficiency. Although deemed efficient, technologies are not perceived as effective unless utilization leads us to do the right things in the right ways. Professional telecommuting is emerging as a satisfactory working arrangement for individuals choosing to reside geographically outside a company's commuting areas. This is also a feasible alternative for professionals with disabilities who have difficulty commuting and working fulltime outside the home. Cost savings, improved time utilization, and increased productivity support the virtues of the virtual workplace.

Bad network connections, older machines, unexpected delays, and participants' ignorance of process are problematic but not insurmountable obstacles. The potential for data exchange and learning are significant. How will technology evolve and change the future? The possibilities are limitless.

Quoting Damer (2001), "The premises of an argument are those statements that together constitute the grounds for affirming the conclusion" (p. 11). To affirm a positive conclusion, concerted efforts were made to present both sides of this argument: the many conflicting and sustaining opinions surrounding the merits and weaknesses of virtual trust.

It is relatively easy to unite individuals, organizations, and institutions into virtual networks.

Purposeful learning communities endorse attentive listening, sharing information, mutual scholarship, and meeting expectations. Virtual teammates work well when all participants stimulate and encourage innovation, search for connections, probe into unknown territories, and become part of something bigger and more powerful than themselves.

To protect the professional integrity of the virtual workplace, drafting a team agreement or contract to clearly define protocols, conduct, responsibilities, and structure provides the foundation for a pleasurable working environment. Anyone feeling uncomfortable, disengaged from the team, or convinced the project cannot and will not work in a virtual environment should immediately request a transfer from the team. This should always be an option.

Virtual teams experience the same triumphs and problems encountered in face-to-face teaming. Individual behaviors follow the person. Regardless of locale or team construction, some persons are simply not team players and prefer independent working conditions.

In closing, consider these words of wisdom:

- The uniqueness of each member must be honored
- It is wise to assign areas of responsibility where persons can excel and build confidence
- Strive for authenticity, focus, and respect
- Release the human potential in all members and encourage them to speak up and share their ideas without fear of ridicule or rejection
- Encourage all members to listen for understanding, to avoid being judgmental, and to seek consensus, enjoyment, and results which exceed early expectations

Participants must practice tenets of shared leadership and co-responsibility. Teammates count on their virtual peers to honor agreements, do their best work while collaborating, avoid all tendencies toward self-serving conduct, and persist until a satisfactory outcome is achieved.

Trustworthiness or lack thereof is prevalent in every professional context. Tompkins (2003) poses the question, "Can we trust the truthfulness of those we never meet" (p. 1). Research suggests that virtual networking is progressively improving in technology, process, function, communication, and relational trust. Whether we establish, build, and maintain trust with others is largely an individual choice. Go for it!

REFERENCES

Andrews, P. (2006). Virtual teams: Establishing trust is key to successful teamwork. *IBM On Demand Business*, 1-3. Retrieved February 25, 2007, from http://www-935.ibm.com/services/us/imc/ondemand/business/trust_building.html

Team building. Theoretical overview of trust building and teamwork in online environments, pp. 1-8. Handout.

Berson, Y., Dan, O., & Yammarino, F.J. (2006a-d). Attachment style and individual differences in leadership perceptions and emergence. *The Journal of Social Psychology, 146*(2), 165-182.

Blackwell, J. (2006). Working with virtual teams. Retrieved February 25, 2007, from http://www.management-issues.com/display_page.asp?section=blackwell&id=2975

Damer, T.E. (2001). *Attacking faulty reasoning: A practical guide to fallacy-free arguments* (4th ed.). Canada: Wadsworth/Thomson Learning, Inc.

Hossain, L., & Wigland, R.T. (2004). ICT enabled virtual collaboration through trust. *Journal of Computer-Mediated Communications, 10*(1), 1-22.

Kanawattanachi, P., & Yoo, Y. (2002). Dynamic nature of trust in virtual teams. *Sprouts: Working Papers on Information, Environments, Systems, and Organizations, 2*(2), 42-58. Retrieved February 25, 2007, from http://sprouts.case.edu/2002/020204.pdf

Kouzes, J.M., & Posner, B.Z. (2002). *The leadership challenge* (3rd ed.). San Francisco: Jossey-Bass.

Livers, A.B., & Caver, K.A. (2003). *Leading in black and white: Working across the racial divide in corporate America*. San Francisco: Jossey-Bass.

Senge, P.M., (1990). *The fifth discipline: The art and practice of the learning organization*. New York: Doubleday Currancy.

Tompkins, P. (2003). *Truth and trust in cyberspace*. Paper presented at the Conference on Communication Ethics and Virtual Reality. Co-sponsored by Brigham Young University, University of Illinois, and WACC.

Walther, J.B. (1995). Relational aspects of computer-mediated communication: Experimental observations over time. *Organization Science, 6*, 186-203.

Walther, J.B., & Bunz, U. (2005a-k). The rules of virtual groups: Trust, liking, and performance in computer-mediated communication. *Journal of Communication, 12*, 828-846.

Chapter V
Media and Familiarity Effects on Assessing Trustworthiness:
"What Did They Mean By That?"

Mark A. Fuller
Washington State University, USA

Roger C. Mayer
University of Akron, USA

Ronald E. Pike
Washington State University, USA

ABSTRACT

This chapter explores the role media effects and familiarity play in the development of trust in CMC environments. As team members interact with one another via technology, each team member assesses information and makes assessments about the trustworthiness of their teammates. Such trustworthiness assessments are known to influence trust, a factor that has been established to have significant effects on the functioning of teams. This research uses media synchronicity theory and the concept of interpersonal familiarity to examine virtual team interactions and the formation of trust. Implications are drawn for researchers and managers as they seek to understand how teams operate in virtual environments.

INTRODUCTION

U.S. corporations commonly use some form of team structure in their organizations. In a global environment, they increasingly have to use more distributed teams using computer-mediated communication (CMC). CMC teams, sometimes referred to as virtual teams, perform substantial portions of organizational work (Townsend, De-Marie, & Hendrickson, 1998, p. 18). They have the potential to draw upon the skills of a widely dispersed workforce and can function better in today's complex and dynamic environment (Townsend et al., 1998, p. 23).

Past research has acknowledged that a crucial determinant of successful team functioning is trust. Trust reduces the uncertainty that permeates technology-mediated environments (Jarvenpaa & Leidner, 1999, p. 792) by, for example, minimizing the effects of misunderstandings that arise from communication delays (Jarvenpaa, Shaw, & Staples, 2004, pp. 262-263). Further, the geographic and cultural diversity often characteristic of CMC-based teams creates additional complexity and enhances the likelihood of misunderstandings, making trust even more important. Overall, trust helps team members cope with uncertainty and maintain positive attitudes about team members in computer-mediated environments.

Trust is most commonly regarded as an interpersonal phenomenon, created (or diminished) as one party forms impressions about the trustworthiness of another (Mayer, Davis, & Schoorman, 1995). In large part these impressions are formed based on the interactions of the parties. In CMC (i.e., virtual) environments, the behaviors that give rise to these impression of trustworthiness are observed through cues conveyed by the media used for communication. The specific capabilities of the communications media used can thus play a major role in determining the types of cues that are conveyed between team members. A medium, in other words, has the potential to play a crucial role in trust formation.

The purpose of this chapter is to explore how media characteristics influence the development of trust and how interpersonal familiarity may influence these media effects. More specifically, employing media synchronicity theory (Dennis & Valacich, 1999), this chapter discusses how different media capabilities, including feedback immediacy, symbol variety, parallelism, rehearsability, and reprocessability, affect the ability of team members to form appropriate levels of trust,

that is, trust based on accurate impressions of how trustworthy other members of the team are. Further, this chapter also explores how media can have potentially detrimental effects on trust formation by *diminishing* the ability to form accurate impressions and creating overconfidence about the accuracy of incorrect impressions. Finally, we also explore how a medium's influence on enhancing or diminishing the accuracy of trustworthiness perceptions depends on the concept of interpersonal familiarity.

As media capabilities continue to evolve and as teams continue to use these media to accomplish tasks, understanding how trust is formed in teams will be an important topic for organizations. In addressing this topic, we first turn our attention to a discussion of trust and trustworthiness, which will serve as a foundation for examining how media capabilities can influence trust formation between teammates in CMC environments.

TRUST

Scholars in the organizational sciences have been examining trust for about half a century (Deutsch, 1958). Researchers have defined the term in a variety of ways, leading to some confusion with the construct. In an attempt to clarify the construct and provide a solid basis for future research, Mayer et al. (1995) developed a model of trust, its causes, and its outcomes (see Figure 1). Mayer et al. defined trust itself as a "willingness to be vulnerable to the actions of another party based on the expectation that the other will perform a particular action important to the trustor, irrespective of the ability to monitor or control that other party" (p. 712). Trust is distinct from the concept of trustworthiness because trustworthiness involves the trustor's perceptions of the trustee's ability, benevolence, and integrity. Trustworthiness thus refers to the sum of the trustee's attributes that are assessed by the trustor in deciding how much trust is warranted.

Figure 1. Trust and trustworthiness

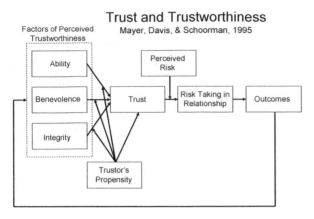

Early on, researchers recognized that trust was important in organizations, suggesting that organizations whose members trust their leaders will perform better than those with less trust (Argyris, 1964). Research by Davis, Schoorman, Mayer, and Tan (2000) found similar results, suggesting that from a financial standpoint, organizations whose general management was more trusted outperformed those whose general management was less trusted.

The factors that drive these performance improvements take many forms. For example, a meta-analysis by Dirks and Ferrin (2002) illustrated that numerous researchers have found that higher levels of trust bring about more organizational citizenship behavior (OCB), that is, helping behaviors directed either at other employees or at the organization itself. Mayer and Gavin (2005) found that when employees trusted their managers, they were more able to focus attention on the organization's work, providing one explanation for increased OCB and organizational performance. A number of scholars have provided evidence that trust fosters cooperation, which is crucial for collaboration (Mayer et al., 1995; McAllister, 1995). Thus, a growing body of evidence supports Argyris's claim that higher levels of trust are helpful to the organization.

Trust is also crucial in business because it helps coworkers and organizations work together

effectively (Bos, Gergle, Olson, & Wright, 2002), particularly when uncertainty is high. When one party cannot effectively control the behavior of another, trust provides a means of coping with the resulting uncertainty. A key issue when examining the effects of trust on performance is the presence of risk, specifically that the trusting party must have something of value at stake. While this risk is often borne by a manager (Mayer & Gavin, 2005) or subordinate, it can also arise from team members being jointly responsible with others to accomplish some task or produce some product or outcome. In such a team environment, one's ability to perform is dependent on cooperation from and performance of others in the team.

In building their model of trust, Mayer et al. (1995) found that three factors stood out as crucial antecedents to the development of trust, and thus parsimoniously captured the concept of a trustee's trustworthiness: ability, benevolence, and integrity (ABI). Ability represents the skills or competencies that enable someone to be helpful to the trustor within a certain domain. Benevolence is the "extent to which a trustee is believed to want to do good to the trustor," and thus represents the trustee's goodwill toward the trustor. Benevolence is more commonly found in well-developed relationships, although even after a lengthy relationship many parties do not develop a perception that the other is benevolent toward them. Finally, integrity is the extent to which the trustee adheres to a set of principles that are acceptable to the trustor. Subsequent empirical research has supported the validity of the model (Mayer & Davis, 1999; Mayer & Gavin, 2005).

For the purposes of examining trust in these different contexts, Mayer et al.'s (1995) model of trust considered not only factors about the trustee (trustworthiness attributes) but also attributes of the trustor. Based on the work of Rotter (1967), they pointed out that a trusting party has some relatively stable level of trust in others in general. Lacking specific information about a trustee, initial trust is primarily driven by this general willingness to trust others, which Mayer et al. (1995) termed propensity to trust. Initial trust is a starting point when there is no (or little) information on which to base perceptions of the trustee's ability, benevolence, and integrity. Thus, a trustor's initial level of trust will be largely dependent on that person's propensity level.

Trust is essential for determining performance in collaborative relationships. As Paul and McDaniel (2004) state, "a direct link between trust and collaborative relationship performance exists; once the need for collaboration is established, trust becomes the salient factor in determining performance" (p. 185). Jones and George (1998) posit that trust promotes seven distinct processes that can foster interpersonal communication and teamwork: flexible and broad role definitions, a communal orientation, high levels of confidence in others, help-seeking behaviors, free exchange of knowledge, subjugation of personal needs and ego, and high levels of involvement in the activities of others.

In virtual collaborations, trust is likely to be particularly important, for collaboration requires both parties to enter into the process with a willingness to open themselves to one another and cooperate in executing a task or solving a problem (Jarvenpaa, Knoll, & Leidner, 1998). Trust thus becomes the glue that binds virtual team collaborators together by engendering faith that members will contribute and not exploit dependencies created in the process of completing tasks (Brown, Poole, & Rodgers, 2004). Given the importance of trust for the effective functioning of virtual teams, understanding how virtual team members form trustworthiness evaluations is critical. Given that such perceptions are formed while participants interact using different types of media, and that these media influence what a team member is (and is not) able to discern about other group members (i.e., cues), understanding how trustworthiness perceptions form in such environments may not be straightforward.

In summary, the formation of trust in a relationship depends on an evaluation of another across the three primary domains of ability, benevolence, and integrity. Initially, to move beyond a level of trust based only on propensity to trust, data for this assessment may come from group membership, credible communications from others about the trustee, and other indirect information sources. Further development of trust thus requires the trustor to collect data directly through experience based on interactions between the parties. When two parties are collocated and interact frequently, contextual cues such as the timing of responses (e.g., immediate vs. after a long pause), facial expressions, tone of voice, posture, and a host of other cues provide a great deal of information to the trustor about how to interpret the person's actual words. Without these cues, it becomes much more difficult not only to interpret the meaning of the communication but also to be confident that the communication was accurately received. Understanding the effect that media has on the presence, or absence, of such cues in virtual groups presents a significant challenge to the process of building trust in a virtual environment. In the section that follows, we consider recent work on identifying the important dimensions along which media vary in their ability to provide cues between communicators.

MEDIA CHARACTERISTICS

Cues that help us form perceptions of other people are influenced by the media being used for communication. In this section, we discuss several theories related to the effects of media on communication, with particular attention to how media may enable or constrain contextual information being passed from person to person. This provides a foundation for our discussion of how various media characteristics affect the development of trustworthiness perceptions, and thus downstream trust.

Media richness theory (Daft & Lengel, 1986) is one of the predominant theories explaining media effects over the last 20 years. Daft and Lengel proposed that different communication media have differing capacities for facilitating communication and understanding. The authors proposed a media richness hierarchy that categorizes different media according to their degrees of richness. Components of media richness include the number of cues available, capacity for immediate feedback, channels utilized, language variety, and personalization. Face-to-face was thought to be the richest medium because it provides body language, tone of voice, and so forth, and uses natural language. Following face-to-face communication in the media richness hierarchy are such media as telephone, electronic mail, letters, reports, and fliers. Media richness theory suggests that people choose media that match the particular task they seek to accomplish. When the medium does not match the situation, the communication may be misinterpreted or ineffective in relation to its intended purpose (Trevino, Lengel, Bodensteiner, Gerloff, & Muir, 1990; Trevino, Lengel, & Daft, 1987). Although media richness has been challenged in the literature, the theory has continued to be influential and its underlying tenets still have support (Kahai & Cooper, 2003).

While a number of researcher have investigated media richness theory (Daft & Lengel, 1986; Kraut, Galegher, Fish, & Chalfonte, 1992), the theory has typically gained support when it is applied in comparing "traditional" media such as face to face, telephone, and letters (Lengel & Daft, 1988; Russ, Daft, & Lengel, 1990). When used to study newer forms of media, such as e-mail or voice-mail, empirical findings have been inconsistent (Dennis & Kinney, 1998; El-Shinnawy & Markus, 1998, p. 244).

These inconsistent findings have stimulated newer approaches to understanding media effects on group communications, including the idea of channel expansion theory (Carlson & Zmud, 1999). Advocates of channel expansion theory (which builds on media richness theory) contend that media capabilities are perceived differently in different situations. Four relevant experiences that affect media perception include "experience with the channel, experience with the message topic, experience with the organizational context, and experience with communication coparticipants" (Carlson & Zmud, 1999, p. 155). As an example, if a user routinely uses e-mail in communicating with a specific individual, then e-mail over time becomes a richer media, allowing the user to have a richer and deeper understanding of the message content than would be predicted by media richness alone.

It has been well established that text-based CMC, such as e-mail or chat systems, prevents the exchange of some nonverbal cues present in face-to-face (FtF) situations. This has led researchers to suggest that associated impression building and relational development may be affected (Kiesler, Siegel, & McGuire, 1984; Siegel, Dubrovsky, Kiesler, & McGuire, 1986). Early research on this topic suggested that people communicating using means other than FtF were limited in their ability to form impressions (Kiesler, 1986; Siegel et al., 1986). Other research, however, revealed that the use of CMC simply changed the temporal frames within which impressions and relationships were formed (Walther, 1993; Walther & Burgoon, 1992). For example, Walther's (1992) social information processing (SIP) theory of CMC argues that communicators adapt their relational behaviors to the cues that a particular CMC context can provide.

More recent work on media has examined the specific capabilities of the specific medium itself. Media synchronicity theory (Dennis & Valacich, 1999) provides the notion of communication effectiveness, which "is influenced by matching the media capabilities to the needs of the fundamental communication process, not aggregate collections of these processes (i.e., tasks) as proposed by media richness theory" (p. 1). Media synchronicity theory argues that group communication processes are composed of two primary processes: conveyance (conveying or sharing information) and convergence (arriving at a mutual understanding). Media synchronicity investigates information-processing capabilities by studying the five key media capabilities: feedback immediacy, symbol variety, parallelism, rehearsability, and reprocessability (the description of each of these capabilities is shown in Table 1). We contend that these five capabilities affect the team member's ability to form accurate perceptions of others. This chapter explores these effects and their implications for understanding the complex interaction between media capabilities and the development of trustworthiness perceptions.

Feedback immediacy can vary greatly depending on the media choice. In terms of feedback immediacy, it could be argued that FtF and telephone conversations are very similar. Both give participants the ability to respond immediately and even break into the middle of a sentence to respond. Postal service, on the other hand, might require a week to get feedback. Other technologies such as e-mail fall between these two. Even within a given technology (e.g., videoconferencing), feedback immediacy may differ as the distance between parties increases. This can easily be seen even during news programs, where live video hookups connecting very distributed participants may have significant delays. These delays have the potential to change the structure of the interac-

tions between communications participants, for example, prompting communicators to ask several questions at once, and at the same time be more tolerant of pauses during the interaction.

Even though telephone and FtF communication are similar in terms of feedback immediacy, they may be dissimilar in terms of other media characteristics, such as symbol variety. While the telephone relies completely on spoken words, FtF conversations also allow the use of facial expressions, posture, hand motions, and other ways of conveying information. Further, within a given medium, symbol variety may vary. As an example, all e-mail is not equal. Depending on the e-mail server and e-mail client, e-mail may or may not facilitate the exchange of sounds, rich text, pictures, and even full-motion video clips. An e-mail message with limited or no visual symbols, for instance, may provides less data for the receiver than a FtF conversation. The reduced data is critical as the receiver has to "process" data into information (Ngwenyama & Lee, 1997).

Parallelism refers to a medium's ability to allow communications to occur in synchronous fashion, for example, by allowing people to effectively "talk" at the same time. Obviously, while parallelism can occur in FtF contacts, it has limits, as people talk over one another and much of the content of the messages may be lost. By contrast, parallelism in computer-mediated communication (e.g., in computer-mediated brainstorming

Table 1. Media synchronicity five media characteristics (Source: Dennis & Valacich, 1999)

Media Capability	Description
Feedback immediacy	Immediacy of feedback is the extent to which a medium enables users to give rapid feedback on the communications they receive. It is the ability of the medium to support rapid bidirectional communication.
Symbol variety	Symbol variety is the number of ways in which information can be communicated—the "height" of the medium—and subsumes Daft and Lengel's (1986) multiplicity of cues and language variety.
Parallelism	Parallelism refers to the number of simultaneous conversations that can exist effectively—the "width" of the medium.
Rehearsability	Rehearsability is the extent to which the media enables the sender to rehearse or fine-tune the message before sending.
Reprocessability	Reprocessability is the extent to which a message can be reexamined or processed again within the context of the communication event.

tools) may allow participants to generate ideas simultaneously but without the loss of the ideas (since they are recorded on the system).

Rehearseability is a medium's ability to allow a communicator to rehearse what he or she wants to say before actually having to send a message. FtF interactions typically limit this ability to rehearse. In contrast, e-mail is generally viewed as a highly rehearseable medium. Many people actually write an e-mail in another application (for example, Microsoft Word) before cutting and pasting the message into their e-mail client. In this way the sender can review and reconsider the message, checking facts and grammar prior to sending it. While the time taken to do these things reduces feedback immediacy, it can also save the sender a great deal of trouble and embarrassment.

Finally, reprocessability refers to a medium's ability to provide the receiver time to review a message, which has long been a significant advantage of written communication over verbal exchanges. In an e-mail conversation, a participant can read and reread the message multiple times in order to ensure full comprehension before replying. Later in the conversation, the participant can refer back directly to text to try to ascertain precisely what was said. FtF and telephone conversations, on the other hand, provide no mechanism for reexamination of the message.

FAMILIARITY AS A MODERATOR OF MEDIA EFFECTS

In the preceding sections we discussed the nature of trust, and how a person's perceptions of trustworthiness in another may evolve as new information is gathered. We also discussed how media may influence the information communicators receive. We now introduce a final factor that influences the effect of media on the interpretation of messages and the subsequent formation of impressions such as trustworthiness about another person.

While much of what is written about the development of trust tends to focus on direct interpersonal interactions between the parties (Lewicki & Bunker, 1996; Mayer et al., 1995), of particular interest in this chapter is the development of trust when communications occur between distributed group members using technology. We contend that the development of trust, like the development of other impressions and aspects of the relationship, is shaped by characteristics of the medium and, of course, by the way participants use the medium. What opportunities and constraints a trustor will have for reevaluating another's trustworthiness depends on how the particular medium allows or inhibits clues and insights about trustworthiness.

However, media characteristics alone do not shape the relationship between the trustor and the trustee. For a trustor to form *accurate* perceptions of the trustworthiness of another person, the trustor needs accurate and sufficient information about the trustee's domain-relevant abilities, benevolence towards the trustor, and integrity. While the characteristics of a communication medium enable or constrain a person's ability to accurately interpret both the meaning of someone's communication and that party's trustworthiness, another factor in play is the level of familiarity between the communicating parties which is shown in Figure 2.

Figure 2. A model of media, familiarity, and trustworthiness

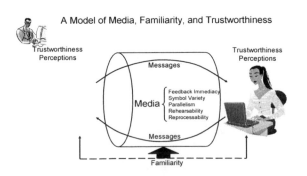

Importantly, familiarity not only refers to a person's personal knowledge about another individual (e.g., background, beliefs, motivations) but also includes broader knowledge about other factors that may enhance the ability to accurately interpret messages (e.g., knowledge about how another person tends to communicate). When familiarity is taken into consideration, media that convey more information and/or cues are not always better. We contend that the relationship between each of the five media characteristics just reviewed and the ability to derive accurate judgments about trustworthiness depends on the extent to which one party understands, or has knowledge regarding, the other. In other words, our familiarity with another person helps us make more accurate attributions about what they mean when they communicate. Thus, the message conveyed (shaped by the media) combined with our level of familiarity with the message's sender should interact to shape the accuracy of our impressions about the meaning behind the message, as well as the sender.

One way to think about familiarity is by employing the ideas presented in self-categorization theory (SCT), which explains how individuals categorize not only themselves but others, at various levels of abstraction (Hogg, 1996; Hogg, Terry & White, 1995; Turner, 1987; Turner, Oakes, Haslam, & McGarty, 1994). For example, an individual might characterize someone else as a co-author, a full professor, an academic, an academic researcher, a Midwesterner, a golfer, and a variety of other characterizations. Each of these characterizations carries certain beliefs and expectations about the message sender, which influence our interpretation of the meaning behind communications. As a particular medium is more or less successful at conveying accurate information about a person, the level of abstraction or reliance upon broad-level referent groups may be replaced by more specific and accurate impressions.

Thought of another way, understanding referent groups may create more "familiarity." This can be seen in cases where both parties are part of a culture, organization, or group that shares a way of seeing and doing things. For instance, Americans are more likely to share an understanding of American customs and approaches to communication and behave somewhat automatically in ways that others within that culture generally understand. For example, idioms or references to classic movies are likely to be appropriately interpreted. This is *less* the case when the communication is between parties from different cultures. It is well accepted that the content of communications from American senders to Chinese receivers is more likely to be misunderstood than within-culture communications. Appropriately interpreting the cues to accurately determine what the communication shares about the trustee's ability, benevolence toward the trustor, or integrity are much more difficult when the parties are from different referent groups, such as cultures.

Because organizations also tend to develop their own idiosyncratic cultures, this reasoning also applies at the organizational level. A message passed between military officers might be more easily interpreted, given the obvious shared referent group, than a message passed from a military officer to non-military personnel. Organizational familiarity also breeds understanding. As an example, one of the authors of this chapter works with (and serves as a faculty advisor for) a PhD student who also happens to be a lieutenant colonel in the military. In a recent meeting of the faculty advisor, the PhD student, and a second faculty member about a new research project, the PhD student offered to do a read-ahead for the faculty advisor. It was only after several e-mails that the advisor eventually understood that the read-ahead was basically a briefing on the second faculty member and his past research, provided by the PhD student so the faculty member would be fully prepared in as efficient a manner as possible. The read-ahead was thus a sign of respect, but one that was not immediately interpreted as such given the difference in organizational backgrounds.

Nationality and culture, as differences that stem from political, sociological, and psychological factors that result from the history, institutions, early life experiences, and education within that nation (Hofstede, 1983, pp. 75-76), have also been shown to affect understanding (Javidan, House, Dorfman, Hanges, & Luque, 2006). High context cultures for instance prefer a communication style that is based upon individuals drawing inferences from non-explicit or implicit information while low context cultures prefer information to be stated directly and exhibit a preference for quantifiable detail (Hall, 1976). IS researchers have recently examined culture in IT and a corresponding IT culture (Leidner & Kayworth, 2006) and cultural impacts on technology adoption (Srite & Karahanna, 2006). In each case the authors found uncertainty avoidance to be a significant issue between cultures. Variance in uncertainty avoidance affects decision-making processes and in turn affects the understanding shared between members of these cultures.

Even if the communicators are not from the same culture, national or organizational, some of the same "familiarity" effect can be derived from one party knowing the culture and traditions of the other. For example, Italians tend to be demonstrative in their communications, while Chinese tend to be more reserved. A Chinese trustor would be less likely to make certain attributions about demonstrative communications from a trustee if he or she knows the trustee is Italian and has dealt with Italians before. While the Chinese trustor is not from the same culture, an awareness of the cultural differences of the trustee can ease the negative attributions made. Without knowledge of relevant aspects of Italian culture, or lacking awareness that the other party was Italian, the trustor is likely to assess trustworthiness less accurately.

Overall, if a particular trustee is familiar to the trustor, the trustor should have a clearer and more specific understanding and framework with which to evaluate the trustee's communications. Such interpersonal familiarity enables a trustor to more accurately interpret the trustee's true meaning. More importantly, it enables the trustor to correctly interpret the extent to which the new communication provides clues that affect the evaluation of the trustee's trustworthiness. Thus, one way that a medium used for communication may affect the formation of impressions relates to the ability of the cues the medium provides to increase familiarity where reliance is less on abstract referent group characterizations and more on the individual themselves. In addition to a particular medium's ability to convey information about another person, a medium may also influence the ability to accurately interpret particular messages. We now turn our attention to this latter point and examine how media characteristics, such as those highlighted in media synchronicity theory, may influence the development of trust.

MEDIA CHARACTERISTICS AND THE DEVELOPMENT OF TRUST

To understand how the level of trust necessary to enable a group of interdependent people to work together effectively develops, we need to understand how team members form accurate perceptions of trustworthiness factors during the course of their work. Serva, Fuller, and Mayer (2005) examined the topic of reciprocal trust formation and found that trust is developed over time based on the interactions (and cues sent) between parties. Reciprocal trust was defined by Serva et al. (2005) as "the trust that results when a party observes the actions of another and reconsiders one's attitudes and subsequent behaviors based on those observations" (p. 627). Thus, accurate perceptions of a party's trustworthiness factors (i.e., ability, benevolence, and integrity) can lead to appropriate perceptions of trustworthiness. It is equally true that inaccurate perceptions of the trustworthiness factors can lead to inappropriate levels of trust. In the next section, we examine how media capabilities—as identified within the media synchronicity theory—influence the accuracy of perceptions of the trustworthiness factors, and thus have the potential to influence trust formation (see Figure 2).

The Effect of Feedback Immediacy on the Formation of Trust

Dennis and Valacich (1999) define feedback immediacy as "the extent to which a medium enables users to give rapid feedback on the communications they receive. It is the ability of the medium to support rapid bi-directional communication" (p. 2). Feedback immediacy is a critical factor in developing trust within virtual teams because developing trust is a reciprocal process that occurs in many incremental steps as people communicate (Serva et al., 2005). Immediate feedback allows for more back-and-forth iterations in the process of developing reciprocal trust. While this is important in any environment, it is particularly important in teams using CMC, for such teams lack the FtF interaction often taken for granted in traditional organizations and, secondarily, because rapid team construction and tight timelines are common within virtual teams (Kasper-Fuehrer & Ashkanasy, 2001). Regardless of the timeline of the project, the absence of feedback immediacy may serve to impede the development of trust, through its affect on trustworthiness impressions. In CMC, for example, the relative speed with which the communicator agrees or disagrees with a statement can be a cue about how to interpret the comment (agreement or disagreement) itself as well as the trustworthiness of the sender (e.g., perceptions of shared values that are by definition an aspect of integrity perceptions).

Jarvenpaa et al. (2004) propose that "initial trustworthiness (trustworthiness perceptions developed before the midpoint of the project) biases the overall view of the other party unless there is distinctive information available that contradicts the view" (p. 254). This would indicate that taking the life of the project to develop accurate perceptions of trustworthiness is inadequate. Accurate trustworthiness perceptions must also be developed very early in the project. Lower levels of feedback immediacy in the communication media may hamper development of these perceptions.

Paradoxically, an increase in feedback immediacy may also, however, have a negative effect on the ability to develop *accurate* perceptions of trustworthiness. Increased feedback immediacy may give the trustor an illusion of understanding that is unwarranted. The trustor may receive immediate feedback, which builds confidence in the trustee, but, depending on other cues a particular media may not be conveying, the trustor may still lack important pieces of information needed to make a well-informed decision. Thus, by relying on feedback immediacy as a surrogate for other cues (e.g., additional cues sent through greater media symbol variety), the trustor may make erroneous conclusions about the sender's trustworthiness.

This relationship, however, is even more complex given our discussion of moderators above. We propose that sharing or understanding the perspective of the trustee (i.e., the concept of familiarity) through the means discussed earlier in this chapter may further influence the actual effects of feedback immediacy on the accuracy of the trustworthiness assessment. For example, returning to our cultural example, the Chinese have a different orientation to time than do people from the United States. Even within the United States, Midwesterners or Southerners have different orientations than do people from the East Coast (cf. the famous "New York minute"). Depending on one's orientation toward time, the trustor will interpret differently a pause or a long pause in a response. Recently one of the co-authors of this chapter was involved in a large-scale virtual team project involving student teams working across continents. One task in a class project involved coming up with an idea for an e-commerce business, an idea-generating task. While the U.S. project team members were actively brainstorming online regarding possible business ideas, Chinese students seemed reluctant to do so, which was interpreted by U.S. counterparts as lack of interest, motivation, or knowledge. After several days of frustration on the part of U.S. students, suddenly a document detailing an e-commerce business idea emerged from their Chinese counterparts. When later asked about their hesitancy to brainstorm, the Chinese students indicated that they were not comfortable with that process, as they believed that ideas perceived to be "silly" would lead to a loss of face. Lacking a full understanding of cultural differences, U.S. students who interpreted the lack of immediate response from Chinese students as indicators that they lacked ability or were uninterested may have formed inaccurate trustworthiness perceptions. From the Chinese students' perspective, the opposite could have been true as well. The immediacy of the U.S. students' responses could well send a negative signal to the Chinese students, who may perceive that their U.S. student counterparts were not careful or thoughtful enough in their responses.

The Effect of Symbol Variety on the Formation of Trust

Symbol variety is "the number of ways in which information can be communicated—the "height" of the medium—and subsumes Daft and Lengel's multiplicity of cues and language variety" (Dennis & Valacich, 1999, p. 2). Increasing symbol variety increases the amount of information that can be transmitted in a given amount of time, and theoretically reduces the amount of ambiguity in the message. By making it possible to send more information, increased symbol variety should increase the receiver's ability to gain accurate perceptions of a sender's ability, benevolence, and integrity.

While symbol variety can increase understanding, and thus more accurate assessments of trustworthiness, one negative consequence of increased symbol variety is that it may give the message receiver a false sense of confidence about the accuracy of their perceptions of a message's meaning and related attributions about the sender. In the case of trust, this can lead to increased confidence in potentially inaccurate perceptions of trustworthiness. As an example, many people are more confident of their ability to "read" a person in face-to-face media environments. However, this confidence may be misplaced if the other party is a practiced liar. In such a case, the message receiver may walk away feeling confident that a person is trustworthy who is in fact not.

Another potential unanticipated problem of increased symbol variety (increasing the amount and type of information a message may contain) relates to forming accurate impressions of another is that a medium that provides more cues may also increase a message's complexity, which may in turn make it more difficult to accurately decode a message's meaning and thus form accurate impressions of another party. An example of this phenomenon might be "mixed" signals, where some cues (e.g., a written message or memo) might speak to a person's trustworthiness (e.g., through expertise displayed in the text), but face-to-face interactions leave a different impression (e.g., where a person may look confused). Increased symbol variety alone, without allowing for other media capabilities such as feedback immediacy, which might allow for probing, may decrease the ability to accurately calibrate the trustworthiness of another party.

Applying the concept of familiarity to our discussion, of the five media characteristics, symbol variety may have the largest interaction with familiarity. For example, without shared or mutually understood cultural reference points, more symbol variety may actually lead to more mixed messages being sent and/or received. Media that carry greater symbol variety where there is not a shared cultural context may provide a greater chance of misattribution of trustworthiness, since more signals will be sent that are apparently contradictory. These contradictory signals may reduce the perception of trustworthiness.

This phenomenon has also been directly experienced by the co-authors. In research collaborations with a Chinese PhD student, a faculty advisor has found that e-mail is a much more effective medium for ensuring understanding and agreement than FtF interactions. During several FtF interactions with the student, the faculty advisor believed that there was a shared understanding about a project, only to later find out that the student was confused. As the faculty member and student continued to interact, c-mail (a medium with typically less symbol variety than FtF interactions) became the preferred medium for ensuring understanding (perhaps because of e-mail's natural advantages in rehearseability and reprocessability, discussed later). This same issue has been documented in cross-cultural negotiations.

The Effect of Parallelism on the Formation of Trust

As discussed earlier, parallelism is the number of conversations that can go on at the same time. An example of a medium that promotes parallelism is an online chat tool, where all parties may submit comments (i.e., "talk") in the same chat room *at the same time*. Essentially, information can be sent to and received from multiple communication partners simultaneously. By allowing for multiple simultaneous transmissions, parallelism reduces some of the process losses caused by turn taking, where ideas may be lost while waiting for others to finish (Dennis, Valacich, Carte, & Garfield, 1997; Valacich, Dennis, & Nunamaker, 1992). Thus parallelism, by increasing information exchange, holds the promise of allowing a trustor to gain a more accurate perception of trustworthiness.

However, as with feedback immediacy and symbol variety, increased parallelism also carries a potential negative influence on accurate impression formation. More information may overwhelm the recipient, leaving the recipient in a state of information overload. Information overload in turn causes anxiety, which reduces cognitive processing ability. This reduction of cognitive processing is critical, as the ability to exchange information with multiple communication partners is cognitively taxing. An increased demand for cognitive processing, coupled with a reduction in cognitive processing ability, may actually lead to information degradation (Potter & Balthazard, 2004), thus decreasing the accuracy of perceptions of trustworthiness.

While a medium's inherent ability to allow parallel conversation promotes information exchange, other perceptions formed during this chat may influence the actual effects of this increased information exchange. For example, let's examine the potential effects of a four-way online chat, where participants can type and respond quickly, but where the fourth member is less adept as a typist. By the time the fourth member gets a comment typed and input into the chat room, the conversation between the other three has moved on. Even though they were relevant when initiated, the slow typist's comments seem out of place by the time they are typed and inserted into the conversation. Thus, while parallelism might be expected to increase the perception of accuracy (through greater information exchange), impressions in this case may not reflect reality (i.e., the fourth member may be seen to be making seemingly incoherent comments, or not keeping up with the conversation). In this way, parallelism, which to a certain extent relies on a person's ability to use the medium at a speed appropriate to the conversation, may reduce perceptions of ability. Applying the concept of familiarity to this example, knowledge about the fourth member's typing limits may temper such perceptions.

The Effect of Rehearsability on the Formation of Trust

Rehearsability is the extent to which a medium provides a mechanism for the sender to rehearse or refine a message before sending (Dennis & Valacich, 1999). A high degree of rehearsability can allow the trustee to craft the message's meaning succinctly and precisely. A clear, well-crafted message allows the trustor to quickly understand the meaning or intent of the trustee and may thereby create the opportunity for the trustor to develop an accurate perception of trustworthiness.

However, as with our other media capabilities, a high degree of rehearsability can also have a paradoxical effect and promote inaccurate perceptions of trustworthiness. Rehearsability allows the trustee to better craft his or her message and potentially even incorporate research or gain assistance from others. This may send signals about a person's level of ability (or knowledge) that may exceed what actually exists, creating a perception of ability-related trustworthiness in excess of what is justified.

As a moderator over these effects, familiarity may influence how we interpret rehearseable messages. In this instance, let us examine how familiarity may be applied, not in relation to whether a receiver knows a *particular* sender, but rather in relation to what the receiver's expectations are regarding a "category" of sender (as discussed earlier). Consider a manager receiving an introductory e-mail or letter from a job applicant. The manager likely expects that such a message will be of high quality, that it will be well crafted and include evidence of ability in the job domain, as well as other attributes. However, the manger also knows that this medium is rehearsable and that the job applicant has likely spent substantial

time on crafting such a message, which might have been sent to multiple recipients. Does the manger, reacting to such a communication, form a perception that the applicant has high ability? Perhaps not in this instance. The level of expectation in a communication of this type, made through a rehearseable medium, may mitigate the potential effect that the expertise exhibited in the letter would have on the manager's perceptions of applicant abilities (and hence trustworthiness on this dimension).

Direct familiarity with the sender may also effect the interpretation of rehearseable messages. For example, if a receiver already has an established relationship with a message sender and has a high opinion about a person's trustworthiness, expectations about how well a message needs to be crafted may be relaxed. In such an instance, there is a shared assumption that the receiver will know how to interpret what is being quickly, casually, or cavalierly communicated. Without this level of familiarity, such as with the new business contact, a mismatch between the level of casualness expected by the receiver and that used by the sender may well influence perceptions of trustworthiness.

The Effect of Reprocessability on the Formation of Trust

Reprocessability is the extent to which a medium allows the message to be reexamined or reprocessed, either during the communication event or after the event has passed. For example, a telephone conversation (assuming it was not recorded) does not allow one to listen again for exactly how a message was worded to verify that it was accurately received and that the nuances of the message were detected and understood. An e-mail, on the other hand, allows one to review what was said and then consider whether the receiver's initial interpretation was accurate.

Media with higher levels of reprocessability may improve the accuracy of trustworthiness perceptions, as the trustor is allowed the opportunity to review the transcript. Reviewing the message as time permits may allow a more complete analysis of the cues that were received, potentially permitting a more accurate perception of the trustee's actual ability, benevolence, and integrity. For example, a more careful post hoc analysis of a message, rather than one based on fallible memory, may reveal subtleties about the message that were initially overlooked. These additional insights might call attention to comments or wording that give cues about trustworthiness. For example, the first time one reads an editor's response letter to a manuscript submitted for publication, the focus of attention is likely to be on the specific content and directions contained in the letter. Only after several subsequent reviews of the letter might the author glean from the editor's choice of words the editor's level of benevolence toward the author. If, however, the words in the letter were not carefully chosen by its writer, perhaps because the letter contained standard boiler-plate phrases, the reader may gain an inaccurate perception of the editor's trustworthiness.

Once again, the extent to which the receiver is familiar with the sender will affect the accuracy of the trustworthiness assessment. If the receiver can read the message repeatedly, the receiver may assume he or she can interpret it better. Why else would one reread? While holding the promise of allowing a receiver to better interpret meaning, if the receiver is not familiar with the sender or how the sender typically communicates, the receiver may actually form a less accurate interpretation of the sender's trustworthiness by reprocessing. For example, a manager may send out a note that everyone expects was carefully framed, but each employee reads it with a different frame of reference. The more the employees reprocess the message, the more dissimilar their individual interpretations of the intent of the message will be, particularly if they do not know the manager well. How accurately they interpret the signals will affect how accurately they interpret the manager's trustworthiness. The likelihood that these additional signals will be correctly interpreted depends on how well the employees each either share or at least understand the manager and his frame of reference.

IMPLICATIONS OF MEDIA AND FAMILIARITY ON VIRTUAL TEAMWORK

Earlier theories on media, for example, media richness theory, posited that the more cues a medium conveyed, the better it might be at promoting understanding in complex tasks. As media theory evolved, for example, with channel expansion theory, such issues as group history and familiarity were added to our understanding of media effects. Recent research on media capabilities, as discussed in media synchronicity theory, point to a medium's capabilities as being more than simply the types of cues it can convey but include issues like feedback speed, message reprocessabiity, and the like. In many cases, expectations about the effects of these capabilities are not as straightforward as one might hope. To fully understand how media capabilities can influence the accuracy of impression formation, we need to once again examine how moderators like familiarity interact with media capabilities to modify these impressions.

While higher levels of each of the five media characteristics in media synchronicity theory can allow message receivers to better understand a message's meaning, and thus form more accurate impressions of the sender, we have demonstrated in this chapter this may not actually occur in practice. Whether or not the potential gains in accuracy of trustworthiness assessments that can come from these five media characteristics are realized is driven by shared or understood perspectives, or by familiarity with the sender. Where the receiver of the information shares a perspective with the sender, understands that sender's perspective, or is familiar with the sender, the accuracy of the interpretation of trustworthiness cues associated with the message is likely to increase. The less the receiver understands the perspective of the sender, the more likely the extra information available from these five characteristics will lead to inaccurate assessments of trustworthiness (see Tables 2 and 3).

Based on this need for understanding, mechanisms for developing familiarity may play a crucial role in successful virtual team functioning, particularly in cases where the formation of trust is important. Many of the traditional team-building

Table 2. The impact and implications of media characteristics

Media Capabilities	Paradoxical Influences on Trustworthiness Perceptions	Implications for Managers
Feedback immediacy	While feedback immediacy might be expected to have a positive impact on trustworthiness perceptions (given that increased ability to have more communication iterations should provide more information about the trustee), immediate feedback may also serve to build unjustified confidence in the trustee. Familiarity (see example on cultural brainstorming) may serve to temper inaccurate negative impressions.	Higher levels of feedback immediacy can certainly be viewed positively and should be a design goal when developing systems. However, managers should be vigilant in training employees to be mindful about how different groups may have different temporal styles or processes of collaboration.
Symbol Variety	While increased symbol variety allows a medium to provide more cues regarding communication intent, thus potentially increasing the accuracy of perceptions, it may also create the opportunity for increased confidence in inaccurate impressions (see example on the practiced liar), or for mixed signals to be sent. Familiarity may have a particularly strong interaction with symbol variety (see example on e-mail as a preferred collaboration medium).	Managers are likely to view increased symbol variety as having a positive influence on the accuracy of perceptions. However, in certain contexts, leaner media may actually by clearer, create more shared understanding, and foster trust in collaborators.
Parallelism	While parallelism allows for interrupted messages to be sent, reducing the effect of certain process losses such as turn-taking, parallelism may also foster cognitive overload and anxiety, reducing the ability to form accurate impressions of others (see example on communication pace). Familiarity may temper those attributions.	As with feedback immediacy, more information does not necessarily lead to enhanced judgment. Employees should be trained to be mindful of differing abilities of collaborators when using such technologies.
Rehearsability	While rehearsability allows a message sender to better craft a message, thus potentially increasing perceptions of the message senders' ability, such attributions may not be accurate. Familiarity, with either the category of communicator (see example on job applicant) or communicator themselves (long-term colleague) should temper such assessments.	Employees should be trained to understand that a high-quality message does not indicate sender ability or competence, and likewise that more casual messages do not symbolize a lack of ability or interest.
Reprocessability	While reprocessability provides a message receiver time to carefully consider a message's content, possibly picking up subtle cues within the message, it can also allow for a receiver's frame of reference to play an even larger role in message interpretation. Familiarity, in this instance, should enhance the accuracy of perceptions formed from such messages.	Reprocessability carries the potential to improve perceptions of trustworthiness. However, it is also time consuming, and may actually create unintended interpretations of a message. Communicators using reprocessable media need to be thoughtful in communications.

exercises organizations use are typically done in face-to-face settings. However, it is precisely the lack of ability to meet face to face that leads to the use of virtual teams. In such instances, training within CMC environments may also be useful. In addition, where the team members are from culturally diverse backgrounds, cultural sensitivity training (e.g., focused on aspects such as the normal pace and style of interactions on certain categories of tasks) may provide great benefits. In the example cited earlier in this chapter, explaining to U.S. students that Chinese students are likely to be uncomfortable with brainstorming will help to inoculate the U.S. students from making inaccurate assessments of the Chinese students' trustworthiness. Likewise, time spent teaching the Chinese students how U.S. students use the approach of brainstorming and sharing of ideas that are not well thought out as a method to enhance creativity is likely to be of value. The Chinese students would be less likely to misinterpret the knee-jerk, fragmented, disjointed thinking reflected in the U.S. students' virtual communications as reflecting low ability and therefore low trustworthiness.

When it is not practical to give virtual team members sufficient exposure to one another so that they can develop interpersonal familiarity, an important step in the right direction is to teach them about broader categories of behavior, for example, behaviors based on cultural differences, organizational work styles, influences from the profession (e.g., engineers vs. advertising executives), or other referent groups. While cultural sensitivity training is a bit more common in modern organizations, our broader concept of familiarity goes beyond this, giving human resource and organizational development specialists a lens through which to reexamine the types of training they use to support virtual team performance. Such a perspective may yield more accurate trustworthiness assessments of team members, enhance trust formation, and result in fewer unpleasant surprises as we increase the use of virtual teams in the workplace.

REFERENCES

Argyris, C. (1964). *Integrating the individual and the organization*. New York: John Wiley & Sons.

Bos, N., Gergle, D., Olson, J., & Wright, Z. (2002). *Effects of four computer-mediated communications channels on trust development*.

Brown, H.G., Poole, M.S., & Rodgers, T.L. (2004). Interpersonal traits, complementarity, and trust in virtual collaboration. *Journal of Management Information Systems, 20*(4), 115.

Carlson, J.R., & Zmud, R.W. (1999). Channel expansion theory and the experiential nature of media richness perceptions. *Academy of Management Journal, 42*(2), 153.

Daft, R.L., & Lengel, R.H. (1986). Organizational information requirements, media richness and structural design. *Management Science, 32*(5), 554.

Davis, J.H., Schoorman, F.D., Mayer, R.C., & Tan, H.H. (2000). The trusted general manager and business unit performance: Empirical evidence of a competitive advantage. *Strategic Management Journal, 21*, 563-576.

Dennis, A.R., & Kinney, S.T. (1998). Testing media richness theory in the new media: The effects of cues, feedback, and task equivocality. *Information Systems Research, 9*(3), 256.

Dennis, A.R., & Valacich, J.S. (1999). *Rethinking media richness: Towards a theory of media synchronicity*. Paper presented at the 32nd Hawaii International Conference on Systems Sciences.

Dennis, A.R., Valacich, J.S., Carte, T.A., & Garfield, M.J. (1997). Research report: The effectiveness of multiple dialogues in electronic brainstorming. *Information Systems Research, 8*(2), 203.

Deutsch, M. (1958). Trust and suspicion. *The Journal of Conflict Resolution (pre-1986), 2*(4), 265.

Dirks, K.T., & Ferrin, D.L. (2002). Trust in leadership: Meta-analytic findings and implications for research and practice. *Journal of Applied Psychology, 87*(4), 611.

El-Shinnawy, M., & Markus, L. (1998). Acceptance of communication media in organizations: Richness or features? *IEEE Transactions on Professional Communication, 41*(4), 242.

Hall, E.T. (1976). *Beyond culture*. Garden City, CA: Anchor.

Hofstede, G. (1983). The cultural relativity of organizational practices and theories. *Journal of International Business Studies, 14*(2), 75.

Hogg, M.A. (1996). Social identity, self-categorization, and the small group. In E. Witte & J.H. Davis (Eds.), *Understanding group behavior: Small group processes and interpersonal relations* (Vol. 2, pp. 227-254).

Hogg, M.A., Terry, D.J., & White, K.M. (1995). A tale of two theories: A critical comparison of identity theory with social identity theory. *Social Psychology Bulletin, 58*(4), 255-269.

Jarvenpaa, S.L., Knoll, K., & Leidner, D.E. (1998). Is anybody out there? Antecedents of trust in global virtual teams. *Journal of Management Information Systems, 14*(4), 29.

Jarvenpaa, S.L., & Leidner, D.E. (1999). Communication and trust in global virtual teams. *Organization Science, 10*(6), 791.

Jarvenpaa, S.L., Shaw, T.R., & Staples, D.S. (2004). Toward contextualized theories of trust: The role of trust in global virtual teams. *Information Systems Research, 15*(3), 250.

Javidan, M., House, R.J., Dorfman, P.W., Hanges, P.J., & Luque, M.S.d. (2006). Conceptualizing and measuring cultures and their consequences: A comparative review of GLOBE's and Hofstede's approaches. *Journal of International Business Studies, 37*(6), 897.

Jones, G.R., & George, J.M. (1998). The experience and evolution of trust: Implications for cooperation and teamwork. *Academy of Management. The Academy of Management Review, 23*(3), 531.

Kahai, S.S., & Cooper, R.B. (2003). Exploring the core concepts of media richness theory: The impact of cue multiplicity and feedback immediacy on decision quality. *Journal of Management Information Systems, 20*(1), 263.

Kasper-Fuehrer, E.C., & Ashkanasy, N. (2001). Communicating trustworthiness and building trust in interorganizational virtual organizations. *Journal of Management, 27*(3), 235.

Kiesler, S. (1986). The hidden messages in computer networks. *Harvard Business Review, 64*(1), 46.

Kiesler, S., Siegel, J., & McGuire, T. (1984). Social psychological aspects of computer-mediated communication. *American Psychologist, 39*, 1123-1134.

Kraut, R.E., Galegher, J., Fish, R., & Chalfonte, B. (1992). Task requirements and media choice in collaborative writing. *Human Computer Interaction, 7*, 375-407.

Leidner, D.E., & Kayworth, T. (2006). Review: A review of culture in information systems research: Toward a theory of information technology culture conflict. *MIS Quarterly, 30*(2), 357.

Lengel, R.H., & Daft, R.L. (1988). The selection of communication media as an executive skill. *The Academy of Management Executive, 2*(3), 225.

Lewicki, R.J., & Bunker, B.B. (1996). Developing and maintaining trust in work relationships. In R.M. Kramer & T.R. Tyler (Eds.), *Trust in organizations: Frontiers of theory and research* (pp. 114-139). Thousand Oaks, CA: Sage.

Mayer, R.C., & Davis, J.H. (1999). The effect of the performance appraisal system on trust for management: A field quasi-experiment. *Journal of Applied Psychology, 84*(1), 123.

Mayer, R.C., Davis, J.H., & Schoorman, F.D. (1995). An integration model of organizational trust. *Academy of Management. The Academy of Management Review, 20*(3), 709.

Mayer, R.C., & Gavin, M.B. (2005). Trust in management and performance: Who minds the shop while the employees watch the boss? *Academy of Management Journal, 48*(5), 874.

McAllister, D.J. (1995). Affect- and cognition-based trust as foundations for interpersonal cooperation in organizations. *Academy of Management Journal, 38*(1), 24.

Ngwenyama, O.K., & Lee, A.S. (1997). Communication richness in electronic mail: Critical social theory and the contextuality of meaning. *MIS Quarterly, 21*(2), 145.

Paul, D.L., & McDaniel, R.R. (2004). A field study of the effect of interpersonal trust on virtual collaborative relationship performance. *MIS Quarterly, 28*(2), 183.

Potter, R.E., & Balthazard, P. (2004). The role of individual memory and attention processes during electronic brainstorming. *MIS Quarterly, 28*(4), 621.

Rotter, J.B. (1967). A new scale for the measurement of interpersonal trust. *Journal of Personality, 35*, 651-665.

Russ, G.S., Daft, R.L., & Lengel, R.H. (1990). Media selection and managerial characteristics in organizational communications. *Management Communication Quarterly: McQ (1986-1998), 4*(2), 151.

Serva, M.A., Fuller, M.A., & Mayer, R.C. (2005). The reciprocal nature of trust: A longitudinal study of interacting teams. *Journal of Organizational Behavior, 26*(6), 625.

Siegel, J., Dubrovsky, V., Kiesler, S., & McGuire, T.W. (1986). Group processes in computer-mediated communication. *Organizational Behavior and Human Decision Processes, 37*(2), 157.

Srite, M., & Karahanna, E. (2006). The role of espoused national cultural values in technology acceptance. *MIS Quarterly, 30*(3), 679.

Townsend, A.M., DeMarie, S.M., & Hendrickson, A.R. (1998). Virtual teams: Technology and the workplace of the future. *The Academy of Management Executive, 12*(3), 17.

Trevino, L.K., Lengel, R.H., Bodensteiner, W., Gerloff, E.A., & Muir, N.K. (1990). The richness imperative and cognitives style: The role of individual differences in media choice behavior. *Management Communication Quarterly: McQ (1986-1998), 4*(2), 176.

Trevino, L.K., Lengel, R.H., & Daft, R.L. (1987). Media symbolism, media richness, and media choice in organizations. *Communication Research*, 553-574.

Turner, J.C. (1987). *Rediscovering the social group: A self categorization theory*. New York: Basil Blackwell Inc.

Turner, J.C., Oakes, P.J., Haslam, S.A., & McGarty, C. (1994). Self and collective: Cognition and social context. *Personality and Social Psychology Bulletin, 20*(5), 454.

Valacich, J.S., Dennis, A.R., & Nunamaker, J.F. (1992). Group size and anonymity effects on computer-mediated idea generation. *Small Group Research, 23*(1), 49-73.

Walther, J.B. (1992). Interpersonal effects in computer-mediated interaction: A relational perpective. *Communication Research, 19*(1), 52.

Walther, J.B. (1993). Impression development in computer-mediated interaction. *Western Journal of Communication, 57,* 381-398.

Walther, J.B., & Burgoon, J.K. (1992). Relational communication in computer-mediated interaction. *Human Communication Research, 19,* 50-88.

Chapter VI
Designing and Assessing Virtual Assurance:
The Role of Computer-Mediated Technologies in Facilitating High Levels of Trust and Distrust

Terry R. Adler
New Mexico State University, USA

Michael Glissmeyer
New Mexico State University, USA

ABSTRACT

This chapter suggests that computer-mediated technologies (CMTs) facilitate organizational trust and distrust by leading to what we introduce as virtual assurance. Through partnering and outsourcing, organizations are exposed to managing simultaneous organizational trust and distrust. For instance, CMTs allow more precise and timely monitoring of organizations in a high trust and high distrust context, a process that leads to virtual assurance. We further describe virtual assurance as a means to manage the fragility of modern interorganizational relationships, especially when high trust and high distrust is present. We also suggest that the presence of virtual assurance will ultimately provide a competitive advantage to firms in making contractual agreements, tracking progress, imposing penalties, and shielding organizations from potential harm.

INTRODUCTION

The role of trust and distrust plays an important role in how computer-mediated technologies are used and the effects of this use on the integration of work between virtual organizations. Computer-mediated technologies (CMTs) like e-mail, group decision support systems, and so forth, provide alternatives that allow organizations to link themselves with their virtual partners, suppliers, and customers. Recent evidence underlying the importance of CMTs in supporting the practices of joint venturing and organizational outsourcing is abundant and indicates that trust is an important factor to consider in interorganizational relationships (The future of outsourcing, 2006; Ratnasingam, 2005). The role of trust is critical to understand when organizations use computer-mediated technologies because attitudes, risk perceptions, and intentions are managed differently than in traditional face-to-face relationships (Wilson, Strauss, & McEvily, 2006). Issues such as time pressures (Ross & Wieland, 1996), group performance (Walther & Bunz, 2005), and collaboration among organizations (Grossman, 2004) have been studied with regard to trust and CMTs.

The role of trust and distrust is even more critical to understand when organizations use CMTs because attitudes, risk perceptions, and intentions are managed differently than in traditional face-to-face relationships (Lee, 1994). For instance, Lee presented detailed information on how poorly conceived e-mail was used to misinform, thereby providing a rich and integrative contextual meaning in a Midwestern, high technology company. Lee and Ngwenyama's (1997) follow-on article indicates that the similar physical artifacts or human action can have different meanings for different subjects as well as for the observing scientist. E-mail provides an open computer-mediated forum to observe communication and meaning from multiple actors and from multiple locations that is quite different than traditional views of face-to-face interaction.

While previous research has investigated and analyzed contextual issues of computer-mediated interaction, we have found a scarcity of research on how trust and distrust affects an integrative and organizational perspective. In this chapter we provide a theoretical model for understanding the effects of organizational trust and distrust through the use of shared CMTs like electronic mail. Rather than looking at antecedents of trust and distrust in computer-mediated relationships, we examine trust and distrust as antecedents of computer-mediated selection and use. We discuss our model relative to how CMTs increase the integration of work between organizational partners. We refer to this particular type of integration due to CMT use as *virtual assurance,* which we define and expand later on in this chapter. Finally, we provide managerial implications of trust and distrust and trends in interorganizational relationships that depend on the use of CMTs. These implications and trends provide further context in understanding the complexities of managing simultaneous trust and distrust in assuring work gets integrated and accomplished that is beneficial to organizational actors.

THEORETICAL MODEL OF SIMULTANEOUS TRUST AND DISTRUST EFFECTS

Trust is a complex organizational as well as interpersonal construct (Tyler & Kramer, 1996). We adapt Boon & Holmes' (1991) conceptualization of trust as having positive expectations while remaining vulnerable in an economic exchange. Distrust is defined similarly as having negative expectations while remaining vulnerable in an economic exchange (Adler, 2005; Kipnis, 1996).

Research and common sense suggest that trust behavior is good while distrust behavior is bad. While trust plays a seemingly wholesome role in business relationships, many authors have formulated interesting theories and results built around concepts related to distrust. Through partnering and outsourcing, organizations are exposed to managing simultaneous trust and distrust (Lewicki, McAllister, & Bies, 1998). Adler (2007, in press), using an adapted model from Lewicki et al. (1998), found that simultaneous trust and distrust is an important and valuable consideration that binds work between partners in an economic exchange. Distrust is similar to the concept of "opportunism," or self-interest seeking with guile, which Williamson (1985) proposes as one of four basic conditions of any business transaction. The more opportunism, or distrust, the more likely an organization will choose to make products and services rather than expose the organization to potential opportunistic behavior with a business partner.

The seminal work by Lewicki et al. (1998), however, suggests that trust and distrust exist simultaneously and that both be managed simultaneously. Lewicki et al. argue that simultaneous high trust and high distrust in business transactions today is the most prevalent form of business transaction. Adler (2005) found support for this claim and many organizational contexts support this claim. For example, on an individual level, employees are given bar-coded security cards which provide access, albeit restricted, to firm operations, resources, and assets. Organizations expect individuals to perform their duties as planned, a high trust situation, but there is a nagging belief that employees will eventually steal information and secrets given the right circumstances, a high distrust situation. Consequently, individuals typically do not have complete access to all corporate facilities, resources, and assets.

From an organizational perspective, this is also quite true. The banking industry provides an example of simultaneous high trust and high distrust between organizations. When organizations like the Federal Deposit Insurance Commission (FDIC) deal with banks, banks are required to provide information to government monitoring organizations like the FDIC. The FDIC trusts banks to carry out their normal financial responsibilities in a competent way. Banks are fundamentally telling the FDIC that they have enough cash to cover the bank's financial transactions. Banks conversely expect the FDIC to accurately review all of their financial information in accordance with current federal policies and law. The FDIC in turn constructs computer-mediated technologies through monitoring software programs to identify when bank data might be inconsistent with the established policies and law. Conversely, the FDIC, as one of the primary government watchdogs, also distrusts banks since banks can make errors, fail to follow policy, or act fraudulently in not having enough cash on hand. Banks also distrust the FDIC for many reasons, none of which compares to the general lack of enthusiasm of being repeatedly monitored and reviewed. Thus, simultaneous trust and distrust exists even at the organizational level of analysis.

TRUST AND DISTRUST EFFECTS IN INTERORGANIZATIONAL RELATIONSHIPS

CMTs allow more precise and timely monitoring of organizations in a high trust and high distrust context. The FDIC example highlights the effects of interorganizational trust and distrust as separate concepts, and, in accordance with Lewicki et al. (1998) and Adler (2005), the degree of trust and distrust affects how decisions will be made. Grossman (2004), for instance, found that with trust between suppliers, distributors, and customers with business partners, computer-information systems were successfully implemented. This study was important in establishing the role of interorganizational trust in the supply-chain literature. Organizations can establish rules and policies on how to interact with their business associates, or partners, using CMTs as the means to haggle and monitor operations. Ratnasingam (2005) provides an example of how e-commerce, as a type of CMT, was also affected by technological and relationship trust.

Figure 1 provides a four-cell model that integrates the model of Lewicki et al. (1998) within a proposed computer-mediated context. We refer to interorganizational relationships as any firm business transaction with another organization. This can be in the form of a contractual basis, joint venture, or business partnership. Interorganizational relationships can be characterized as either high to low in trust and distrust with their partner firms. We provide an explanation of each cell of the Lewicki et al. model with regard to how trust and distrust affects the use of computer-mediated technologies to manage these interorganizational relationships.

Low Trust and Low Distrust

A low trust and low distrust environment is typical where partner firms do not have a lot of information about each other. New business ventures and low reputational cues lead firms to use bounded, arms-length contracts that include standard boilerplate terms and conditions. Superficial respect and communication occurs as firms know little about each other. Thus, since there is little integration between organizations, interorganizational relationships tend to be simplistic and CMTs tend to be stand-alone systems with little interdependence between work done in each firm. This type of organizational interrelationship is probably not likely given the amount of information on potential or current partners available today through the Internet.

High Trust and Low Distrust

Business partners that have positive expectations, with little evidence of opportunistic behavior about their trading partner(s) are said to have a high trust and low distrust relationship. The

Figure 1. Simultaneous trust and distrust through CMTs

	Low Trust	High Trust

High Distrust	• Undesirable events feared
	  o Research through CMT indicates reputational effects indicate dishonesty
	• Interdependence managed
	  o CMT used to severely restrict access to information and resources
	• Contractual preemption
	  o CMT used to communicate penalties for abuse to contractual terms
	• Paranoia
	  o Worst case scenarios dominate your organization supported by CMTs
		• Trust but verify
		  o Virtual assurance is sought by relying on CMT to verify work accomplishment
		• Relationships highly segmented
		  o Firm resources segmented and bounded to protect key products, intellectual property, and decisions
		• Opportunities pursued/risks monitored
		  o CMT provides data and information regarding potential investors and regulators
		  o CMT monitors work completion, timeliness, and accuracy

Low Distrust	• Casual acquaintances
	  o Limited previous history so a new opportunity for everyone
	• Limited integration
	  o Inputs and outputs between organization's processes are independent and not integrated
	• Bounded, arms-length
	  o Emotional relationship and transaction attachment between organizational actors is not a factor to consider
	• Professional courtesy
	  o Respect for each other is superficial
		• High-value congruence
		  o Perceptions between organizational actors align
		• Interdependence promoted
		  o Sharing of work is expected
		• Opportunities pursued
		  o Creativity and innovation is an outcome of the CMT
		• New initiatives
		  o New strategies for maximizing the CMT are discussed openly and pursued by both parties

dimensions of this cell are described in Figure 1. Since values between organizations tend to be congruent in this scenario, firms pay more attention to how they can share work to take advantage of mutual opportunities. CMTs help link organizations together as common software, hardware, and systems are designed and shared and information is exploited between organizations. Shared CMTs are encouraged for use between partners to facilitate information exchange. As Adler (in press) discovered in the development of organizational contracts, organizations can use interorganizational relationships to promote skills like joint conflict resolution, communication exchange, and shared opportunity seeking that promotes new initiatives. High trust and low distrust conditions tend to garner respect and mutual support between partners since goal compatibility and positive expectations are open and verified.

Low Trust and High Distrust

Figure 1 also describes a low trust and high distrust situation where firms live in paranoia because reputational information indicates a less than honest business partnership. Many ask why a firm would even do business with an opportunistic firm like this. The answer is always not so clear but there are many reasons why this might occur. A firm may be the sole provider of a product or service. This would put the buyer or partner firm in a disadvantageous position, as described by Porter (1980), since little can be done to reduce the firm's economic power over their potential business partner. Organizations may also be forced to do business with a firm due to governmental oversight and pressure, top management direction, or just plain necessity to keep abreast of market changes. Global ventures also suggest that firms enter into business partnerships that appear one-sided because of the necessity to enter new markets in other countries and cultures.

If firms are in this condition, high distrust tends to make undesirable events anticipated and expected. Any interdependence is managed through CMTs. CMTs can be used to monitor organizational behaviors and actors to limit access to corporate resources and assets. Management tends to focus on how to restrict information that might be sensitive and proprietary. Penalties for abuse of partner privileges and security are set and reviewed periodically to prohibit any future occurrences of distrust in the business transaction. Organizations learn that there are clear delineations between what is acceptable and what is not acceptable, how to prevent potential opportunism by setting rules and conditions, and how to preempt a negative experience.

The last cell of high trust and high distrust as suggested by Lewicki et al. (1998) is a common form of interorganizational relationship found today. We propose that this simultaneous trust and distrust, when combined with the capabilities that CMTs can provide, leads to what we refer to as *virtual assurance*. We next provide a definition and model of virtual assurance.

VIRTUAL ASSURANCE

While Lewicki et al. (1998) provided a model of simultaneous high trust and distrust, we suggest that CMTs are the conduit for simultaneous trust and distrust in that information can be simultaneously restricted and provided in an interorganizational relationship. The permission between two or more organizations to simultaneously allow and restrict access to proprietary information and resources through the use of CMTs is what we refer to as virtual assurance.

As background, Williamson (1985) and Ouchi (1981) state that for trust to exist, organizations must be able to haggle with and monitor one another in order to adequately integrate the work of both (or multiple) partners in a business transaction. Computer-mediated technologies provide a basis for monitoring and haggling that allow business partners to get what they want in the transaction while at the same time attempting to safeguard their own assets from their business partner. The concept of virtual assurance is ultimately based on the fundamental and underlying concepts of high trust and high distrust (see Figure 2).

CMTs allow business partners to blend their resources with their partners without losing these resources or information. Compartmentalization, or segmentation, of information and resources allows interaction and interdependence between key partner representatives all the while restricting access to those not allowed to have certain types of information. Given the immense amount of information available that can be shared, the advent and use of CMTs truly provides an advantage to firms that may have never existed before. CMTs provide avenues for information gathering, summarizing, and sharing. As new technologies like radio frequency identification (RFID) and biomaterial clothing (Clothes with a silver lining, 2006) continue to be developed, so will the ability

of firms to haggle and monitor continue to grow, and firms will have even more options to provide and restrict information simultaneously.

Virtual assurance can truly enhance future interorganizational relationships even when reputational factors suggest that there might be a downside to doing business in traditional ways. Virtual assurance is different than other constructs like virtual collaboration because the focus is on providing and restricting access simultaneously whereas collaboration suggests the absence of opportunism in a business transaction. While contract management views the business transaction from a litigious viewpoint, the use of CMTs provides flexibility through the use of common information system tools to address daily business operations. As another comparison, even though the concept of quality assurance attempts to provide confidence that a transaction will satisfy relevant quality standards, virtual assurance allows firms to develop individualized processes that provide confidence in managing restricted and shared information.

Figure 2. Theoretical framework leading to virtual assurance

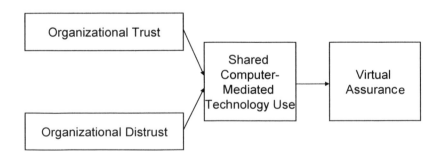

Since information has many classifications and forms, the concept of virtual assurance is much more generalized and practical to any aspect of a business operation. Virtual assurance, through a CMT, adds some interesting strategic tension where organizational boundaries become blurred and partnerships become mature and complex. The roles of trust and distrust play an important part in how computer-mediated technologies are used. As organizations collaborate, they are confronted with the challenge of managing the roles of trust and distrust simultaneously. The next section discusses the managerial implications of how organizations collaborate and achieve virtual assurance.

MANAGERIAL IMPLICATIONS IN ACHIEVING VIRTUAL ASSURANCE

Using the framework described in Figures 1 and 2, we offer insight into how computer-mediated technologies allow organizations to simultaneously manage organizational trust and distrust in a virtual collaboration—the most likely business scenario for industry today. Implications relative to virtual collaborations might include topics like outsourcing, franchising, licensing, or joint venturing. The assumption of our model in Figure 2 is that most, if not all, organizations want assurance that the transaction they are creating with a future business partner will proceed in a manner that meets both (or multiple) partner expectations and needs. Interorganizational trust and distrust is the basis for generating this assumption and computer-mediated technologies are the tools for ensuring that expectations and needs are met. More specifically, we will discuss four managerial implications in the following section to provide further context to the term virtual assurance:

- Documentation of transactions through e-mail adds to the social context of virtual assurance between business partners. E-mail

is a powerful medium that is affected by trust or distrust. We will discuss how an e-mail's content and timing is an outcome of trust and distrust and how the use of e-mail affects the concept of virtual assurance.

- The scope of a partnership is more open for review since details can be easily provided. CMTs allow business partners to do detailed analysis and monitoring as discussed in the bank and FDIC example. How these CMTs affect the development of virtual assurance will be discussed with regard to honesty in haggling and monitoring.

- Reputation of business partners is generally more available since sources exist to research past and monitor current performance. We discuss these implications in interorganizational relationships with regard to information sharing and decision making through the use of CMTs.

- Restriction of data and proprietary information can be easily accomplished through CMTs through the use of passwords and security nets. Business partners can maintain and safeguard resources more easily and we discuss the implications of these practices with regard to virtual assurance.

E-MAIL AS A CONTRIBUTOR TO VIRTUAL ASSURANCE

An article in *The New York Times* (Dillon, 2004) discusses the situation of Kathy Keenan, a legal proofreader who teaches business communication at the University of California, Santa Cruz extension. One e-mail received by Kathy read: "hi KATHY I am sending u the assignmnet again, I had sent u the assignment earlier but I didn't get a respond. If you get this assignment could u please respond". Mrs. Kennan said most of her students are midcareer professionals in high-tech industries. In the same article Dr. Hogan, a retired professor who heads an online school for business, suggests "e-mail is a party to which English teachers have not been invited" and "e-mail has just erupted like a weed, and instead of considering what to say when they write, people now just let their thought drool out on the screen". In another example from business, a systems analyst wrote this e-mail to her supervisor, "I updated the Status report for the four discprepancies Lennie forwarded us via e-mail (they in Barry file).. to make sure my logic was correct It seems we provide Murray with incorrect information…However after verifying controls on JBL—JBL has the indicator as B????—I wanted to make sure with the recent changes—I processed today—before Murray make the changes again on the mainframe to C". The supervisor had no idea what the analyst was saying. With communication between business partners essential for success of a venture, it is critical that e-mail content be of the highest quality and timing be monitored to virtually assure partners to a transaction that expectations and needs will be met. Can you imagine receiving an e-mail like the business example above in a business transaction where you may have never met the individual sending the e-mail to you?

Due to availability of information through e-mail, business transactions can come together swiftly and end abruptly. In the example above with the analyst, social context would be highly affected if the company had come highly recommended by another business partner who you trusted (Rotter, 1971; Stack, 1988). Similarly, timing also affects the trust and distrust relationships (Straus, 1997). One of the conveniences of e-mail is that it identifies senders, receivers, and time the message was sent through the server (Dubrovsky, Kiesler, & Sethna, 1991). While timing issues and the impact of those issues seem to be well understood, to be clear, timing is important in regards to contracts, deadlines, and so forth, all of which influence the concept of virtual assurance. CMTs allow business partners to view not only content but timing of messages sent via e-mail to gauge if and when communication took place. This might be very important when deadlines are at stake and when business partners are required to share information.

An e-mail with carelessly crafted content or poorly timed transmittal, whether intentional or not, can be the result of high trust and high distrust depending on how that interorganizational relationship is framed. E-mail can be both a factor that leverages interorganizational relationships while at the same time, if poorly conceived, can cultivate the seeds of destruction powerful enough to end a partnership. Many organizations, thus, believe it necessary to have control over the design, development, and implementation of their own information systems, including the development of unique software programs necessary to manage their work. The degree that business partners live up to their own commitments in the partnership can be monitored easily using CMTs. The use of CMTs to deter anticipated ineffective or opportunistic behavior and indicate performance of commitments by their business partner(s) leads to high trust and high distrust scenario which we call virtual assurance.

CMTs AND THE MANAGEMENT OF INTERORGANIZATIONAL CONTRACTS

There continue to be advances in the design of an infrastructure of software programs tailored to improve security standards and monitoring (Itani & Kayssi, 2004; Lam, Chung, Gu, & Sun, 2003). Software programs can be something as simple as counting inventories or as complex as monitoring the daily activities of employees like in RFID wristbands or X-Static clothing as discussed previously. CMTs that are used to monitor contracts with suppliers in tracking the cumulative costs on a contract or in assisting managers in the development of plans, agreements, or contracts with a future partner are relatively simple. The following example shows how a video teleconferencing technology could be used to monitor an interorganizational contract. This example shows how a manager of a distribution facility hypothetically had to interface with a key supplier:

In an effort to stay on the job site to improve productivity and cut down on wasted time, all of our meetings with suppliers are done by videoconferencing. At a recent meeting with our supplier we had issues with some contract requirements. However, through use of an overhead scanner, we were able to have the supplier make adjustments to the contract on the scanner as we watched. The contract was completed to our satisfaction and faxed over within minutes. We got the required signatures and faxed the contract back and had our shipment later that day.

We enjoy doing business with this supplier because of the ease and speed of the contracting issues. We are at a point now where the contracting issues are resolved in a matter of minutes because of the relationship we have.

The key feature to having an open interorganizational relationship in this example was the ability to utilize videoconferencing software and hardware to manage a contract management system. As more and more firms utilize outsourcing to accomplish work, the need to fully utilize CMTs becomes imperative.

While high trust leads to the use of this particular type of CMT, videoconferencing, a more robust CMT, could have been chosen to manage a high trust and high distrust situation. For instance, if the supplier had a nefarious background, CMTs could have been employed to assess completion of work as it was being accomplished by this particular supplier. Many service firms log hours per client, or contract, and these hours become the basis for charges to the client. CMTs could

be used to track and monitor hours provided and gauge whether work was being accomplished in accordance with the contract, potential for cost overruns, and other types of performance measurement. CMTs could also be established to restrict supplier access to information, say contractual information relative to the other suppliers of this particular organization. This would have provided what we refer to as virtual assurance in that information is both provided and restricted in the interorganizational relationship.

REPUTATION VERIFICATION AND INTERORGANIZATIONAL DECISION-MAKING

The reputation of a future business partner is a fundamental issue in virtual collaboration (Shmatikov & Talcott, 2005). The economic perspective of reputation is based on the past performance of a company signaling its true intentions (Clark & Montgomery, 1998). For example, when a bank issues a credit card, they check the credit history (i.e., reputation) of the consumer. This is an example of an inherent high distrust scenario. Provided the reputation of the consumer is clean, the bank will issue a credit card with the expectation that the consumer will act in accordance with the card's conditions, a high trust scenario.

In most individual transactions, utilization of a virtual partner with a poor reputation is rare. Consider the case of eBay. A seller's reputation as documented by past buyers' satisfaction with the transaction is a valuable asset (Rietjens, 2006). When buying on eBay, the seller's satisfaction rating is transparent. With several sellers offering the same desired item, a buyer can easily determine who the best seller is, based on satisfaction and usage, and buy from that individual or company. Why buy from a seller with a poor rating, unless, as mentioned above, the seller is the only one offering the desired item? Recent research on reputation found that prominence of a reputation had a significant effect on price premium (Rindova, Williamson, Petkova, & Sever, 2005).

The obvious implication in the context of reputation is that if an organization has a more prominent reputation, it can charge more for services. In partnering, the firm with more prominence will give up less to get more. The partnership may be mutually beneficial, but the firm with a better reputation (and willingness and ability to leverage its reputation) should ultimately hold the stronger position in an agreement.

Both partners to the contract are better able to understand what a future business partner will provide and thus are more assured their expectations will be met. CMTs allow for better decisions, either based on reputation of a future partner, or in how information is exchanged in general between organizations. For instance, research shows that groups communicating electronically displayed more extreme or unconventional decisions than when the same groups meet face-to-face (McGuire, Kiesler & Siegel, 1987). Additionally, the impact of status and expertise on credibility is reduced when groups communicate in some form of information system (Dubrovsky et al., 1991). Linde (1988) discusses this phenomenon in the aviation industry where several accidents could have been prevented if lower ranking crewmembers had been allowed or felt empowered to voice concerns regarding the safety and security of the flights.

CMTs allow richer information exchanges that could indicate bad or good reputation of potential future business partners. Research indicates that group decision-making is often superior to individual decision making (Witt, Andrews, & Kacmar, 2000) and group decision-making allows for organization-wide sharing of information. This leads to better organizational flexibility and creativity taking place (Cotton, Vollrath, Lengnick-Hall, & Froggatt, 1990). CMTs break down status barriers and allow for more honest

and open communication (Dubrovsky et al., 1991; Lee, 1997).

CMTs, thus, based on trust and distrust perceptions, become a conduit for more diverse information sharing between business partners to allow monitoring and haggling more freely. While this diversity of information has not been studied, several studies have found individuals using CMTs give more unusual and less socially acceptable responses (McGuire et al., 1987; Sproull & Kiesler, 1986). We believe more unusual and less socially acceptable responses lead to more diverse information being shared as opposed to less diverse information. In an interorganizational relationship, all members can freely communicate without fear of evaluation anxiety for high status members, and without fear of rejection from low status members (Dubrovsky et al., 1991). With the social attention given in face-to-face meetings, both ends of the status spectrum would attend to their respective roles more closely, thereby integrating work between partners better and increasing the potential for meeting partnership expectations and needs. Virtual assurance occurs because both sides are able to identify what they need more accurately and monitor how progress is being achieved.

MAINTAINING AND SAFEGUARDING RESOURCES

The partitioning of work relies heavily on the framework of simultaneous high trust and high distrust of Lewicki et al. (1998). Trust behaviors in business partnerships include issues such as information sharing, transparency, openness, and honest communication. As explained earlier, when high trust is combined with high distrust, managers must "trust but verify." CMTs can assist managers with technologies like firewalls, protective passwords that allow for monitoring and safeguarding of resources, and software that tracks user actions. A hypothetical bank manager might relate this story with regard to CMT design:

The bank had been working on developing a company intranet that would allow employees access to company briefings, success stories, and best practices. Additionally there was a section for managers where they could request help from the legal department, the human resource department in the form of confidential documents, and the cash vaults.

There was a glitch in the system that was caught almost immediately where employees had access to the "managers" section of the intranet. The firewall in place failed to distinguish between employees. Some employees found out what other employees were making and what medical and dental benefit coverage they had. It was fixed the same day the intranet was rolled out with no litigation or other problems stemming from the release of information. It could have been a huge problem, but was averted quickly and easily through an adjustment of the firewall.

This story illustrates both the potential for problems if CMTs are not implemented properly, as well as the necessity to redesign security features in CMTs to build trust in business operations. The monitoring of the new firewall was constant and as soon as the technology support center realized the problem of unauthorized access, they moved quickly to shut down the intranet access until the firewall was corrected. CMTs are designed and redesigned based on the perceptions of trust and distrust. The intranet developers trusted the employees to only access areas that were available to them, but were verifying and monitoring the access just-in-case.

In interorganizational relationships, the same concepts apply. CMTs provide robust and flexible means by which providers, suppliers, customers,

and vendors can limit their exposure to a future business partner. The partitioning of work is characteristic of most business firms today. Even employee badges do not usually allow complete access to the company records. As organizations increase the amount of work they outsource, as organizations diversify their supplier base, and as organizations grow globally interacting with multifaceted governments, CMTs provide avenues for protecting proprietary information and resources. The protection of key information and resources provides another example of virtual assurance that business partners will meet their expectations and needs in the business transaction.

TRENDS AND DIRECTIONS

Our proposed model will change as CMTs change. More robust software packages will most likely provide an accurate means to measure past performance based on any number of criteria. For instance, the following questions will most likely be developed into an on-going database regarding a firm's reputational factors: (a) How has an organization has invested its resources? (b) Where have organizations ventured by country and for what purposes? (c) Where have firms conducted new product development? (d) Have organizations obeyed all local, regional, state, and federal laws with regard to environmental, health, and other social concerns? (e) and probably some form of measurement of how suppliers, vendors, and partners rate a firm as a business partner on some form of satisfaction scale.

Another trend will be the maturation of users within organizations on how to gather, assess, and improve interorganizational relationships through the use of CMTs. As firms continue to expand in the use of outsourcing, especially on a global scale (The future of outsourcing, 2006), the development and use of CMTs will provide a competitive advantage to firms in making contractual agreements, tracking progress, imposing penalties, and shielding organizations from potential harm.

Finally, this model could be refined by considering the timing of information in the relationship between trust and distrust perceptions, the use of CMTs, and virtual assurance. Meyerson, Weick, and Kramer (1996) suggest that organizational members develop "swift trust" perceptions based on limited time together to process information about an individual, team, or organization. We suggest that limited, or constrained, time together may in fact limit the development and use of CMTs thereby limiting the usefulness of virtual assurance as an organizational concept. For instance, maybe an organization employs a temporary team to manage a supplier. The temporary team uses their own collective perceptions to develop CMTs to manage the supplier and then after this temporary team is disbanded, all information regarding the supplier is lost. In other words, swift trust and distrust scenarios create interesting questions with regard to how trust and distrust perceptions are developed into information systems and then retained so that organization's actually learn from these experiences. While we do not downplay the importance of virtual assurance in current business practices, the idea of "swift trust and distrust" certainly has some kind of role to play in the use of CMTs in interorganizational relationships.

CONCLUSION

We suggest that trust and distrust occur simultaneously and this situation creates a fragile partnership that must be managed with caution. CMTs provide a means to manage the fragility of modern interorganizational relationships, especially when high trust and high distrust is present. As organizations expand into other markets, the amount, accuracy, and timeliness of information about potential business partners becomes acute.

CMTs continue to provide a means for firms to parlay their trust and distrust perceptions into a manageable and workable business solution.

We conclude that most organizations want predictability in their transactions with other organizations. This is necessary so that an organization's expectations and needs are more likely met rather than failed. We have discussed several managerial implications of using CMTs to facilitate high trust and high distrust. We have introduced the concept of virtual assurance that embodies previous research and adds to our understanding of how interorganizational relationships can be improved through CMTs. The proposed model in this chapter highlights how work is changing in organizations today, thereby transforming today's modern marketplace. As more organizations continue to find success outsourcing and as more firms find global partners, we suggest a thorough knowledge of CMTs relative to trust and distrust will lead to a competitive advantage for firms relying on interorganizational relationships in the future.

REFERENCES

Adler, T.R. (2005). The swift trust partnership: A project management exercise investigating the effects of trust and distrust in outsourcing relationships. *Journal of Management Education, 29*, 714-737.

Adler, T.R. (2007, in press). Trust and distrust in long-term contracting. *Journal of Business Strategies.*

Boon, S., & Holmes, J. (1991). The dynamics of interpersonal trust: Resolving uncertainty in the face of risk. In R.A. Hinde & J. Grobel (Eds.), *Cooperation and prosocial behavior* (pp. 190-213). Cambridge, UK: Cambridge University Press.

Clark, B.H., & Montgomery, D.B. (1998). Deterrence, reputations, and competitive cognition. *Management Science, 44*, 62-82.

Clothes with a silver lining. (2006, Fall). *Military Officer*, p. 28.

Cotton, J.L., Vollrath, D.A., Lengnick-Hall, M.L., & Froggatt, K.L. (1990). Fact: The form of participation does matter – a rebuttal to Leana, Locke, & Schweiger. *Academy of Management Review, 15*, 147-153.

Dillon, S. (2004, December 7). What corporate America cannot build: A sentence. *The New York Times*, 23.

Dubrovsky, V.J., Kiesler, S., & Sethna, B.N. (1991). The equalization phenomenon: Status effects in computer-mediated and face-to-face decision-making groups. *Human-Computer Interaction, 6*, 119-146.

The future of outsourcing. (2006, January 30). *Business Week*, 50-61.

Grossman, M. (2004, September). The role of trust and collaboration in the Internet-enabled supply chain. *The Journal of American Academy of Business*, pp. 391-396.

Itani, W., & Kayssi, A. (2004). J23ME application-layer end-to-end security for m-commerce. *Journal of Network and Computer Applications, 1*, 13-33.

Kipnis, D.(1996). Trust and technology. In R.M. Kramer & T.R. Tyler (Eds.), *Trust in organizations: Frontiers of theory and research* (pp. 39-50). Thousand Oaks, CA: Sage Publications.

Lam, K.Y., Chung, S.L., Gu, M., & Sun, J.G. (2003). Lightweight security for mobile commerce transactions. *Computer Communications, 26*, 2052-2061.

Lee, A. (1994). Electronic mail as a medium for rich communication: An empirical investigation using hermenuetic interpretation. *MIS Quarterly, 18*(2), 143-157.

Lee, A., & Ngwenyama, O. (1997). Communication richness in electronic mail: Critical social theory and the contextuality of meaning. *MIS Quarterly, 21*(2), 145-167.

Lewicki, R.J., McAllister, D.J., & Bies, R.J (1998). Trust and distrust: New relationships and realities. *Academy of Management Review, 23*(3), 438-458.

Linde, C. (1988). The quantitative study of communicative success: Politeness and accidents in aviation discourse. *Language and Society, 17,* 375-399.

McGuire, T.W., Kiesler, S., & Siegel, J. (1987). Group and computer-mediated discussion effects in risk decision-making. *Journal of Personality and Social Psychology, 52*, 917-930.

Meyerson, D., Weick, K., & Kramer, R. (1996). Swift trust and temporary groups. In R.M. Kramer & T.R. Tyler (Eds.), *Trust in organizations: Frontiers of theory and research* (pp. 166-195). Thousand Oaks, CA: Sage.

Ouchi, W. (1981). Theory Z: *How American business can meet the Japanese challenge.* Reading, MA: Addison-Wesley.

Porter, M.E. (1980). *Competitive strategy: Techniques for analyzing industries and competitors.* New York: Free Press.

Ratnasingam, P. (2005). E-commerce relationships: The impact of trust on relationship continuity. *International Journal of Commerce and Management, 15*, 1-16.

Rietjens, B. (2006). Trust and reputation on eBay: Towards a legal framework for feedback intermediaries. *Information & Communications Technology Law, 15*, 55-78.

Rindova, V.P., Williamson, I.O., Petkova, A.P., & Sever, J.M. (2005). Being good or being known: An empirical examination of the dimensions, antecedents, and consequences of organizational reputation. *Academy of Management Journal, 48*, 1033-1049.

Ross, W.H., & Wieland, C. (1996). Effects of interpersonal trust and time pressure on managerial mediation strategy in a simulated organizational dispute. *Journal of Applied Psychology, 81*(3), 228-248.

Rotter, J.B. (1971). Generalized expectancies for interpersonal trust. *American Psychologist, 35*, 1-7.

Shmatikov, V., & Talcott, C. (2005). Reputation-based trust management. *Journal of Computer Security, 13*, 167-190.

Stack, L.C. (1988). Trust. In H. London & J.E. Exner, Jr. (Eds.), *Dimensionality of personality* (pp. 561-599). New York: Wiley.

Sproull, L., & Kiesler, S. (1986). Reducing social context cues: Electronic mail in organizational communication. *Management Science, 32*, 1492-1512.

Strauss, S.G. (1997). Technology, group processes, and group outcomes: Testing the connections in computer-mediated and face-to-face groups. *Human-Computer Interaction, 12*, 227-266.

Tyler, T.R., & Kramer, R.M. (1996). Whither trust? In R.M. Kramer & T.R. Tyler (Eds.), *Trust in organizations: Frontiers of theory and research* (pp. 1-15). Thousand Oaks, CA: Sage Publications.

Walther, J.B., & Bunz, U. (2005, December). The roles of virtual groups: Trust, liking, and performance in computer-mediated communication. *Journal of Communication, 55*, 828-846.

Williamson, O.E. (1985). *The economic institutions of capitalism.* New York: Free Press.

Wilson, J.M., Strauss, S.G., & McEvily, B. (2006). All in due time: The development of trust in computer-mediated and face-to-face teams. *Organizational Behavior and Human Decision Processes, 99*(1), 16-33.

Witt, L.A., Andrews, M.C., & Kacmar, K.M. (2000). The role of participation in decision-making in the organizational politics-job satisfaction relationship. *Human Relations, 53*, 341-358.

Chapter VII
Trusting Remote Workers

Beverly Leeds
University of Central Lancashire, UK

ABSTRACT

This chapter examines the nature of trust from a number of theoretical bases, with reference to remote workers more often referred to as teleworkers or telecommuters. It examines the relationship between a manager and the remote worker (teleworker). It is concerned with the nature and conditions of trust rather than an examination of the importance of trust or how trust can be created. As well as examining the bases of trust, the chapter examines different levels of trust that can support a teleworking relationship. It draws a distinction between individual and organizational trust and between the conscious and unconscious states of trust. It concludes with a conceptual model that provides a framework to explain some of the anomalies and confusion in the debate regarding the nature of trust in teleworking arrangements. The chapter also suggests how the model may be used to analyze trust in these remote working arrangements and as a framework on which to build trust using different bases and at different levels.

INTRODUCTION

As we move further into the 21st century, technology is providing the opportunity to work anytime and anywhere. There has been an increase in remote working (teleworkers) and virtual teams, which brings new challenges and opportunities for managers and employees. This phenomenon highlights and exacerbates existing tensions and issues. Trust is an issue that managers have wrestled with for decades but as location independent working becomes the norm, the issue of

trust needs to be re-addressed. The importance of trust amongst a particular group of remote workers, known as teleworkers, has been noted by many authors (Handy, 1996; Huws, Korte, & Robinson, 1990; Olson, 1988). These authors suggest that the successful management of these remote workers requires trust and new forms of supervision and means trusting and empowering employees to complete the work when and where it suits them best (Korte, Steinle, & Robinson, 1988). Despite the fact that there are many who argue that trust is required for the successful management of remote teleworkers, there has been very little discussion regarding the nature and condition of trust. Other studies concerned with trust and telework have not examined the nature of trust and have not clearly defined the type of trust that is necessary for remote working to be effective. Some authors have failed to recognize both the multidimensional nature of trust and the different levels of analysis of trust, referring to trust as if there is only one definition, one source, one type, and that trust has only one state. This chapter seeks to address this by examining the nature of trust from a number of theoretical bases including economics, sociology, philosophy, and psychology. It is not concerned with an examination of the importance of trust or how trust can be created but seeks to explore the nature and bases of trust between a manager and a remote teleworker. It draws a distinction between the conscious and unconscious states of trust and between individual and organizational levels of trust. The chapter finishes with a theoretical model that provides an explanatory framework for some of the confusion and inconsistencies in the trust debate regarding remote working arrangements. The model illustrates both the awareness level of trust and the type of trust that can exist in a remote working relationship. It suggests that the model may be used as a framework on which to build trust using different bases and at different cognitive levels as well as offering an analytical framework to examine trust in remote working.

TRUST AND TELEWORK

One of the difficulties in examining trust in telework is that not only are there a number of different opinions regarding the fundamental nature of trust but there are also a plethora of definitions of telework. Nilles, Carson, Gray, and Hanneman (1976) considered telework as working from home with electronic support. The time-space distantiation incorporated into telework practices provides a number of alternative work forms such as satellite centers, neighborhood work centers, and mobile work as well as home-based telework (Jackson & van der Weilen, 1998; Kurland & Bailyn, 1999). Many recent studies, (see, for example, Felstead et al., 2001) distinguish between two types of remote workers: the home based teleworker and the mobile teleworker. This chapter is concerned with both of these types of remote workers, as they are both located independently from the organizational office. However, for the purposes of discussion, the UK's Labour Force Survey (LFS) definition of teleworkers as "people who do some paid or unpaid work in their own home and who use both a telephone and computer" will be used in this chapter. In addition, the employment relationship examined is that of an employee rather than a self employed teleworker.

The successful management of teleworkers requires trust and new forms of supervision that may be contrary to current practice (Huws et al., 1990; Olson, 1988). Most organizations tend to be arranged on the assumption that people cannot be relied upon or trusted (Handy, 1995) and this lack of trust by managers is frequently seen as a constraint on the development of telework, as trust requires a change in the way managers have traditionally managed. Remote working requires a new style of management and organizational culture that relies more on trust between the manager and employee and calls for productivity to be judged in terms of output or service delivery rather than the number of hours worked.

Trust can be seen by some managers as a contradiction in terms as it is both necessary and risky. It can be demonstrated that it is cheaper and more efficient to trust employees rather than regulate and control them, but it carries an element of risk. Furthermore, empowering employees by trusting them with more responsibility also requires them to be supplied with information that will allow them to make decisions. Handy (1995) argues that flexible working requires that management adopt a service role, supporting employees and not acting as an authority over them. If the teleworker is provided with an objective and left to get on with the task this will demonstrate a confidence in the teleworker's competence and commitment. Results can be assessed after the work is completed and thus control occurs after and not before or during the task set. If a remote worker cannot be relied on to complete a task, then some system of control which is incompatible with distance working (Huws et al., 1990) will need to be established. Trust in employees is necessary for the support and development of remote working arrangements. Without trust, telework will require control systems for managers to monitor the work being undertaken. In the long term, the absence of trust and necessity for control may prove to be too expensive and therefore detrimental to the continuation or expansion of remote working arrangements.

THE NATURE OF TRUST

Over the years researchers have studied trust from several disciplinary perspectives—anthropology, economics, psychology, sociology, and political science, among others. As can be expected with such a diversity of disciplines, not only have researchers from different disciplines addressed the same problem from different approaches and with differing methods, but they also have different opinions over the fundamental nature of trust. Most definitions of trust involve a belief that one partner will act in the best interest of the other partner. Researchers have defined trust in a variety of ways and at various levels of abstraction, with some authors endeavoring to organize the trust literature according to theory types (i.e., Hosmer, 1995; Lewicki & Bunker, 1995; Mishra, 1996; Sitkin & Roth, 1993). An investigation into the work of these authors reveals that the theoretical diversity among trust constructs is substantial. Rotter (1980), for example, views trust as a personality characteristic; Gambetta (1988) suggests

it is a rational decision; Zucker (1986) maintains it is a preconscious expectation; and Shapiro (1987) equates the principal-agent relation to trust. The considerable degree of diversity in the literature prevents a useful universal definition from being used. The only common ground of trust research appears to be the theme of actor vulnerability resulting from the acceptance of risk or uncertainty that an individual will not be taken advantage of by another in the relationship (Lane, 1998).

Basis of Trust

There is a considerable degree of disagreement concerning the grounds or social bases for trust. These differences appear to center on the model of human behavior underlying the different theories and differences in both the object of trust and the context of the relationship. In general, economists tend to argue that trust is based on calculation, whereas sociologists and organizational theorists identify the basis of trust as being concerned with common values or moral considerations (Lane, 1998). There is, however, some consensus in that the grounds for trust between employee and employer will vary depending upon the social context, the object of trust, and the stage of the relationship. For some theorists, trust is seen as multidimensional with more than one basis. Some economists, for example, highlight the combination of calculative trust with either cognitive or morally based trust (Dasgupta, 1988).

Calculative Trust

The calculative or rational choice view of trust involves expectations about another, based on calculations evaluating the costs and benefits of the trustor or trustee. This view of trust is based on the theory of the individual as a rational actor, where a course of action is chosen that will provide an individual with the maximum utility (Dasgupta, 1988). The rational actor only bestows trust if a calculation suggests the gain from reciprocated trust is higher than the loss threatened by a betrayal of trust (Preisendörfer, 1995 quoted in Lane, 1998). A manager might weigh up the risk of the teleworker not working while away from the office, against the possible gains in productivity to be made while the teleworker works uninterrupted. Hence a manager may calculate the costs and benefits and, on balance, decide to trust or not to trust the remote worker. This fear by management that teleworkers will not work unless under the watchful eye of a manager has been reported as the principal barrier to the growth of telework (Nilles et al., 1976; Olson, 1988). Coleman (1990) maintains that the making of a precommitment in trusting behavior equates to the issue of "social credit slips". In addition, teleworkers tend to be high trust employees (Handy, 1996). These teleworkers have built up their stock of credit slips to earn their place as a trustee and granted the opportunity to telework.

Transaction cost economists also recognize that actors are limited rationally and are influenced by opportunistic behavior. As opportunistic behavior is dealt with by control mechanisms, it incurs a cost. Thus, a calculative or rational choice manager of remote workers will only be willing to trust if there is an expectation that the balance of costs and benefits will favor co-operative behavior. Furthermore, a manager may deny an employee the opportunity to telework as the cost of monitoring and surveillance to ensure control is perceived as a cost that outweighs the financial gains of having offsite employees.

Value- or Norm-Based Trust

Some theorists object to the calculative or rational-choice view of trust, stating that this view fails to consider the social nature of action. Weber (1978) has argued that this notion of man as a rational egoist is far too narrow. Refuting the notion of calculative trust, Parsons (1951) argues that trust cannot exist unless individuals share common values. He perceives trust as being based on the expectation by the trustor, that the trustee (particularly if in a position of power) will meet social obligations and responsibilities. Thus trust is morally based and collectively orientated. Teleworkers, because no-one is watching them, can be seen in a position of power. Value-based trust can be seen to occur if the manager expects the remote workers to meet their work obligations and commitments without constant supervision. The teleworker shares this value that the manager trusts the teleworker to meet the work obligations. Fukuyama (1995) supports this in stating that "trust comes out of shared values". However, the assumption that trust is solely based on shared values is as partisan as the unilateral belief in calculative trust. The concept of value based trust is adopted by many authors concerned with trust in the fields of economics and organizational and management studies (Lane, 1998). A number of writers have adopted a more limited idea of value-based trust, arguing that values and norms are applicable in specific circumstances or contexts (Bradach & Eccles, 1989; Granovetter, 1985; Sako, 1998).

System Trust

System trust is identified by Luhmann (1979) as being based on the functioning of the system and is, in essence, confidence in an abstract system. Lewis and Weigert (1985), quoted in Seal (1998) further develop this concept, arguing that system trust does not derive from emotion, but rather has a presentational base which "is activated by the appearance that everything seems in proper order" (p. 974). They argue that this is necessary to ensure the organization or system functions effectively. Luhmann (1988) also argues that this trust is symbolic and that it is sensitive to symbolic events; therefore, system trust can be created or destroyed. An organization, therefore, which has remote working arrangements, can create trust by creating confidence in the system. Trust in the teleworking arrangement can be derived from symbolic representation. By accepting telework as the norm and by removing artificial barriers to its growth, an organization may be able to create system trust.

Contractual Trust

Williamson highlights the fact that some economists refer to the presence or absence of contractual safeguards, rather than the presence or absence of trust (1993). Therefore, an employer could interpret such safeguards as a means of minimizing or eliminating vulnerability, risk, or reliance on "the word of another" (Rotter, 1967, p. 651). Contractual safeguards can be seen as removing the necessity to trust the individual teleworker. However, although formal mechanisms such as contracts may act as a substitute for trust, they are often used in a symbolic rather than heavy-handed, legalistic manner. Sitkin and Roth (1993) highlight that attempts to resolve trust violations legalistically are usually unsuccessful because, ironically, they reduce the level of trust rather than producing it. This contractual arrangement produces a psychological barrier between the two parties that reinforces the formality and hence the need for more rules (Sitkin & Roth, 1993); thus, legal remedies and tight monitoring may reduce tendency of teleworkers to be unreliable but may create a "distance" between the teleworker and the organization that causes trust to diminish.

LEVELS OF TRUST

It is without doubt necessary for trust to exist between a manager and teleworker in telework arrangements. However, it is naïve to assume that trust is only necessary at the individual level. As well as having different bases, trust can and does exist at different levels. Some authors have defined trust as personal and impersonal (Luhmann, 1979; Zucker, 1986), others have distinguished between trust as an individual attribute, trust as a behavior,

and trust as an institutional arrangement (Sitkin & Roth, 1993). Luhmann (1979) distinguishes between personal trust, based on familiarity and system trust, based on the functioning of the system. Zucker (1986) distinguishes between three sources of trust (process, characteristic, and institutional) by the way they assume unity of expectation to be present. Process and characteristic trust are seen as sources of personal trust, whereas institutional trust that cannot rely upon commonality of characteristics or past history is impersonal. Trust in telework can be seen as an individual attribute between the individual teleworker and manager or as an organizational or institutional arrangement between individuals and the organization.

Individual Trust

Where trust is considered as an individual attribute, the relevance of expectations concerning the trustworthiness of another individual is only evident when there is an element of dependence on the prior action or cooperation of another person (Dasgupta, 1988; Luhmann, 1979), and is where the manager is dependent on the teleworker's cooperation and vice versa. Lewis and Weigart (quoted in Lane, 1998) go further, arguing that, except for social relationships, there would be no necessity for individuals to trust if there was no dependence or vulnerability. This individual trust may be enhanced by personal interaction between manager and remote worker. Trust may be based on personal knowledge of the teleworker, but it goes beyond just possessing information about the teleworker, as possession of information does not directly correspond to trusting that person. Luhmann (1979) argues that trust "needs history as a reliable background" (p. 20) and that without it, trust cannot be conferred. Furthermore, trust is not merely an inference from the past but defines the future. Hence, a manager needs information regarding the past experience and performance of the teleworker. Similarly, the teleworker needs knowledge and experience of how a manager has responded in the past. This past experience and knowledge can help frame or hinder the future teleworking relationship.

Zucker (1986) identifies two sources of personal or individual trust: process-based and characteristic-based trust. Process-based trust she perceives as being concerned with past or expected exchanges, which presumes a degree of stability. This type of trust is seen as being built incrementally through the gradual accumulation of direct or indirect knowledge about another. Sydow (1998) highlights that process-based trust may eventually end in creating a reputation for the person, or system, based on past practices. However, process-based trust by its nature cannot be created. A teleworker's or manager's past experiences can only be used to define future expectations.

Characteristic-based trust, however, is one where trustor and trustee are socially similar or have some cultural unity (Zucker, 1986). The global nature of remote work may exacerbate the difficulty of enhancing or creating this type of trust. Characteristic or process based trust is also difficult to produce due to the increasingly complex nature of background expectations. Although both process and characteristic based trust are important in a teleworking relationship, they are almost impossible to create. Furthermore, Sydow (1998) argues that process- and characteristic-based trust can only be developed through face-to-face interactions, reinforcing the argument that this type of trust cannot be used to enhance trust in teleworking relationships.

Organizational Trust

Shapiro (1987) identifies impersonal trust as common in developed societies where individuals act as trustees. Zucker (1986) identifies this impersonal trust as institutional trust, where there is a great reliance on formal structures which guarantee trust. This type of trust becomes part of "the external world known in common" (Zucker, p. 63); it is, in effect, institutionalized. However, unlike process- and characteristic-based trust, Zuker argues that both individuals and organizations can intentionally construct institutional trust. Furthermore as organizations become more impersonal through geographical remoteness, individual trust may be replaced or supplemented by institutional trust (Zucker). Individual trust between a teleworker and a manager may be created or reinforced by trust created within the organization as institutional based trust. Formal structures may be put in place to ensure trust in the telework arrangement can exist.

Conscious and Unconscious Conditions of Trust

Garfinkel (1967, quoted in Lane, 1998) distinguishes between expectancies that exist in the background and those expectancies that are developed. He identifies the unreflective quality of trust, drawing attention to trust as "expectancies of persistence, regularity, order and stability in the everyday and moral world" (Garfinkel, 1967, p. 173, quoted in Lane, 1998). Thus trust enables the manager to take for granted, that is, to take on trust, many routine actions in a teleworking relationship. Zucker (1986) builds on Garfinkel's supposition that expectations may be background and/or constitutive and that trust is developed more easily where more background expectations exist. Therefore if a manager works with a teleworker of good reputation, trust becomes habitual and unconscious rather than explicit. This habitual state is obviously beneficial to both employer and teleworker, allowing both to concentrate on other things. Thus habit, by enabling a predictable event to be managed with very little effort, allows an individual to concentrate on the unpredictable (Young, 1988). Once an employer has agreed that an employee work as a teleworker, then the relationship becomes mainly habitual. The activities that make up the transaction are not evaluated in detail. Therefore, the manager is confident that work will be undertaken and completed and the teleworker is confident payment will be received. This unconscious trust will continue until there is any variation in the teleworker's or manager's behavior. If a teleworker undertakes the work given and delivers work correctly and on time, then a habitual relationship will be formed between the employer and employee. This habitual state with unconscious trust can become conscious trust or, worse, conscious distrust.

TOWARDS A CONCEPTUAL MODEL

From the preceding discussion it can be concluded that trust in a teleworking arrangement operates in a number of dimensions and at a different levels. Not only can trust be an individual attribute (Luhmann, 1979; Zucker, 1986) it can also be an institutional arrangement (Sitkin & Roth, 1993). At the individual level, trust is personal, based on both process (past performance, etc.) and characteristics (Zucker, 1986). This personal trust may be calculative (Dasgupta, 1988), based on a subjective

cost benefit analysis based on past performance (process-based) and it may also be based on shared values or norms (characteristic-based). At the institutional level, trust is symbolic, reliant on abstract system features such as reputation, and so forth (Luhmann, 1979). Contractual trust is a form of trust that may occur at a specific event or for a specific purpose in the employment relationship. In addition trust may not always be at a state of consciousness. Moreover, in a relationship, trust may be habitual and unconscious.

Consequently, the complex and multidimensional nature of trust offers an explanation for the diversity of opinions as to its nature. Trust can be individual or institutional (Zucker, 1986) and conscious or unconscious. These can be seen not as definitive states but positions along two continua. An individual manager may need to trust the individual teleworker but she also needs to trust the organization to support the teleworking arrangement. This trust may be a deliberate and conscious effort to trust, or it may be habitual and unconscious. Figure 1 illustrates the two continua along which trust can be placed. It is these two continua that provide an opportunity

Figure 1. Conceptual model of the nature of trust in teleworking relationships

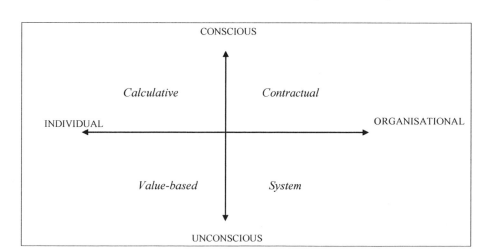

to explain other competing theoretical models of trust. At any point in time whether at the start of a teleworking arrangement, after a key event, or while teleworking, trust may exist in a conscious or unconscious state and at the individual or institutional level. When trust moves to and from a conscious/unconscious state and to and from an individual/institutional level the nature of trust can be seen to change.

At the individual level, a conscious state of trust requires the individual manager to decide whether to trust the teleworker by evaluating the risks involved. Thus, trust at the conscious state and individual level can be seen as calculative. When this trust becomes habitual and slips into the unconscious level, it becomes trust based on values or norms. It is in this quadrant where blind trust would be placed - trust placed by an individual manager based on shared values alone without thinking. Similarly, trust at the organizational level at a conscious state requires a decision or encouragement to trust. This can be seen as contractual-based trust, a formalization of organizational trust. If this organizational trust is or becomes unconscious and habitual, then it becomes a trust based on the system.

This model also offers an explanation for the conflicting opinions regarding whether trust can be created. While each of the four quadrants of trust are commensurate and trust may occur in one or all quadrants at the same time, there is a fundamental difference between trust at the individual and institutional levels. This key difference is the degree to which trust can be created. It has been argued that trust at the individual level cannot be brought about at will (Luhmann, 1979). Individuals cannot will others to trust them; neither may individuals will themselves to trust others. To do this would destroy trust itself. However, at the institutional level trust can be instilled through the building of a reputation (unconscious) or through contractual and regulatory standards (conscious).

CONCLUSION

The conceptual model presented in this chapter provides a framework to explain some of the anomalies and confusion in the debate regarding the nature of trust in teleworking arrangements. In addition, it also provides a model to illustrate both the level of trust and the condition of trust that can exist in teleworking relationships. Trust may be perceived as the property of a relationship and hence the distinction between personal (individual) and system (organizational) trust. The object of personal trust is the individual teleworker whereas the objects of system trust are technical or social systems of teleworking organizations. In addition to trust existing at the individual and organizational levels, it is found as conscious and unconscious states. A manager who has evaluated the reputation of a teleworker supplier will form a habitual relationship, expecting work to be undertaken and completed as requested. This habitual and unconscious trust will continue until there is any variation in the teleworker's behavior that may change this passive expectation of continuity into conscious trust or conscious distrust (Blois, 1998). Hence, trust in telework, whether individual or organizational, can be seen as both fragile and emotional and hence can easily be transformed into suspicion and doubt.

The implications of the conceptual framework for telework are: it provides a framework on which to examine trust in telework arrangements; it highlights where and in what contexts trust can be created; it defines the nature of trust that can help support teleworking relationships; and it draws attention to the bases on which trust should be built for successful teleworking relationships. The model highlights that in teleworking arrangements individual trust based on calculation is time consuming, and furthermore, this trust is easily destroyed. In addition, organizational trust based on contractual arrangements may be costly to arrange and may also serve to damage trust in the relationship. Calculative or contractual based trust can be seen as acting as a deterrent to the growth of telework. It could be argued that on these bases of trust, a conflict exists between the costs and benefits calculated for trust and the costs and benefits calculated for operational efficiency. Consequently, limiting telework to high trust employees may be difficult to justify against cost efficiencies brought about by transferring workforce to offsite employees. It is clear that there is a need to consider other bases of trust other than, or in conjunction with, the notion of contractual or calculative trust. This does not mean that it can be assumed that trust within a teleworking relationship should be completely free of calculation, but that it should not be based solely on contractual or calculative relationships.

Value or norm based trust can be seen to help drive and increase teleworking arrangements and system based trust can be seen to support and promote teleworking arrangements. An organization should look to create values and norms that can assist with trust at the individual level and a system within the organization that will reinforce trust at the organizational level. Overall, the implication for teleworking organizations and teleworkers is that trust can and should be created at both the individual and organizational level. Furthermore, this trust should become habitual and unconscious and not contractual or calculative.

The role of trust has been examined by a number of authors and there is widespread agreement that trust is vital for successful teleworking arrangements. Despite this, there is very little empirical evidence that exists regarding the nature of trust in telework. This conceptual model has only sought to examine the nature and conditions of trust in teleworking arrangements from a theoretical perspective. Empirical evidence should be sought to assess the viability of the model in teleworking organizations. Furthermore, the framework does not consider the impact of national culture or different demographic profiles

of teleworkers. There are many opportunities for further developing the model to include these factors. Empirical evidence is needed to test the model and also to provide practical guidance for a teleworking organization to increase trust at the individual and organizational levels. In addition, cross-cultural studies of trust development in teleworking arrangements would be beneficial, particularly given the global nature of telework. Furthermore, the model should be examined using teleworkers with differing demographic profiles in order to examine the impact of gender and other factors.

REFERENCES

Blois, K. (1998). A trust interpretation of business to business relationships: A case-based discussion. *Management Decision, 36,* 5.

Bradach, J. L., & Eccles, R. G. (1989). Price, authority and trust: From ideal types to plural forms. *Annual Review of Sociology, 15,* 97-118.

Coleman, J. S. (1990). *The foundations of social theory.* Cambridge, MA: Harvard University Press.

Dasgupta, P. (1988). Trust as a commodity. In D. Gambetta (Ed.), *Trust: Making and breaking of co-operative relations.* (pp. 49-72). Oxford: Blackwell.

Felstead, A., Jewson, N., Phizacklea, A., & Walters, S. (2001). Working at home: statistical evidence of seven key hypotheses. *Work, Employment and Society, 15*(2), 215-31.

Fukuyama, F. (1995). *Trust: The social virtues and the creation of prosperity.* London: Hamish Hamilton.

Gambetta, D. (1988). Can we trust trust? In D. Gambetta (Ed.), *Trust: Making and breaking co-operative relations.* (pp. 12-25). Oxford:

Blackwell.

Granovetter, M. (1985). Economic action and social structure: A theory of embeddedness. *American Journal of Sociology, 91,* 481-510.

Handy, C. (1995, May-June). Trust and the virtual organization, *Harvard Business Review, 73*(3), 40-50.

Hosmer, L. T. (1995). Trust: The connecting link between organizational theory and philosophical ethics. *Academy of Management Review, 20*(2), 379-403.

Hotopp, U. (2002). Teleworking in the UK: The trends and characteristics of teleworking in the UK and comparisons with other Western countries. *Labor Market Trends, 110*(6).

Huws, U., Korte, W. B., & Robinson, S. (1990). *Telework: Towards the elusive office.* New York: John Wiley.

Jackson, P. J., & van der Wielen, J. M. (1998). *Teleworking: International Perspectives: From Telecommuting to the Virtual Organisation.* Routeledge, London.

Korte, W. B., Steinle, W. J., & Robinson S. (1988). *Telework: Present situation and further development of a new form of work.* Bonn: Elsevier Science.

Kurland, N., & Bailyn, L. (1999). Telework: the advantages and challenges of working here, there, anywhere, and anytime. *Organizational Dynamics, Autumn,* 53-68.

Lane, C. (1998). Theories and issues in the study of trust. In C. Lane & R. Bachman (Eds.), *Trust within and between organizations.* (pp. 1-30). New York: Oxford University Press.

Lewicki, R. J., & Bunker, B. B. (1995). Developing and maintaining trust in work relationships. In R. M. Kramer & T. R. Tyler (Eds.), *Trust in organizations: Frontiers of theory and research* (pp. 114-139). Thousand Oaks, CA: Sage.

Luhmann, N. (1979). *Trust and power.* Chichester: John Wiley.

Mishra, A. K. (1996). Organizational responses to crisis: The centrality of trust. In R. M. Kramer & T. R. Tyler (Eds.), *Trust in organizations: Frontiers of theory and research* (pp. 261-287). Thousand Oaks, CA: Sage.

Nilles, M., Carlson, F., Gray, P., & Hanneman, G. (1976). *The telecommunications-transportation tradeoff.* New York: John Wiley & Sons.

Olson, M. H. (1988). Organizational barriers to telework in telework. In W. B. Korte, W. J. Steinle & S. Robinson (Eds.), *Present situation and further development of a new form of work* (pp. 135-151). Bonn: Elsevier Science.

Parsons, T. (1951). *The social system.* London: Routledge Keegan & Paul.

Rotter, J. B. (1967). A new scale for the measurement of interpersonal trust. *Journal of Personality, 35*(4), 651-65.

Rotter, J. B. (1980). Interpersonal trust, trustworthiness, and gullibility. *American Psychologist, 35,* 1-7.

Sako, M. (1998). Does trust improve business performance? In C. Lane & R. Bachman (Eds.), *Trust within and between organizations.* (pp. 88-117). New York: Oxford University Press.

Seal, W. B. (1998). Relationship banking and the management of organizational trust. *International Journal of Bank Marketing, 16*(3), 102-108.

Shapiro, S. P. (1987). The social control of impersonal trust. *American Journal of Sociology, 93,* 623-658.

Sitkin, S., & Roth, N. (1993). Explaining the limited effectiveness of legalistic remedies for trust/distrust. *Organization Science, 4,* 367-392.

Sydow, J. (1998). Understanding the constitution of interorganizational trust. In C. Lane &

R. Bachman (Eds.), *Trust within and between organizations.* (pp. 31-63). New York: Oxford University Press.

Weber, M. (1978). *Economy and society.* Berkeley: University of California Press.

Williamson, O. E. (1993, April). Calculativeness, trust and economic organization. *Journal of Law and Economics, 36*, 453-86.

Young, M. (1988). *The metronomic society: Natural rhythms and human time keeping.* London: Thames and Hudson.

Zucker, L. G. (1986). Production of trust: Institutional sources of economic structure, 1840-1920. *Research in Organizational Behaviour, 8*, 53-111.

Chapter VIII
Trust in Virtual Multicultural Teams

Iris C. Fischlmayr
Johannes Kepler University, Austria

Werner Auer-Rizzi
Johannes Kepler University, Austria

ABSTRACT

This chapter analyzes the phenomenon of trust with regard to its significance for virtual teams. Guided by the existing literature on trust, this chapter presents different kinds of trust and the development of trust over time. The challenges inherent to virtual multicultural teams, thus to working teams, which are geographically dispersed and communicate with the help of electronic media, raise the questions of their consequences on trust. As virtual teams are mostly used in companies operating in different countries all over the world, the different cultural backgrounds of the team members are taken into account as well. To give an example for the relevance of this issue in practice, an illustrative case study on experiences international business students have made during virtual team projects is presented.

INTRODUCTION

Trust is a broadly discussed issue in all disciplines. In the field of organizational behavior, trust is, for example, reflected in the relationship between leader and subordinate and among employees, but also in groups and teams. Working teams, which are geographically dispersed and communicate with the help of electronic media (referred to as virtual teams), are dominating today's international business area. In this contribution we

analyze the phenomenon of trust with regard to its significance for virtual teams.

Guided by the existing literature on trust, we start the chapter by presenting different kinds of trust, such as affective and cognitive trust, as well as factors influencing trust. Given that trust is not a stable component, we also give insight into the development of trust over time. By discussing the nature of virtual teams, we will introduce some challenges that go hand in hand with that modern form of collaboration. As virtual teams are mostly used in companies operating in different countries all over the world, we need to take into account the different cultural backgrounds of the team members.

The challenges inherent to virtual multicultural teams raise the question of their consequences on trust. With the help of the existing literature on trust and on virtual teams, we will first clarify how and in which form trust is appearing in virtual teams and how these forms evolve over time.

To give an example for the relevance of existing literature in practice, we present an illustrative case study on experiences international business students have made during virtual team projects. We use the observation of those students, their individual reflections on that experience, and questionnaires as a tool to give an example of the practical appearance of trust in virtual multicultural teams. Based on these results we assume that in virtual multicultural teams, affective and cognitive trust have a different development.

THE PHENOMENON OF TRUST

Different Facets of Trust

Luhmann (2000) regards trust as an efficacious mechanism to reduce social complexity and as a possibility to enlarge one's scope of action. This approach is based on the idea that in a complex world a person is only able to perceive and process a marginal part of all possible information and therefore only has a rather limited basis for rational decision making. If the person was able to rely on future actions of another person, then the complexity of the world would be reduced because a certain part of the other person's possibilities to act could be excluded from her behavioral repertoire. In Luhmann's terms, trust simplifies one's life through taking risk (2000, p. 93).

The following paragraphs are based on the major distinction between trust and distrust, which are interrelated, yet separate phenomena. In the area of trust, the literature is referring to two different levels, the interpersonal one on the one hand, and the abstract system on the other.

Interpersonal trust is based on the relationship of two parties, the trustor and the trustee. It is the willingness of the trustor to make himself vulnerable to the actions of the trustee. The willingness is based on the trustor's expectation that the trustee will act in a way that is important to the trustor and is independent of the possibility to control the trustee. From this perspective, Luhmann (2000) defines trust as a previous engagement on the part of the trustor, which involves uncertainty and risk. In addition, according to Deutsch (1952), trust is only possible in situations where the possible damage of a breach of trust is bigger then the possible advantages gained when trust has been proven (Luhmann, 2000). Trust, therefore, goes beyond a mere rational calculus. The deliberate acceptance of negative consequences is not equal to trust as long as the risks remain within acceptable limits. Trust is only required if a bad outcome would make you regret the decision (Luhmann, 2000). Trust is always associated with a positive attitude towards the trustee, which to a certain part stays irrational and is not based on risk, but reduces or substitutes it.

According to Lewis and Weigert (1985), interpersonal trust has cognitive and affective foundations. McAllister (1995) found empirical evidence in the sociological and social psychological literature for the distinction between cognition- and affect-based trust. *Cognition-based trust* is grounded in the trustee's competence and responsibility, as well as reliability and dependability. It is the trustor's perception of the trustee's ability to deliver as promised or expected. *Affect-based trust* consists of the emotional bonds between individuals and demonstrates goodwill and care. "People make emotional investments in trust relationships, express genuine care and concern for the welfare of partners, believe in the intrinsic virtue of such relationships, and believe that these sentiments are reciprocated" (McAllister, 1995, p. 26). McAllister found in his empirical investigation that in his sample, the general levels of cognition based trust were higher than affect based trust. "This is consistent with the understanding that some level of cognition-based trust is necessary for affect-based trust to develop" (McAllister, p. 51). However, although cognition- and affect-based trust may be causally connected, each form of trust functions in a unique manner and has a distinct pattern of association to antecedent and consequent variables.

System trust is based on the thought that things will develop according to norms and regulations in a usual manner and is independent of persons. It refers to social institutions and systems, such as the free market economy, money, legal system, and so forth, whose representatives are unknown. Two components of system based trust are discussed in the literature: (a) an individual may "believe in a normal situation," that is the individual relies on the situation being not exceptional but adequately ordered and therefore is able to successfully interact within the system, and (b) the second component of system based trust is the belief in sufficient structural measures like rules and procedures, guaranties, legal recourse, and so forth. System based trust has some influence on interpersonal trust, especially at the beginning of a trust relationship with still a relatively low level of information about the other person who is perceived as a representative of a functioning system.

Distrust has a strong interrelationship with trust. Recent findings in social psychological research indicate that trust and distrust are inter-related, yet different phenomena (Lewicki, McAllister, & Bies, 1998). Luhmann (2000) refers to distrust as a "functional equivalent" of trust. He argues that trust reduces social complexity and makes the person capable of acting in the first place. If trust is reduced, then full complexity and incapability of action is merely restored. According to Luhmann (2000), the person needs to find a different strategy in order to be able to define a reasonable situation: Distrust in terms of raising expectations to the negative can also serve as a function of reducing complexity. It is reflected in high cautiousness, scepticism, even to the point of paranoia and is characterized by defensive and vigilant behavior, whereas trustful relationships are characterized by good faith and benevolence. According to Lewicki et al. (1998), individuals do not assess others in a relationship only on one criterion but use more concurrently. Therefore trust and distrust can co-exist at the same time.

Development of Trust

1. **Gradual development of trust:** Traditionally researchers on trust implicitly assumed that trust between persons starts at a relatively low level and is then nurtured by mutual corresponding actions and experiences. Shapiro, Sheppard, and Cheraskin (1992) describe the development of trust as a linear sequence with three different kinds of trust coming into play, each one is in the foreground at a different stage depending on the maturity of the relationship. In the beginning of a relationship there is *calculus-based trust* (deterrence based trust), which is grounded in the consistency of the trustee's behavior with promises. It is kept alive through the threat of punishment in the case of breach of trust (i.e., loss of relationship, loss of good reputation, etc.) or rewards in the case of meeting the expectations. In this way it constitutes a sheer economic market oriented calculus in which the costs of building and deepening the relationship are contrasted to the costs of endangering the relationship. *Knowledge based trust* at the next level refers to the predictability of behavior. One knows the other person sufficiently well in order to be able to judge future behavior. Knowledge based trust is developed over time through ongoing interactions and regular communication and observations. The third level is *identification based trust*. At this stage trust develops through identification with the beliefs, values, and intentions of the other person. It emerges from knowing as well as sharing other persons´ needs and preferences. According to Shapiro et al. (1992), identification based trust enables vicarious acting and makes control unnecessary in the first place.

2. **High initial trust:** Recent research has suggested that parties at the start of a social encounter more often than not display high levels of trust (Jones & George, 1998; McKnight, Cummings, & Chervany, 1998). Empirical studies by Berg, Dickhaut, and McCabe (1995) and Kramer (1994) showed that despite the lack of incentives and knowledge about the other party, subjects displayed high levels of initial trust. Luhmann (2000) even argues that people prefer to trust someone and to assume that the other person is within the limits of one's own value system, rather than to show initial distrust. It would cost too much time and effort to investigate the real nature of the other person's value system.

Meyerson, Weick, and Kramer (1996) have argued that especially for temporary work teams it is necessary to start with a high level of trust which they called *swift trust*. The kind of situations for example film teams, theatre and architectural groups, presidential commissions, senate select committees, and cockpit crews are confronted with, are characterized by tight deadlines, team members having diverse skills and a limited history of working together and low probability of working together in the future. There is no time for relationship building and incremental gradual knowledge acquisition about the behavior of others. According to Meyerson et al. (1996), trust in these situations is mainly imported rather than developed and is initially based on broad categorical social structures and later on action and refers to the cognitive rather than the affective dimension.

3. **Factors influencing trust development:** One factor which influences the development of trust is the trustor's propensity to trust which constitutes a relatively stable personal trait and has been developed since earliest childhood. It is the general disposition to trust others and can be characterized through the common belief in humanity; thereby one takes for granted that people in general are trustworthy and benevolent. Another possibility for a high propensity to trust is that a person has the attitude that trust as a virtue per se is good and advantageous, independent of what one thinks about other persons.

Characteristics of the trustee also influence the level of trust. According to McKnight et al. (1998), the trustor's perception of the factors ability, integrity, and benevolence explains the most part of the trustee's trustworthiness. The inclusion of perceived ability makes the concept of trustworthiness task and situation specific. One can think much of someone in one area of competence, whereas at the same time not think much in a different area. Benevolence means that the trustee needs to have a positive orientation towards the trustor and does not pursue egocentric motives. Integrity refers to the trustee's adherence to principles which are seen as good and right by the trustor.

VIRTUAL MULTICULTURAL TEAMS

Working across borders without being limited by geographical distances is becoming increasingly popular and necessary in the modern business world. One manifestation of this trend is the use of so-called "virtual teams". According to Cohen and Gibson (2003, p. 4), virtual teams can be defined as functioning teams whose members are geographically dispersed and their communication is technology-mediated rather than face-to-face. As geographical distance is one of their key features, teams are characterized by the cultural diversity of their team members and we therefore term them "virtual multicultural teams".

For almost two decades, researchers have shown increasing interest in this modern form of collaboration. Many of those studies focus on the particular characteristics of virtual teams such as technological tools (Duarte & Snyder, 2001; Riopelle et al., 2003) or communication (Pottler

& Balthazard, 2002). Others deal with team processes and focus on issues such as team building (Hart & McLeod, 2003; Huang, Wei, Watson, & Tan, 2002) or team performance (Driskell et al., 2003; Lawler, 2003; Levenson & Cohen, 2003). Still others simply provide "best practices" (Kirkman, Rosen, Gibson, Tesluk, & McPherson, 2002; Lurey & Raisinghani, 2000). So far, little attention has been paid to social factors influencing work in virtual teams. Some authors have started to point out the relevance of these factors based on their importance in "normal" face-to-face teams. Among these, leadership (Davis, 2003; Duarte & Snyder, 2001; Tyran, Dennis, Vogel, & Nunamaker, 2003; Zigurs, 2003), conflict (Griffith, Mannix, & Neale, 2003), influence (Elron & Vigoda, 2003) and commitment (Workman, Kahnweiler, & Bommer, 2003) are dealt with most frequently.

Due to their geographical dispersion, most of the virtual teams used in multinational companies are composed of members of different cultures; thus, the phenomenon of culture has a crucial influence on their collaboration and also on trust. Although research on the influence of culture on teams is still in its infancy (Maznevski, Davison, & Barmeyer, 2005), several studies have shown that cultural diversity has an impact team processes (Cox, 1993; DiStefano, & Maznevski, 2000; Konradt & Hertel, 2002; Lipnack & Stamps, 2000; Thomas, 1999; Watson, Johnson, & Zgourides, 2002). However, researchers can be divided according to their opinions of whether the multicultural composition of a team lowers (Kirchmeyer & Cohen, 1992; Watson, Kumar, & Michaelsen, 1993) or enhances group performance (Driver, 2003; Konradt & Hertel, 2002; Richard, 2000; Stumpf & Alexander, 1999; Thomas & Ravlin, 1996).

Basic differences among cultures such as the perception of time, individualistic vs. collectivistic orientation, power distribution, the attitude towards risk or feminine vs. masculine life style have been formulated in different models on cultural dimensions (Hall, 1976; Hall & Hall, 1990; Hampden-Turner & Trompenaars, 2000; Hofstede, 1980, 2001; Kluckhohn & Strodtbeck,

1961; Trompenaars, 1993). From those, especially the questions of relationship building, commitment, taking over responsibility, decision making processes, and issues related to communication such as context, communication style, use of communication tools seem to be the crucial factors influencing multicultural virtual teams. Particular problems might stem from the fact that more collectivistically oriented cultures strive for relationship building first and serve as a basis for doing business, whereas individualistic cultures are highly task oriented and personal relationships are of lower order. For the former, face-to-face meetings are crucial for relationship building, because building successful relationships is very difficult using only technological media relationships. Additionally, whereas collectivistic cultures see group works as being completed together, individualistic ones tend to split tasks and value individual contributions higher.

Further, hierarchy oriented cultures prefer autocratic and powerful leaders and will perhaps have their problems in virtual teams where more democratic and egalitarian behavior has been revealed to exist. Another crucial cultural dimension influencing virtual multicultural teams is the context orientation of cultures in communication (Hall, 1976; Hall & Hall, 1990). Low context cultures where direct and clear communication dominates will have less problems with electronic media and communication tools than high context cultures where people need more non-verbal communication and more context to transfer their messages (Konradt & Hertel, 2002; Maznevski et al., 2005). Summarizing, we can state that team members often have been showing sensitivity concerning the attitude and perception towards foreign cultures, and do not understand the different perspectives. One reason is a rather ethnocentric orientation which means that behavior, cultural patterns and values from one's own culture are regarded as leading and are thus supposed to be valid for the other culture as well (Konradt & Hertel, 2002).

REVIEW OF EXISTING STUDIES ON TRUST IN VIRTUAL TEAMS

When it comes to the issue of trust, only a few academic publications can be found. Most of them try to draw conclusions from "traditional" face-to-face teams to virtual ones. The influence of different cultural backgrounds is more or less neglected. As the first empirical studies on trust in virtual teams date back to late 1990s, we can clearly classify this field of research as a rather young one. By trying to group the existing articles and studies on trust in virtual teams, we can define different categories: (a) contributions on different factors influencing trust, (b) studies on the development of trust over time, and (c) articles dealing with the effects of trust on performance in virtual teams. In the following paragraphs we intend to give an overview on the existing literature regarding trust in virtual teams, focusing on empirical studies contributing to our knowledge about the phenomenon.

Initial Influencing Factors

In some articles we could find issues that turned out to influence initial trust in virtual teams to a certain extent; thus, we named this category of studies "initial influencing factors." As such, we have identified the following: behavioral control (Piccoli & Ives, 2001), process management (Pauleen & Young, 2001), interpersonal traits (Brown, Poole, & Rodgers, 2004), perceived trustworthiness (Zolin, Hinds, Fruchter, & Levit, 2004) and parts of the study conducted by Aubert and Kelsey (2003). As the findings of those studies are not interrelated, we will only give a short overview on each single article respectively factor in chronological order.

Pauleen and Young (2001) show that the effects of crossing organizational and cultural as well as time and distance barriers have a great impact on building relationships in virtual teams. Although their article on relationship building is only indirectly related to trust, we see strong parallels and overlaps with affect-based trust. With the qualitative method of grounded theory, they aim not to test any hypotheses but show how facilitators use and might use information and communication technology (ICT) across borders for building and maintaining relationships between the team members. Their data collection took place over three years and was done with professional business people involved in an action-learning virtually based training program. Pauleen and Young's results give the impression that stronger relationships between team members go hand in hand with higher task performance and the effectiveness of information exchange. Further, they assume that strong ties are connected to increased creativity, motivation, morale, better decisions, and fewer process losses. Thus, the authors recommend that facilitators should support the building of stronger relational ties by carefully selecting formal and informal communication channels. Without any personal relationships before the task in question, using only formal communication media (such as e-mail, telephone, or desktop video conferencing) establishing solid relationships and trust-building is problematic. They see it as problematic for the establishment. Providing team members with the opportunity to use informal communication media (e.g., ICQ) spontaneously as well facilitates the socialization process and enables them in a certain way to exchange feelings and emotions as well. Organizational, cultural, and time and distance barriers render the situation even more complex and difficult as more time is needed for relationship building. Pauleen and Young seem to be the only authors also taking cultural diversity under consideration, at least by mentioning that the facilitators need to find out the required degree of personal relationships necessary to render the team functioning. All in all, according to the authors, a clever mix of formal and informal as well as synchronous and asynchronous communication channels helps a lot to manage the process of relationship and thus, also trust building.

In their longitudinal study on the effects of behavior control on trust, Piccoli and Ives (2003) conducted an experiment with 51 virtual student teams, half of them using control mechanisms known from traditional teams, the other half being self-directed. With a combination of a quantitative study and the in-depth analysis of communication logs, they could show that traditional control mechanisms have a negative effect on trust in virtual teams. Through behavior control mechanisms (e.g., defining detailed work assignments, specifying regulations and procedures, delivering continuous reports, or sticking closely to project plans) salience and vigilance increased and enhanced the opportunity that their failure was detected. Thus, the decline in trust is mainly rooted in team members not meeting the required obligations. According to Piccoli and Ives (2003), especially close to deadlines, failure is most likely to come to the surface and has then the strongest impact on the decline in trust. In those situations, the attention is more on individual contribution and performance than on mutual responsibility and duty. The self-directed teams have selected similar control mechanisms (such as over-structuring the team's interaction, fixing deadlines, proposing individual tasks, and demanding weekly feedback) but different combinations of them. The only difference that could be shown is between externally imposed and internally chosen regulations. But also in the self-regulated teams, the control mechanisms have lead to enhanced salience and vigilance and thus, to the probability that team members´ failure will be detected, which again contributed to trust decline.

Zolin et al. (2004) concentrated more on cross-functional virtual teams and tried to find out more on interpersonal trust in those by studying virtual long-term projects among architecture and construction students. With the help of surveys used together with interviews, they could show that above all perceived trustworthiness (defined according to Mayer, Davis, & Schoorman, (1995)) is crucial for the development and character of trust. The initial perception of this trustworthiness plays a significant role for the commitment and the stability of trust. Interestingly, members of one's own professional culture are seen as more trustworthy. When it comes to cultural differences, they could show that the more diverse the culture, the more fragile the trust. Zolin et al. concluded that in cross-functional teams it is harder to evalu-

ate the other one and thus, initial impressions as well as perceived trustworthiness and perceived follow-through influence to a high degree the character and development of trust.

A purely theoretical contribution is the study of Brown et al. (2004) on interpersonal traits and their effect on trust. They assume that the type of personality has an impact on: one's disposition to trust; the perceived trustworthiness, communication, and willingness to cooperate. According to Brown et al., trust is a "function of a constellation of attributes grounded in the individual's core personality: interpersonal traits influence individual interaction styles, which in turn shape the experiences that build or undermine trust and expectations about trust" (p. 133). In their remarks they come up with different propositions, for example, they assume that hostility is associated with distrust—individuals who are low in affiliation and high in hostility are characterized as rather competitive, cold, hostile, and mistrusting. These dominance-oriented, hostile attitudes cause these people to tend to distrust their relationship partners. Contrarily, people who are high in affiliation are named to be assured, sociable, friendly, warm, and open and are thus believed to be high trusters in virtual teams. Those extraverted, open individuals who take the initiative are said to be well-accepted and positively taken by the other team members. Further, Brown et al. conclude that these personality types in combination with the person's attitudes towards trust affect an individual's perceived usefulness of information technology (IT) tools as well. Hostile and thus impatient and criticizing people are said to expect a lower level of use of IT and tend to stop using these media.

An aspect of the study conducted by Aubert and Kelsey (2003), which will be described in more detail under the next category, refers to antecedents of trust. Based on the work of Mayer et al. (1995), they found that perceptions of integrity and ability had an impact on trust levels, both for the evaluation of local and remote team members. Trustor´s perceptions of benevolence only had a small influence on the initial trust evaluation of local members, and also the trustor's propensity to trust only slightly influenced final trust levels towards remote members. Aubert and Kelsey (2003) carefully conclude that:

... members with high propensity to trust will be more likely to trust people whom they perceive have strong abilities, whereas members with low propensity to trust will tend to trust others with high perceived integrity. However, all this is relative because in both groups, perceived ability and integrity are the main factors explaining trust. It is their relative importance that changes from one group to the other. (p. 595f)

Development Over Time and Performance

The second category of studies on trust in virtual teams deals with trust development over time (i.e., over the period of the collaboration, over the project duration) and its relation to performance. The studies by Jarvenpaa and Leidner (1999), Kanawattanachai and Yoo (2002), Aubert and Kelsay (2003), and Jarvenpaa, Shaw, and Staples (2004) fit into this category.

Jarvenpaa and Leidner (1999) found in their study of four to six person student teams whose members were spread around the world that in about one half of the teams a high initial level of trust existed from the beginning of their projects. This result shows that swift trust (Meyerson et al., 1996) in virtual teams can develop. Two-thirds of the teams in Jarvenpaa and Leidner's study that developed swift trust were able to maintain high levels of trust until the end of the project. Moreover, Jarvenpaa and Leidner (1999) show that teams that started and ended with high levels of trust showed an outstanding performance. On the other hand, only 4 of 14 teams in Jarvenpaa and Leidner's study that did not develop swift trust were able to get to higher levels of trust towards the end of their projects. The teams with low initial trust engaged in unproductive behaviors (e.g., nonresponding members, withholding participation, lack of initiative to put the project

forward, little social communication). However, the teams who were able to overcome their initial lack of trust and developed trust throughout the project also performed well.

As a result of their in-depth case analysis, Jarvenpaa and Leidner (1999) put forward major characteristics of communication behaviors and member actions that facilitate trust throughout the lifespan of a virtual team. In early stages, it seems to be important to concentrate on social communication. However, Jarvenpaa and Leidner (1999) point out, that "extensive social discussion appeared to foster trust in the beginning of the project but was insufficient in maintaining trust over the longer term" (p. 807). Social communication leads to higher level of trust as long as it is not at the expense of task focus at later stages of the project. Another factor that enhances trust in the beginning of a project is communicating enthusiasm and optimism. In contrary to low trust teams, high trust teams were able to cope with technical and task uncertainties and showed more individual initiative instead of demanding initiative from others.

At later stages, Jarvenpaa and Leidner (1999) point out the importance of maintaining predictable communication patterns and the need for situational leadership roles for the development and maintenance of trust. In their high trust teams they found that leadership rotated among members, depending on exhibited skills, ability, or interest critical for the task to be accomplished. At later stages of a project Jarvenpaa and Leidner state that for high levels of trust it is important for teams to make a transition from a social and/or procedural focus to a task orientation.

After Jarvenpaa and Leidner's (1999) first contribution on trust in 1999, she, along with other authors (Jarvenpaa et al., 2004) developed an understanding furthered by trust the influence of the situation's structure on trust and the effects of an individual's trust in virtual teams in different situations. In student virtual teams (one with and one without intervention) with 94 students from 11 universities in eight countries the participants were measured by quantitative analysis as well as their individual and group contributions concerning a business project. The authors conclude that in situations with weak structures, like in the starting phases of a team, trust has the strongest impact on a virtual team. Like Zolin et al. (2004), they believe that initial trusting beliefs (i.e., trustworthiness) have a positive and direct interrelation to a member's trust. After transition, the structure of the situations becomes stronger. Jarvenpaa et al. (2004) expect that in teams with high initial trust, the ties between communication among team members and a member's attitudes (e.g., of task quality) are quite weak as the concentration is more on task execution than on communication. Contrarily, in teams with low initial trust, members are more reflecting about the others' commitment and will evaluate the team due to communication levels. Thus, more communication ends up in more positive attitudes, higher levels of satisfaction and a stronger (perceived) cohesion of the group. As structures become stronger as a consequence of good and intense communication, the effects of trust on the individual are expected to weaken as concerns can be put aside and trust will not affect the effort anymore. All in all, the authors clearly show that trust is not linear and often has no direct effects and as structure varies with time, it can be considered as an influencing factor as well.

Kanawattanachai and Yoo (2002) conducted a study on the dynamic nature of trust in virtual teams using a student sample of 30 teams. The members represented ten different nationalities and were in four different countries. Overall, the teams in their study developed higher levels of cognition-based trust than affect-based trust. Kanawattanachai and Yoo (2002) also studied the possible relationship between trust and virtual team performance. They found that the presence of high levels of initial trust, swift trust (Meyerson et al., 1996), in the cognitive dimension was related to team performance, but they did not find such a relationship for affect-based trust. Their findings are in accordance with the proposition made by Meyerson et al. that in temporary work teams the cognitive element of trust is more important than the affective one.

In their third research question, Kanawattanachai and Yoo (2002) looked at differences between high- and low-performing teams regarding changing patterns of trust levels over time. They found that high-performing teams not only quickly established trust at the beginning of the project but also managed to maintain it at a high level throughout the project. This result is consistent with the findings of Jarvenpaa and Leidner (1999). Kanawattanachai and Yoo (2002) conclude that although they:

... do not attempt to draw any conclusive causal relationship between swift trust and team performance, it at least seems safe to say that the presence of swift trust in the cognitive dimension can be used as an early predictor of team performance. (p. 205)

Kanawattanachai and Yoo (2002) also found that over time high-performing teams were more likely to maintain their levels of cognition- and affect-based trust.

Aubert and Kelsey (2003) hypothesized a relationship between levels of trust and virtual team performance. They measured trust at the beginning and at the end for both, local and remote members. Their analyses show, however, that none of the trust measurements except final trust towards remote members had a significant influence on team performance. The other variable in their study that had an impact on team performance was team members' individual performances. To answer the question as to why there was no influence of trust on team performance, Aubert and Kelsey (2003) take arguments from the group process literature on process losses and gains. In a post hoc analysis Aubert and Kelsey note that interestingly "teams showing process losses had lower levels of trust at all times (both within local and remote teams) than teams showing process gains. However, none of these differences was significant at the $p < .05$ level" (p. 601).

In an analysis of subteam reports Aubert and Kelsey (2003) found out that low performing teams had different perceived sub-team goals, communication problems (e.g., complaints about responsiveness, lack of punctuality, absenteeism, insufficient feedback), and recognized performance problems but did not effectively resolve them. On the other hand, high performing teams who agreed on work ethics and norms, and had high levels of transparency.

Aubert and Kelsey (2003) conclude that although according to their results:

... trust did not directly influence performance, many elements in the sub-team reports suggest that the overall process effort was lower in teams who also trusted their team mates. They experienced reduced information asymmetry and increased understanding of the progress of each member's work. They also reduced the amount of time required to coordinate efforts by constantly communicating their ideas and progress, hence, eliminating the need for additional messages to monitor each other. All these elements suggest lower agency costs. (p. 604)

This could explain why final remote trust had a significant influence on performance. Members on high trusting teams were able to devote their effort more towards the task and had less need of monitoring and procedural actions.

ILLUSTRATIVE CASE STUDY

In the following section, we present an illustrative case study on experiences international business students had during virtual team projects. Together with students of a U.S. university, Austrian and international students were part of a course on cross-cultural virtual teams. The first (theory) part of the course was held at each university separately. For the second part, four person-teams of each university joined a team from the other university in order to build an eight person virtual team that had to work on a common group project. Starting with a videoconference, the project lasted a bit more than six weeks. The groups had to fulfill different tasks by communicating via electronic media. At four different points of time (start of the project, after videoconference, after first task, after second task, after project), students answered a questionnaire with both closed and open-ended questions related to trust in virtual teams, which provided the possibility of illustrating an example of how trust developed in a multicultural virtual team setting.

The attitude towards trust, respectively the perceived trustworthiness is viewed differently by the participants, although the majority is confident that the other members of their teams would give their best. They based their confidence on positive impressions after the videoconference. Interestingly, the U.S. students revealed a more open and positive attitude than the Austrians who also saw problems concerning trust development and the fact that physical meetings were missing. This skeptical approach about the consequences of virtuality on teams might stem from cultural differences—perhaps the very individualistic American culture makes people feel more confident in virtual settings than less individualistic ones such as Austria (Konradt & Hertel, 2002). Further, virtual collaboration has a component of uncertainty. As Austrians show a higher degree of uncertainty avoidance than the Americans (Hofstede, 1980), they feel perhaps uncomfortable with this risk factor; however, detailed research does not exist and would be required to confirm this assumption.

From the students' reflections, we can conclude that a form of swift trust has played a certain role at the beginning of the project. Only a few participants expected problems in developing trust with their remote partners. Interestingly, questioning the trust towards one's own team's members has never been an issue and it was taken as granted. Possible reasons for the very positive impressions of video-conferencing may be experiences with former groups, and/or impressions of preparation due to preceding theory sessions of the course. Also Jarvenpaa and Leidner (1999) mention that experiences have a positive impact on forthcoming teamwork as communication patterns have already been learned and practiced. Here, no cultural differences could be stated.

In this case, system trust and the project organization of the teachers and the project organization of the teachers did not seem to significantly impact the team work in our case. Although the students anticipated different challenges (such as language barriers, time difference, clear task distribution, or team leadership), they did not believe that these challenges would influence their attitude towards trust.

During the collaboration the frequency of communication increased, especially close to deadlines. Students who were at the same place have also used the possibility of face-to-face meetings in their sub-teams. Consequently, a higher solidarity with team members at the same physical site as well as clear differentiation between U.S. and Austrian members or "on-site and off-site group" could be observed in the teams. Surely, a more intense communication and the closer collaboration between on-site team members lead to increased knowledge about the others, and, thus, accelerated trust building among on-site members. Between the members of the on-site teams, both affective and cognitive trust was developed, and members of the off-site team were obviously excluded. Hand-in-hand with these closer ties, on-site team

members' trust in the performance of "the others" decreased. Because formality dominated the conversation between the two sub-teams, a bridge providing frequent information exchange was not built (Ripperger, 1998). Wong and Burton (2001) even see some danger in such weak ties between sub-teams as reciprocity and mutual trust are much weaker than between members with strong ties. Further, missing rules for communication or weak communication were named to be the reasons for a stable and existing, but not very high level of trust. One more contributing factor to the general low level of trust was that communication took place only between sub-team leaders or single members. Again, cultural differences about the common understanding of a team's purpose, the role of a team leader or different attitudes about efficient team work might play a major role.

Overall, our teams are characterized by a strong task-orientation and little exchange of social and personal information. This may be due to the rather masculine orientation ("live in order to work") of both cultures (Hofstede, 1980). Low social exchange, according to Zaccaro and Bader (2003), is in the long run always associated with low trust levels. But although our team

Figure 1. Development of cognition-based and affect-based trust over project time

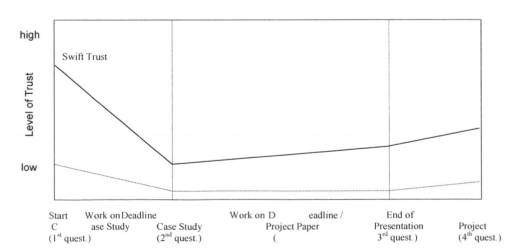

members raised their doubts about "the others", they reported to count on the others and valued their integrity and friendliness. These observed challenges or problems during the team work were slightly ameliorated between the second and the third questionnaire. The perceived reliability and the satisfying fulfillment of tasks were appraised as sources of trust.

Trust was mostly associated with task fulfillment and results—the students' reflections indicated that cognition-based trust was the dominating form in the virtual teams. This is in accordance with Kranawattanachai and Yoo's (2002) results confirm that cognition-based trust is prevalent in temporary and also virtual teams.

Towards the end of the project, participants indicated a general increase in the trust level, which might also have been a consequence of the more frequent communication and the perceived fulfilment and quality of the individual tasks delivered.

Although our teams were not given any recommendations concerning behavior control, different mechanisms revealed: setting deadlines, distributing individual tasks among team members for supervision, or the exchange of current versions of documents in order to reach high agreement among all team members. Also Piccoli and Ives (2003) report that their self-directed teams came up with similar behavior control mechanisms, even if not given any guidelines. Again, this strong task-oriented relation and the setting of control mechanisms might stem from the fact that both cultures are rather masculine, and Austria shows a high degree of uncertainty avoidance (Konradt & Hertel, 2002).

Summarizing, we can draw the picture on different forms of trust over the project time, as shown in Figure 1.

At the beginning of the project, we see the existence of cognition-based swift trust (high initial level). This is probably due to the positive expectations and positive impressions after the video-conference. During the collaboration of the first task, the level of cognition-based trust was quite high as a consequence of a smooth start,

first successful subtask completions, and face-to-face meetings of on-site members. Before the completion of the first task difficulties occurred: suboptimal communication, geographic distance, quality lower than expected, little transparency, and so forth. This led to a decrease in already existing cognition-based trust. Some team members indicated an absence of trust in their team members after the first task. During the work on the second task, the level of trust ameliorated slightly and at the end of the project a quite acceptable and satisfying level of cognition-based trust was observable. Affect-based trust seemed to have only little significance but also declined after the first enthusiasm and correction of initial sympathies.

The study indicated several factors which support and some which hinder the development of trust (see Table 1). Again, most of those factors contribute to the development or non-development of cognition-based trust and only little to affect-based trust.

PRACTICAL IMPLICATIONS

Our example shows that the issue of trust in a virtual multicultural team is a current and crucial one. For practitioners we can thus formulate some recommendations stemming from the literature review and the insights from the illustrative case study:

- Most important is to provide enough time to team members for to get to know each other, to exchange personal information, to get familiar with the different cultures involved and to establish social ties, which goes hand in hand with the establishment of trust. But also during the working process, there should be possibilities for personal communication and relationship development. Team leaders should actively seek to provide time and space for communication and try to motivate people to exchange personal information as well.
- As collectivistic oriented cultures rely more on relationship building before working on

Table 1. Factors supporting and hindering the development of trust

Factors supporting the development of trust	Factors hindering the development of trust
Sticking to self-established deadlines	Communication only among single people
Reliability (task fulfilment)	Not letting others to speak as well
Integrity	Distance
Punctuality	Missing communication
No big discussions	Missing personal relationships
Effort	No face-to-face contact
No tricks	Different working styles
Face-to-face meetings with local members	Unexpected behavior of others
Intense and good communication (especially among sub-team)	Delay in responding to e-mails
Good results, good performance	Dissatisfaction with tasks of others (especially from remote partners)
Personal and social information	Little commitment

a task, a variety of appropriate communication tools for establishing these relationships (e.g., video-conference, Web cam, etc.) is also necessary. As cognitive-based trust dominates virtual teams, its maintenance is of utmost importance. This can be ensured by keeping deadlines, answering e-mails and other messages, handing in tasks with the required level of quality, instances of absences or delays, and so forth. Additionally, technical features should function and be reliable. Thus, some behavioral tools which might already be implemented by the project leader are seen as helpful. Support from the team leader and well-organized, clearly defined tasks also play a crucial role in the establishment of trust among team members.

However, in the long run it is important for the team members to develop some affect-based trust in order to utilize the full group potential. For example, Fryxell, Dooley, and Vryza (2002) show in the domain of international joint ventures that young ventures were more successful when

establishing formal control mechanisms. But as ventures mature and are not able to establish affect-based trust between the partners, their performance will suffer. For virtual multicultural teams this means that not only cognition-based trust but also affect-based trust is necessary in a long term perspective, especially when the tasks are increasingly more complex and situations require single team members to make decisions without consulting the others. This is only possible when affect-based trust has been established.

FUTURE TRENDS AND CONCLUSION

These existing studies clearly show that research on trust in virtual teams is still in its infancy. Most of the articles only take one or the other aspect and test the authors' propositions related to their expectations on a particular issue. The majority of studies have used student teams for their empirical investigations. Using students for studying social phenomena certainly gives a first good insight at early stages. Furthermore, behavioral factors turned out to be quite similar between business practitioners and students (Reber & Berry, 1999). Nevertheless, we assume that there are more influencing factors in the business world context which might have an impact on trust, such as power issues, the different organizational functions of team members, hierarchies, resources, department loyalties, connections, or networks. Thus, studies in the business world would complete the picture on trust in virtual multicultural teams.

One factor that is extremely is the cultural diversity of team members in a virtual team and its effect on trust. We assume that the issue of trust becomes again more complex when we deal with differing attitudes and the handling of trust among members with different cultural backgrounds Huff and Kelley (2005), for example, expect that individuals of individualistic societies have a higher propensity to trust, and trust their partners more than those of collectivistic cultures. The reason for this rather astonishing assumption is the strict differentiation of the latter between in- and out-group members. Whereas they would give their lives for members of the in-group, they exclude members of out-groups and do not tend to trust them. In fact, the cultural influence on trust issues in virtual teams remains purely normative. Studies simply do not exist so far, and only some few researchers give slight attention to a potential interrelation between cultural dimensions and trust (Huff & Kelley, 2005; Konradt & Hertel, 2002; Maznevski et al., 2005). Maznevski et al. (2005, p. 101), who are known as pioneers concerning current issues in cross-cultural management, clearly state that research about cultural influences on virtual teams (and consequently also on trust in virtual teams) is still in its infancy.

REFERENCES

Aubert, B.A., & Kelsey, B.L. (2003). Further understanding of trust and performance in virtual teams. *Small Group Research*, *34*(5), 575-618.

Apfelthaler, G. (1999). *Interkulturelles Management*. Wien: Manz Schulbuch GmbH.

Berg, J., Dickhaut, J., & McCabe, K. (1995). *Trust, reciprocity, and social history* (Unpublished Working Paper). University of Minnesota, Minneapolis.

Brown, H.G., Poole, M.S., & Rodgers, T.L. (2004). Interpersonal traits, complementarity and trust in virtual collaboration. *Journal of Management Information Systems*, *20*(4), 115-137.

Cohen, S.G., & Gibson, C.B. (2003). In the beginning: Introduction and framework. In C.B. Gibson & S.G. Cohen (Eds.), *Virtual teams that work* (pp. 1-13). San Francisco: Jossey-Bass.

Cox, T. (1993) *Cultural diversity in organizations*. San Francisco: Berrett-Koehler.

Davis, D.D. (2003). The Tao of leadership in virtual teams. *Organizational Dynamics, 33*(1), 47-62.

Deutsch, M. (1952). Trust and suspicion. *Journal of Conflict Resolution*, *2*, 265-279.

DiStefano, J.J., & Maznevski, M. (2000). Creating value with diverse teams in global management. *Organizational Dynamics, 29*(1), 45-63.

Driskell, J.E., Radtke, P.H., & Salas, E. (2003). Virtual teams: Effects of technological mediation on team performance. *Group Dynamics, 7*(4), 297-323.

Driver, M. (2003). Diversity and learning in groups. *The Learning Organization, 10*(3), 149-166.

Duarte, D.L., & Snyder, N.T. (2001). *Mastering virtual teams: Strategies, tools and techniques that succeed* (2nd ed.). New York: Jossey-Bass.

Elron, E., & Vigoda, E. (2003). Influence and political processes in virtual teams. In C.B. Gibson & S.G. Cohen (Eds.), *Virtual teams that work* (pp. 317-334). San Francisco: Jossey-Bass.

Fryxell, G.E., Dooley, R.S., & Vryza, M. (2002). After the ink dries: The interaction of trust and control in US-based international joint ventures. *Journal of Management Studies, 39,* 865-886.

Griffith, T.L., Mannix, E.A., & Neale, M.A. (2003). Conflict and virtual teams. In C.B. Gibson & S.G. Cohen (Eds.), *Virtual teams that work* (pp. 335-352). San Francisco: Jossey-Bass.

Hall, E.T. (1976) *Beyond culture.* New York: Anchor Books.

Hall, E.T., & Hall, M. (1990). *Understanding cultural differences.* Yarmouth, Maine: Intercultural Press.

Hart, R.K., & McLeod, P.L. (2003). Rethinking team building in geographically dispersed teams: One message at a time. *Organizational Dynamics, 31*(4), 352-361.

Hampden-Turner, C., & Trompenaars, F. (2000). *Building cross-cultural competence: How to create wealth from conflicting values.* Chichester, UK: John Wiley & Sons.

Hofstede, G. (1980). *Culture's consequences: International differences in work related values.* Beverly Hills, CA: Sage.

Hofstede, G. (2001) *Culture's consequences: Comparing values, behaviours, institutions and organizations across nations* (2nd ed.). Beverly Hills, CA: Sage.

Huang, W.W., Wei, K.-K., Watson, R.T., & Tan, B.C. (2002). Supporting virtual team-building with a GSS: An empirical investigation. *Decision Support Systems, 34,* 359-367.

Huff, L., & Kelley, L. (2005). Is collectivism a liability? The impact of culture on organizational trust and customer orientation. A seven nation study. *Journal of Business Research, 58*(1), 96-102.

Jarvenpaa, S.L., & Leidner, Dorothy E. (1999). Communication and trust in global virtual teams. *Organization Science, 10*(6), 791-815.

Jarvenpaa, S., Shaw, T.R., & Staples, D.S. (2004). Toward contextualized theories of trust: The role of trust in global virtual teams. *Information Systems Research, 15*(3), 250-267.

Jones, G., & George, J. (1998). The experience and evolution of trust: Implications for cooperation and teamwork. *Academy of Management Review, 23*(3), 531-546.

Kirchmeyer, C., & Cohen, A. (1992). Multicultural groups, their performance and reactions with constructive conflict. *Group and Organization Management, 17,* 153-170.

Kirkman, B.L., Rosen, B., Gibson, C.B., Tesluk, P.E., & McPherson, S.O. (2002). Five challenges to virtual team success: Lessons from Sabre. *Academy of Management Executive, 16*(3), 67-79.

Kluckhohn, F., & Strodtbeck, F.L. (1961). *Variations in value orientations.* Evanston, IL: Peterson.

Kramer, R. (1994). The sinister attribution error: Paranoid cognition and collective distrust in organizations. *Motivation and Emotion, 18*(2), 199-230.

Konradt, U., & Hertel, G. (2002). *Management virtueller teams.* Weinheim & Basel: Beltz Verlag.

Kranawattanachai, P., & Yoo, Y. (2002). Dynamic nature of trust in virtual teams. *Journal of Strategic Information Systems, 11,* 187-213.

Lawler, E.E. (2003). Pay systems for virtual teams. In C.B. Gibson & S.G. Cohen (Eds.), *Virtual teams that work* (pp. 121-144). San Francisco: Jossey-Bass.

Levenson, A.R., & Cohen, S.C. (2003). Meeting the performance challenge: Calculating return in investment for virtual teams. In C.B. Gibson & S.G. Cohen (Eds.), *Virtual teams that work* (pp. 145-174). San Francisco: Jossey-Bass.

Lewicki, R., McAllister, D., & Bies, R. (1998). Trust and distrust: New relationships and realities. *Academy of Management Review, 23*(3), 438-458.

Lewis, J.D., & Weigert, A.J. (1985). Trust as a social reality. *Social Forces, 63*(4), 967-985.

Lipnack, J., & Stamps, J. (2000). *Virtual teams: Reaching across space, time and organizations with technology.* New York, Chichester: John Wiley & Sons.

Luhmann, N. (2000). *Vertrauen* (4th ed.). Stuttgart: Lucius & Lucius.

Lurey, J.S., & Raisinghani, M.S. (2001). An empirical study of best practices in virtual teams. *Information & Management, 38,* 523-544.

Mayer, R., Davis, J., & Schoorman, F.D. (1995). An integrative model of organizational trust. *Academy of Management Review, 20,* 709-734.

Maznevski, M., Davison, S.C., & Barmeyer, C. (2005). Management virtueller teams. In G. Stahl, W. Mayrhofer & T. Kühlmann (Eds.), *Internationales personalmanagement* (pp. 91-114). Mering: Rainer Hampp Verlag.

McAllister, D. (1995). Affect- and cognition-based trust as foundations for interpersonal cooperation in organizations. *Academy of Management Journal, 38*(1), 24-59.

McKnight, D., Cummings, L., & Chervany, N. (1998). Initial trust formation in new organizational relationships. *Academy of Management Review, 23*(3) 473-491.

Meyerson, D., Weick, K.E., & Kramer, R.M (1996). Swift trust and temporary groups. In R.M. Kramer & T.R. Tyler (Eds.), *Trust in organizations: Frontiers of theory and research* (pp. 166-195). Thousand Oaks, CA: Sage Publications

Pauleen, D.J., & Yoong, P. (2001). Relationship building and the use of ICT in boundary-crossing virtual teams: A facilitator's perspective. *Journal of Information Technology, 16,* 205-220.

Piccoli, G., & Blake, I. (2003). Trust and the unintended effects of behaviour control in virtual teams. *MIS Quarterly, 27*(3), 365-395.

Pottler, R.E., & Balthazard, P.A. (2002). Virtual team interaction styles: Assessment and effects. *International Journal of Human-Computer Studies, 56,* 423-443.

Reber, G., & Berry, M. (1999). A role of social and intercultural communication competence in international human resource development. In S. Lähteenmäki, L. Holden, & I. Roberts (Eds.), *HRM and the learning organisation* (pp. 313-343). Publications of the Turku School of Economics and Business Administration, Series A-2.

Richard, O.C. (2000). Racial diversity, business strategy, and firm performance: A resource-based view. *Academy of Management Journal, 43*(2), 164-177.

Riopelle, K., Gluesing, J.C., Alcordo, T.C., Baba, M.L., Britt, D., McKether, W., et al. (2003). Context, task and the evolution of technology use in global virtual teams. In C.B. Gibson & S.G. Cohen (Eds.), *Virtual teams that work* (pp. 239-264). San Francisco: Jossey-Bass.

Ripperger, T. (1998). *Ökonomik des Vertrauens: Analyse eines Organisationsprinzips.* Tübingen: Moor Siebeck.

Shapiro, D., Sheppard, B., & Cheraskin, L. (1992). Business on a handshake. *Negotiation Journal, 8*(4), 365-377.

Stumpf, S., & Alexander, T. (1999) Management von Heterogenität und Homogenität in Gruppen. *Personalführung, 5*(99), 36-44.

Thomas, A., & Ravlin, E.C. (1996). Effect of cultural diversity in work groups. In P.A. Bamberger, M. Erez, & S.B. Bacharach (Eds.), *Research in the sociology of organizations* (pp. 1-33). Greenwich: JAI Press.

Thomas, A. (1999). Cultural diversity and work group effectiveness. *Journal of Cross-Cultural Psychology, 30*(2), 242-263.

Trompenaars, F. (1993) *Riding the waves of culture.* London: N. Brealey.

Tyran, C.K., Dennis, A.R., Vogel, D.R., & Nunamaker, J.F. (1992). The application of electronic meeting technology to support strategic management. *MIS Quarterly, 16*(3), 313-334.

Watson, W.E., Kumar, K., & Michaelsen, L.K. (1993). Cultural diversity's impact on interaction process and performance: Comparing homogeneous and diverse task groups. *Academy of Management Journal, 36*(3), 590-602.

Watson, W.E., Johnson, L., & Zgourides, G.D. (2002). The influence of ethnic diversity on leadership. Group process, and performance: An examination of learning teams. *International Journal of Intercultural Relations, 26*, 1-16.

Wong, S.-S., & Burton, R.M. (2000). Virtual teams: What are their characteristics and impact on team performance? *Computational & Mathematical Organization Theory, 6*, 339-260.

Workman, M., Kahnweiler, W., & Bommer, W. (2003). The effects of cognitive style and media richness in commitment to telework and virtual teams. *Journal of Vocational Behaviour, 63*, 199-219.

Zaccaro, S.J., & Bader, P. (2003). E-leadership and the challenger of leading e-teams: Minimizing the bad, maximizing the good. *Organizational Dynamics, 31*(4), 377-387.

Zigurs, I. (2003). Leadership in virtual teams: Oxymoron or opportunity? *Organizational Dynamics, 31*(4), 339-351.

Zolin, R., Hinds, P.J., Fruchter, R., & Levit, R.E. (2004). Interpersonal trust in cross-functional geographically distributed work: A longitudinal study. *Information & Organization, 14*(1), 1-26.

Chapter IX
Building Trust in Networked Environments:
Understanding the Importance of Trust Brokers

Tom E. Julsrud
Norwegian University of Science and Technology, Norway

John W. Bakke
Telenor Research & Innovation, Norway

ABSTRACT

As organizations grow and become multi-national, distributed work, that is, work where members are located in different sites, cities, or countries usually follows (Meyerson, Weick, & Kramer, 1996; Jarvenpaa & Leidner 1999; Zolin & Hinds 2002; Hossain & Wigand 2004; Panteli 2005). Yet such teams and groups have fewer opportunities to build social networks as is common in traditional groups, such as time spent together and frequent informal interaction. The "paradox of trust" in distributed work then, is that while trust is a need-to-have asset for distributed work groups, in particular for knowledge work, it is also difficult to foster due to the lack of physical co-location (Handy, 1995). This chapter argues that one way to deal with the paradox is to recognize the importance of trust as generated through individuals that have trustful ties that cross central boundaries, that is, trust brokers. Based on a relational approach to trust in groups as well as empirical studies of distributed work groups, we argue that trust brokers can help to establish trust quickly and make the group operate in more robust and sustainable ways.

INTRODUCTION

Over the last two decades, a rich stream of research has emphasized the importance of *trust* for large scale organizational processes as well as individual employees. As organizations become more and more knowledge-oriented, trust has moved to the center of attention as a supplement and also as a corrective for control as a coordinative mechanism. As recently argued by Adler and Heckscher (2006), this seems to be especially important for organizations that are engaged in innovations and knowledge-based work:

Knowledge work ... requires that each party offer something with no guarantee that they will get anything specific back in return. They must trust that the other has useful competence and knowledge that will help in their joint effort; that the other can understand her own ideas well enough to engage them productively. (p. 30)

Another aspect of modern organizations that may make trust even more critical for the functioning of organizations is the increase of more geographically dispersed physical structures. As organizations grow and become multi-national, *distributed work,*[1] that is, work where members are located in different sites, cities, or countries, usually follows. According to a recent Nordic study, every third Nordic manager in knowledge intensive businesses plans to reorganize their workplaces, and over 50% of these managers considered "distributed and mobile work" as a relevant option (Julsrud & Bakke, 2004).

There are several reasons for establishing and upholding distributed organizations: In addition to having distributed work as an instrument for establishing presence in different regions and markets, as in the case of regional offices, distributed work may also be a way of saving facilities costs and costs related to work travels. Setting up distributed work groups may also help organizations save expenses, as compared to the collocation of groups and employees. Distributed organizations may also be part of a strategy for developing new knowledge in teams by including people from various organizational units. Distributed groups by definition represent groups with participants situated in different physical settings and organizational and national cultures. To the extent that these people also include differences in knowledge and points of view, distributed work groups can be hubs for development of knowledge and in-

novations (Cummings, 2004). The challenge is to get such groups working together with a limited amount of physical contact, although supported by a diverse set of communication tools.

The Paradox of Trust in Distributed Work Groups

At a general level, the phenomenon of trust can be described as, "a willingness of a party to be vulnerable to actions of another party based on the expectations that the other will perform a particular action important to the trustor, irrespective of the ability to monitor or control that other party" (Mayer & Davis, 1995, p. 712). Whereas collaborating in distributed work groups is emerging as a common way of working, the ability to monitor or control the other party is drastically reduced, and, in essence, this is what makes trust a core asset for organizations practicing distributed work. There is a risk that distributed work may become fragmented if people cannot work together with a sense of comfort or if they feel that they must constantly use time and efforts on controlling the distant colleagues or employees. The "paradox of distributed work" is that while, in general, trust is a "need to have" asset for distributed work groups, in particular, for knowledge work, it is also difficult to foster due to the lack of physical co-location (Handy, 1995). Distance reduces the abilities to interact and to gradually develop trust over time. Even if interaction on Web-based infrastructures and software applications like e-mail and instant messaging (IM), as well as mobile communication provides rich opportunities for instant communication, it often lacks the differentiating cues that influences judgments about trustfulness[2] (Nissenbaum, 2004).

We will in this chapter argue that one way to deal with the paradox of trust in distributed work is to focus on the role of *trust brokers*. Based on a relational approach to trust in groups, we argue that trust can be enhanced by centrally located trust brokers that establish and sustain ties over distances and across boundaries. We will first clarify the concept of trust brokers, drawing on literature in the broad fields of social network analysis and organizational trust. We propose that trust brokering should be understood as an activity involving persistent elaboration of relations based on position in a social network. Next, we will describe trust-broker activities based on a case study of distributed workers within a large Nordic ICT-company. Deploying a combination of qualitative analysis and social network data, we found that trust brokers were important for the positive development of trust within this group. In the last section we will discuss how trust brokering mechanisms can be used strategically by organizations as a way of enhancing the development of trust in distributed groups.

The purpose of this study, then, is to demonstrate how certain qualities of the relations between actors play important roles in the establishment of trust in computer-mediated work environments and other forms of distributed work. The concept of trust brokering, we argue, is a key to understanding the construction of trust across distance.

A Note on the Methodology

This chapter is based on an empirical field study of distributed work groups in a Nordic ICT-company. Over a period of 15 months, a sample of five groups were followed closely. These groups worked in established, distributed work groups with employees situated in different places and countries, and they were also working together with people in other organizational units.

This study has been guided by an inductive approach, trying to understand how trust was built up in the groups over time (Eisenhardt, 1989; Ragin, 1994). In this process, in-depth interviews of participants were combined with formal questionnaires. The network techniques were applied to assist us in building an understanding of both the roles individuals had in the distributed social

networks and of the flow of information within the networks. Social networks were mapped by distributing a list of collaborators to each participant so that adjacency matrixes could be constructed. This approach contrasts and supplements much of the former research in this area, which, to a large extent has, had a focus on testing selected theoretical hypotheses.

One of the core findings from this inductive approach was that individual employees figured as important "nodes" active in the process of developing trust across the boundaries. We will here label this as *trust brokering*, and we will in this chapter explain further the mechanisms and activities involved with trust brokering.

TRUST BROKERING: CONCEPT AND DIMENSIONS

Trust brokering can be described as an activity, informally or formally, targeted at creating trustful relations between two or more groups[3]. As a working definition, we will here describe trust brokering as the active building of trust across distinct groups and/or subgroups, through the development of social relations. Trust brokering thereby refers to an activity within an organization, whereas the term trust broker refers to the corresponding role.

Reflecting the definition of trust cited above, trust brokering may be seen as an activity aiming at increasing positive expectations and reducing negative expectations about other parties in particular groupings. As indicated by the definition, trust brokering relates to trust building as an activity in the development of relations across distance between distinct social groups. In cases where distributed work is based on collaboration between employees belonging to multiple organizations, departments, or locations, the integration of such units becomes an important challenge. We will in this section explain how trust can be understood as a relational concept with cognitive and affective aspects and that trust brokering can be analyzed from its relational and positional aspects.

Cognitive and Affective Dimensions of Trust

Trust may be seen as a multidimensional construct with both cognitive and affective dimensions (Lewis & Weigert, 1995). The *cognitive dimension* refers to the calculative and rational characteristics demonstrated by trustees, such as reliability, integrity, competence, and responsibility. *Affect-based trust*, on the other hand, involves emotional elements and social skills of trustees.

The affective aspects of trust have in particular been studied in close relationships, but they have also been found to be important in work-related relationships (Boon & Holmes, 1991; McAllister, 1995). It has also been argued that in temporary and distributed groups the cognitive aspects are most important because there are fewer opportunities to develop affective ties (Jarvenpaa & Leidner, 1999; Kanawattanachai & Yoo, 2002; Meyerson et al., 1996). Yet recent studies of trust in organizations tends to emphasize the importance of also capturing the affective side of the concept (Kramer & Tyler, 1996). Hence the term trust brokering should strive to capture both cognitive and affective dimensions and we will in this article include both these dimensions.

A Relational Approach to Trust

When trust is defined as "a willingness of a party to be vulnerable to actions of another party based on the expectations that the other will perform a particular action important to the trustor, irrespective of the ability to monitor or control that other party" (Mayer, Davis, & Schoorman, 1995, p. 3), trust is defined as a relational concept, referring to characteristics of both the trustor and the trustee.

In actual studies, trust is nevertheless often seen as a characteristic of the trustee alone: Measures of individuals' trust levels may then be compared, or aggregated as a group characteristic, for example, when groups are rank-ordered according to the dimension of high trust/low trust (Jarvenpaa & Leidner, 1999; Kanawattanachai & Yoo, 2002; Piccoli & Ives, 2003).

In this article, where we investigate how trust-based relations develop within a group of distributed workers, we will deploy the relation-based approach to trust, also on the methodological level. This approach gives the benefits of exploring in depth the structure of relations within a group and the roles that are related to position in these networks. To reflect the cognitive and the affective aspects of trust, this chapter explores relations based on preferred collaboration partners when it comes to solving difficult work issues, as well as relations based on discussing a potential change of job situation. The affective and cognitive trust relations will be combined with relations based on both mediated and face-to-face daily interaction.

Two Aspects of Trust Brokering

The concept of trust brokering, as defined above, addresses two central issues: the establishment of trustful relationships and the "bridging" of formerly weakly connected groups or sub-groups within a larger structured network. While the first issue mainly has been elaborated by psychologically oriented studies of organizational trust (Kramer & Tyler, 1996; Lewicki & Bunker, 1996; Mayer et al., 1995; McKnight, Cummings et al., 1995), the latter has been discussed in particular within social network oriented approaches (Burt, 2005; Coleman, 1988; Granovetter, 1973; Kilduff & Tsai, 2003; Krackhardt & Kilduff, 2002). The "relational" and "positional" aspects of trust brokering, will be discussed briefly below.

Relational Aspects of Trust Brokering

A trust broker may be seen as an individual that actively seeks to establish trustful ties across groups with low levels of trust, whereas trust brokerage may be seen as the outcome of trust brokering activities or of activities that have the establishment of trust brokerage as a by-product.[4] In traditional network terms, trustful relations are usually described as "strong ties" (Granovetter, 1973; Krackhardt, 1992; Krackhardt & Brass, 1994). Strong ties are often found in denser social units like in families and between close friends or partners, while weaker ties exist between acquaintances. A strong tie is usually seen as a provider of more trustful relationships than a weak one. As argued by Mark Granovetter, the strength of ties is the outcome of "the combination of the amount of time, the emotional intensity, the intimacy (mutual confiding), and the reciprocal services that characterize the tie" (Granovetter, 1973, p. 1361). A wide range of research has indicated the value of having a broad network of weak ties. There are also studies exploring the more obvious phenomenon, that strong ties are also important. According to David Krackhardt (1992), the "strength of the strong ties" is that they help reduce risks in insecure environments and predict the behavior of others. This indicates a close conceptual relation between strong ties and trustful relation, and empirical studies corroborate that stronger ties usually are more trustful than weaker ties (Burt & Knez, 1996).

Few studies in the social network tradition have explored the activities that are involved in the development of trust and trustfulness between individuals. Although this issue has been developed and discussed within general studies of trust within organizations (Dirks & Ferrin, 2001; Kramer & Cook, 2004; Lewicki & Bunker, 1996; Mayer et al., 1995). Summing up different studies, Mayer and his colleagues proposes three central factors that influence the general trustworthiness

of a person: ability, benevolence, and integrity (Mayer et al.). Ability refers to the competence and skills the party is believed to have or display on a certain task. If people are believed to have certain skills their trustworthiness is usually high. This is probably particularly important in situations involving knowledge-based work. Benevolence refers more directly to the expected motivation the trustee has to help or support the other party. In certain situations the relationship between the parties is of a kind that supports benevolence, such as between teacher and pupil. Thus benevolence refers to the particular role a party has and his relations to the trustor (i.e., the person that is to be trusted). And finally, the integrity of the trustor is believed to be important for the trustworthiness of a person. If the party is believed to adhere to a set of principles that has acceptance for the trustor, this affects the perceived integrity. But also knowledge about earlier achievements and actions may affect perceived integrity. Thus, the trustworthiness of a certain person builds on how a trustor understands the particular person's competence, intentions, and personal integrity.[5]

It is, however, important to note that these forms of understanding are not evolving in a social vacuum; they are affected by the particular context and the situation within which the relationships take place. Particular qualities of institutional systems like organizations and states will in most cases affect the willingness and possibilities to trust the other part (Mishira, 1996). Sudden changes in organizations can, for instance, create power differences and destabilize trust between individuals. Similarly, duration of interaction over time is believed to be important for the emergence of trustful relationships. Based on these three core concepts, one may say that contextual factors and interaction over time is likely to affect the understanding of the other part's ability, benevolence, and integrity.

Positional Aspects of Trust Brokering

Trust brokering is not only about developing trust between individuals but in particular about connecting individuals with low trust across boundaries. Social network studies have traditionally used the term "brokers" and "brokerage" to describe individuals who actively profit from connecting information and/or people belonging to different groups or networks (Boissevain, 1974; Burt, 2005; Cross & Prusak, 2002). Brokers are described as individuals who try to get personal advantages from negotiating information between parties. As described by Boissevain: "A broker is a professional manipulator of people and information who brings about communication for profit" (Boissevain, 1974).

In technical terms, the *information broker,* then, can be described as a person having an active transmitter role, mediating information between to two other roles; sources, and destinations. The information broker gets information or messages from one "source-node" and transmits it over to a "destination-node". Based on the position within these groups, the information broker can act as a coordinator, consultant, gatekeeper, representative, or liaison (Fernandez & Gould, 1994). Table 1 presents these different positions.

In all these positions, the information broker is active in transmitting or trading information between actors across the boundaries of two or more groups (or within a group). A high level of brokerage activities indicates a central position between two or three groups, which is fundamental for the exploitation of opportunities provided by the "structural holes," understood as gaps in the social worlds across which there are no current connections. According to Burt, these holes in the networks can be connected by savvy entrepreneurs who thereby gain control over the flow of information across these gaps (Burt, 2002, 2005).

A *trust broker* may in principle be located in every one of Fernandez and Gould's (1994) positions. Nevertheless, information brokerage

Table 1. Information broker positions (Based on Fernandez & Gould, 1994)

ROLE TYPE	DESCRIPTION
Coordinator	*Indicates brokerage within the same group*
Consultant	*Indicates brokerages where the broker belongs to one group, and the other two belong to a different group*
Gatekeeper	*The source node belongs to a different group than the broker and the destination node*
	Indicates that the destination node belongs to a different group than the broker and the source node
Liaison	*Indicates that each node belongs to a different group*

and trust brokerage are in principle distinctively different since the latter is less focused on getting access to information and more oriented towards developing ties and relations across distances. This implies a difference of relational quality, as well as a difference of network structure; information brokerage in terms of self-interest is best achieved when there is only one connection between two network components (or groups) and the tension between these groups can be exploited at the maximum (Burt, 2005). Trust brokerage, on the other hand, will seek to develop more relations, and move towards a "closure" of networks. There is also an important difference related to motivation: The goal of trust brokering is to develop trustful relations, not to exploit information from different sources. Thereby, it is more driven by a motivation of creating a common understanding and identity within a group. The trust broker can, similar to the information broker, be positioned differently between groups, but the difference between source and destination is less important in trust brokerage, since it is always a question of brokering in both directions, since brokering is a bi-directional activity[6].

The trust broker then, as described, is a role in a network that is directed towards develop stronger relations between distant units, and to develop more cohesive structures within the group. An important element in the development of trustful relations in network theories may be the use of third parties, that is, individuals outside the dyad that can ensure the trustfulness of the other (Coleman, 1988; Granovetter, 1973). If persons B and C have a strong relationship, this can be used as a platform to develop trust further. If C also has a strong tie to A, C may display a middleman position between B and A that opens for trust brokering (See Figure 1). Given that A has an interest to establish or develop a trustful relation to B, person C can be used as transmitter or mediator of trust, ensuring that A is trustful and has good intentions. The trustworthiness B has to C then spills over to A. Related to the relational qualities described above, we can say that brokering involves the mediation of trustful relations in a network by acting as a middleman between more weakly connected nodes. It is in particular the integrity that can be affected by trust brokering; ensuring that the new person is trustworthy may affect the person's integrity.

An important point is that even though the role as a middleman can be performed in a passive way, there is an opportunity for C to act purposeful, as a trust-connector, when he is aware of the needs and capabilities of A and B. He will then not only act as a guarantor for the relationship, but will also create the new "triadic" unit, ABC. Trust brokers can enhance the denser network structures that are usually perceived as important for the establishment of common norms and security (Coleman, 1988). Compared to the two dyadic relationships AC and CB, the triad ABC will in most cases appear as a social unit with other properties than the dyad, which would more likely induce trust. According to general network theory, a triad is usually more likely to induce trust than a dyadic relationship (Krackhardt, 1999; Krackhardt & Kilduff, 2002; Wolff, 1950).

Summing Up

The discussion above demonstrates that trust brokering involves both relational and positional aspects. On the one hand, the performance of particular actions and communication help to build up trustfulness across boundaries. Central elements here are exposure and demonstration of individual integrity, ability, and/or benevolence. On the other hand, trust brokering involves the connecting of stronger ties within the group and, in particular, across boundaries. This could be done directly by elaborating on relations or indirectly by involving third parties. In addition, we have noted that relational trust in general involves both cognitive and affective aspects.

Figure 1. Inclusion of a third party (C) in a dyadic relation (A & B)

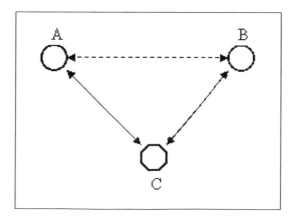

This general outline of trust brokering then suggests recognizing this as a position in network, but also as a position that requires particular actions to enhance relations and ties. One implication of this general attempt is to go beyond the strong structural approach that often is associated with network theories and to bring the individual back in using social network analyses in organizations (Kilduff & Corley, 2000).

DEVELOPMENT OF TRUST-BASED RELATIONS IN OMEGA

The case company, NOMO, is a Scandinavian ICT company with several thousand employees in more than 10 countries.[7] The company has experienced a significant growth in the last years, and investments, mergers, and acquisitions have made it one of the largest European companies within its business area.

Previously, the different national activities of NOMO were relatively independent, but when central divisions of NOMO were merged with ICT companies in Sweden and Denmark, closely interlinked forms of distributed work were initiated across both national and organizational boundaries.[8] A key motivation for the merger was to create synergies across the former divisions, while still keeping contacts with the respective local markets. The transformation from a national ICT company towards a larger multinational company created new challenges for the company. One manager in NOMO told us that "the main challenge for NOMO now is to get the different units work together as one company, not to keep on starting 'national wars' to get local advantages every time there is a potential conflict" (John, Norwegian HR-manager).

To understand more about collaboration within the multinational, distributed groups, a study of distributed work across the former organizational boundaries was launched. Five different distributed work groups were studied in depth over a period of 18 months.[9] We used evidence from one group of product developers, Omega, to illustrate how trust brokerage was important for the development of trust. The study started 15 months after the merger and involved structured analysis of interaction within the group, as well as qualitative interviews with the employees and managers involved.

We will first describe the development within the group during the study period before we turn to a closer description of the networks of trust we found within the group. We will then move on to discuss further some of the most essential nodes and relations within these networks; thus, we try to capture both the positional as well as the relational dimensions of trust brokering, as described in the former chapter

From Crisis to the Re-Establishment of Trust

The core task for the group of 17 developers on Omega was to develop new products for users of computer related services. They were not only located in two of the countries, Norway and Denmark, but they were also at different physical locations within the two countries. In total, people in the group were situated at four different locations (see Figure 2).

The interviews showed that the merged group had experienced a tough initial phase, characterized by numerous intrigues and conflicts. There were underlying conflicts about which product lines that were to be continued in the future. Many of the Danish employees felt their products were rejected in favor of the Norwegian product lines. The challenges were, however, not due to the increased distance between the product developers, but rather to a more complex organizational model where the local marketing units had been given more control of the product development. The product developers needed to establish relations

Figure 2. Location of employees in Omega (number of employees inside boxes)

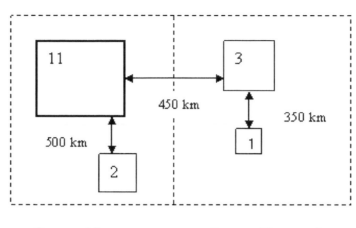

Omega-Norway   Omega-Denmark

with employees in market units in three countries to get resources for developing products. This proved to be difficult as long as the group did not manage to develop a common understanding.

The reorganization initially created a situation that seemed to make the group drift towards mistrust, rather than trust. Underlying much of the conflicts were changes of tasks in Denmark due to the merger. For many of the Danish employees, this was perceived as unpleasant changes, involving a lot of uncertainty. The product development group, which used to be a highly independent and strong unit within the former Danish firm, now experienced problems with being integrated in the larger and more complex NOMO. The understanding of the goals of the group as well as their individual task was low in the first period. In particular the Danish employees reported differences in understanding the new organizational model as well as their role in it.

We simply did not know what to do. All the old was taken away, and projects were closed. I will call this chaos, and very close to an untenable situation. Satisfaction surveys confirmed our problems, and all the "warning-lamps" were blinking. (Ronny, Danish employee)

The situation called for action and 12 months after the merger it was decided to reorganize the group by establishing minor, more specialized units within the groups. A new Norwegian leader (Torhild) was recruited from another division in NOMO, with an objective of facilitating the integration of the groups of developers in Denmark and Norway. When we conducted the interviews, the degree of satisfaction with the new structure was high. The reorganization of Omega was accompanied by changes in the larger NOMO group, involving clearer assignments of tasks, both within the Danish and the Norwegian group of product developers, and to the market units. Although problems with the market units persisted, most of the interviewees emphasized that the group was now moving in a more positive direction than before. Thus, 15 months after the merger, most employees expressed positive attitudes to the new Omega group.

There has been a dramatic improvement in our group during the last couple of months. We have now better people in our management group, and the motivation within the group is much higher. The roles and the responsibilities for the various tasks and assignments are now more clearly defined. (Kai, Norwegian employee)

According to Kai, this attitude was shared by most employees: The group had managed to re-orient their collaboration in a more positive direction.

Positional Aspects of Trust Brokering

In order to better understand the collaboration patterns and the relations within the group, a social network survey was conducted. The following two questions were used to capture cognitive and affective aspects of trust (C-trust and A-trust):

- Who in your group would you talk to if you needed a professional advice in your daily work?
- If you were planning to apply for a job similar to the one you have today, but in another company, whom would you prefer to discuss this with?

In addition, questions that captured the general daily and weekly interaction was used, including face to face communication, as well as the use of e-mails, telephone conversations, and text messages (SMS) on mobile phones (the enterprise deployed mobile phones as the primary work telephone):

- How often have you sent /received e-mails to/from this person the last 7 days?
- How often have you sent/received SMS to/from this person the last 7 days?
- How many mobile phone calls have you had with this person the last 7 days?
- How often have you been in contact with this person during the last 7 days?

All network data was gathered through retrospective reports of the frequency of communication.[10] The data was then coded as regular 1-mode social network data in sociomatrices for valued data. The data was used to conduct differ-

ent analysis, using UCINET software to further explore the trust network vis-a-vis other relational networks.[11] We will here refer to some of the findings and use the directed graphs to illustrate how certain persons in Omega were central in the two trust-based relational networks. We will, also use some simple measures on centrality and density of the networks. *Indegree* centrality indicates the number of incoming lines for each node in a node-by-node network, while *outdegree* centrality indicates the number of outgoing lines (Freeman, 1979). This is a frequently-sed indicator on prestige and popularity in valued networks and in this particular study it indicates whom the other in the group tends to trust. The *density* of a network is measured as the number of actual connections as a proportion of the maximal possible connection, going from 0 to 1.

The cognitive trust network had a dense structure, with connections criss-crossing the group, whereas the affective network was looser: For the C-trust network the density was 0.169, while for the affective trust network, the density was only 0.0542, showing that the general level of cognitive trust was much higher than the level of affective trust. This finding corroborates much former research on trust in distributed group, finding that across distance, cognitive trust is easier established than affective trust.

The head of the department, Torhild, proved to be central in both the trust networks and in the interaction-based network (Table 2 provides data on the degree of centrality for C-trust, A-trust, and daily interaction.) In the interviews, she was acknowledged for playing an important role in connecting the local units. The material also showed that a small group of other individuals with no formal positions proved to be central in these networks. In particular Kai and Martin figured as central in both the C-trust network and the interaction network. All the participants in the group knew someone whom they would trust to give them professional advice, indicating a certain amount of coherence in the group. Yet, when it

Table 2. Indegree and outdegree centrality indicators for position in the cognitive trust network (C-trust), affective trust network (A-trust) and the general interaction network in Omega

	C-trust		A-trust		Interaction	
	Indegree	Outdegree	Indegree	Outdegree	Indegree	Outdegree
Kai	12,00	3,00	0,00	0,00	6,00	4,00
Torhild	10,00	3,00	2,00	0,00	8,00	3,00
Martin	6,00	3,00	3,00	0,00	7,00	4,00
Knut	5,00	3,00	0,00	0,00	1,00	5,00
Kari	4,00	2,00	0,00	2,00	5,00	6,00
Marianne	3,00	2,00	2,00	2,00	3,00	1,00
Ronny	2,00	2,00	0,00	1,00	0,00	1,00
Daniel	1,00	3,00	1,00	0,00	1,00	5,00
Jørgen	1,00	3,00	0,00	0,00	1,00	1,00
Andreas	1,00	3,00	0,00	0,00	1,00	0,00
Emil	1,00	2,00	2,00	2,00	1,00	0,00
Erika	0,00	3,00	0,00	3,00	3,00	7,00
Heidi	0,00	3,00	1,00	1,00	3,00	1,00
Sissel	0,00	3,00	1,00	0,00	3,00	2,00
Simon	0,00	3,00	1,00	2,00	2,00	1,00
Liv	0,00	3,00	0,00	0,00	2,00	3,00
Mathias	0,00	2,00	-	-	2,00	7,00
MEAN	*2,71*	*2,56*	*0,813*	*0,813*	*3.00*	*3*
SD	3,54	0,76	0.950	1.014	2.223	2,301

comes to affective trust, 9 of the 17 employee did not consider anyone in the group as "trustworthy". In addition the nodes that tended to be central in the C-trust network did not appear as highly central in the A-trust network; Kai, for instance, was highly central in the cognitive network, but not included in the affective trust network. Emil, on the other hand, was trusted by two individuals in the group on the affective dimension, but only by one in the cognitive network. Other employees, like Heidi, only had indegree ties in the affective network.

Table 3. Degree of centrality for interaction through e-mail, mobile dialogues and SMS in Omega

Node	E-mail	Mobile	SMS	SUM
Knut	19	15	11	45
Martin	20	10	11	41
Torhild	24	8	7	39
Kai	17	12	9	38
Kari	19	8	5	32
Marianne	13	10	5	28
Erika	14	6	6	26
Jørgen	8	8	6	22
Sissel	14	3	3	20
Mathias	7	6	7	20
Liv	10	4	5	19
Simon	9	5	4	18
Ronny	8	6	4	18
Heidi	10	5	2	17
Daniel	9	3	5	17
Andreas	6	6	3	15
Emil	8	3	1	12
MEAN	11,889	6,941	5,529	
SD	5,801	3,244	2,746	

The indicators for daily interaction showed that Torhild, Martin, and Kai were the most central partners for communication within the group, as well as for the cognitive trust network. Of these three persons, Torhild and Martin were also central in the affective trust network (Table 2). An analysis of communication patterns through mediated channels of communication indicates that the affective trust network follows the cognitive trust networks closely.

A rough measure of the centrality of the network members can be established by looking at the aggregate level of communication, established by adding the incoming and outgoing lines for each partner in the network, while ignoring the direction of communication (Freeman, 1979).[12] Table 3 presents this measure of centrality for all three communication channels. The material shows interesting differences between the networks, based on e-mail, telephone conversations and text messages (SMS) on mobile phones: The manager, Torhild, was most central in the e-mail network, indicating that this perhaps was a more formal medium. Knut was active in the mobile communication interaction, including the use of SMS, even though he had very low centrality in the affective trust network. This is an indicator that interaction frequency is not necessarily closely linked to centrality in trust networks.

Figure 3. Affective and cognitive trust relations in the Omega-network (Danish employees white, Norwegian colored)

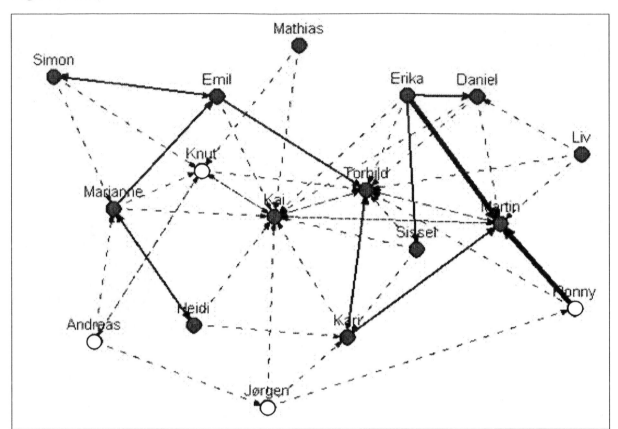

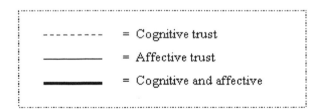

The network survey indicated that the manager, Torhild, as well as Kai and Martin, were most central in the cognitive trust network. Knut, on the other hand, was central in the mediated information flow, but not in particular as a cognitive trust partner. The outdegree interaction table also suggested that he was a sender more than a receiver of information and messages. As illustrated in Figure 3, the centrality of Kai, Torhild, and Martin was based on their relations to both Danes and Norwegians. Two of the 12 persons

seeking advice from Kai were from the Danish part of the group and three of the Norwegians would ask Knut for advice, even though he came from the Danish part of NOMO. As such, these could be considered as trust brokers along the cognitive dimension. When it comes to the affective trust relations, only Martin displayed ties that crossed the national boundary. He was the only person that filled the role as an affective trust broker in this group.

Relational Aspects of Trust Brokering

The network measures and the accompanying diagram showed clearly that some nodes were more central in the interconnected networks of Omega. This finding confirms several earlier studies of social networks of teams and groups, where individual variations in centrality is common (Cross & Prusak, 2002; Cummings & Cross, 2003). Further, the study indicated that the cognitive and the affective dimensions of trust followed rather different tracks. While the cognitive and task-oriented type of trust was present among almost all the employees, the affective trust relation was more sparsely distributed. Interestingly we found that individuals who were central in the cognitive trust network were in some cases not included in the affective trust network. This suggests that we might have individuals that connected along affective or cognitive dimensions only, or along both dimensions.

To give a closer understanding of the relational dimensions of trust brokering, we will here focus on the activities of Kai and Martin, two Norwegian employees who appeared as central players having several connections to the distant units. Martin appeared as the most important trust broker in this group, as he also had affective bonds that crossed the organizational boundary. His close collaborator in Denmark, Ronny, expressed that the development of a strong relationship with Martin was something of a turning point for him, stating:

The fact that Martin now has joined the group with his high level of competence really makes me believe in this. He actually is the first Norwegian that I can say that I really trust. (Ronny, Danish employee)

The relations between these two employees had become an important tie that strengthened the relations not only between two employees but between different geographical units within Omega. It is worth noting, however, that Ronny emphasized Martin's competence and abilities as main reasons for trusting him. For Martin, frequent visits to Denmark, together with frequent communication by electronic media, appeared to be part of a deliberate effort to create a better climate of collaboration within the group:

I use much of my time on communication and on the establishment of a common understanding within the group. I must establish agreement, not by dictates but by communication. Our organization has not done enough to foster this type of understanding across the national boundaries. (Martin, Norwegian employee)

Kai had a particular central role in the cognitive trust network. It turned out that he had a significant advantage by speaking both languages fluently. He had lived in Denmark for long periods of time and he used his insights into culture and language actively to avoid conflicts and misunderstandings. He considered that he had a special responsibility to act as a mediator in the group, due to his ability to detect language-based misunderstandings:

I speak Danish with my collaborators in Denmark, and Norwegian with the collaborators in Norway. In many situations I become a mediator between the environments, and frequently I must change into a role of an "interpreter" in situations where I suspect that people misunderstand each other. (Kai, Norwegian employee)

Thus, Kai's bilingualism helped him to detect misunderstandings but perhaps also to strengthen his own integrity across the nationalities. The deliberate development of relations across the boundaries also involved active use of communication tools.

Kai told us that he had made a routine of calling his colleagues regularly just to hear "how things were going". One of these distant colleagues had recently experienced a critical conflict. He argued that the frequent telephone calls were important to better understand the colleagues' feelings:

I call the other colleagues in my groups often to hear how things are going. I want them to feel that there is interest for what they are doing. When I do not sit beside them and see their faces, I need to call them up and hear how things are going. You must 'read between the lines' to know how their actually are doing in their work.... Sometimes I also talk to others to get information about these issues. (Kai, Norwegian employee)

Martin and Kai were not only developing relations, they were also actively surveying and following up on the others' work within the group. Interestingly, Kai expressed that he actively used third parties to get a better understanding of other colleague's situation. The concern was, on the one hand, that of work-related control, since he was in the position of being the manager of a sub-unit. On the other hand, it was also related to concern about the well-being of his colleagues and an interest in "sorting out" problems in the group. Thus, aspects of control seemed to be intertwined with establishment of trustful relation in this case.[13] All in all, however, Kai and Martin had more interest for the group's activities and their colleagues work than most of the others in Omega. In addition to having an active attitude regarding the connection of ties across the local units, Kai and Martin also seemed to deliberately make use of existing relations on a broader scale. Both were employees who not only had longest records of working in the company, but also of working in different parts of the organization. This was important as Omega was highly dependent on collaboration with other groups within the larger NOMO system. Access to a wide network, then, was also clearly seen as an advantage by the others in the group. "Martin has experience from working in the market units. This gives him access to very rich networks of

contacts that is really useful to us now" (Erika, Norwegian employee).

Kai and Martin enjoyed high levels of trust, at least partly based on their experiences and wide network of contacts within the company. As far as we discovered, this was not used to keep the others at a distance, or to take credit of having exclusive access to central information and resources. Kai expressed that he tried to use help his Danish colleagues to develop their own network within the Norwegian part of the organization. In this way he, implicitly, saw himself as a stepping stone for Danish colleagues in order to develop relations in the Norwegian part of NOMO.

Collaboration across the two countries is difficult. One of my colleagues is coming to me on Thursday, and he has not been here for six months. He needs to get help to develop his networks of contacts in the Norwegian part of the organization. (Kai, Norwegian employee)

This indicates that mediation of relations, and potentially trustfulness, actually took place in the group.

Martin and Kai both reported being involved in trying to solve or moderate conflicts within the group as well as with partners outside the group. Kai emphasized that many conflicts seemed to be based on misunderstanding due to cultural and language differences. Martin, however, said that Norwegians in some situations had been complaining to him about others in the Danish part of the group, recognizing that he had stronger relations here than others. This situation also indicated that Martin operated as a "bridgehead" between the Danish and the Norwegian part of Omega, moderating conflicts.

It is noteworthy that Kai and Martin (as well as Torhild and Knut) developed different types of relations within the group. In a way they might be considered as a "team" of trust brokers, creating a common platform to develop trust across the group. The reorientation of Omega into smaller groups probably also helped the brokers to develop trust within the group based on a common set of tasks and common professional ideas and norms.

Summing Up

Our investigation of Omega found that some employees in the group were important for integrating the two former weakly connected sub-units and building trust within the group. These employees did not only play roles as central connectors, but also acted as trust builders in a network initially suffering from low trust relations. While there were several that figured as trust brokers along the cognitive dimension, affective trust brokers were more infrequent. In Omega only one employee had such a position.

Our qualitative inquiry provided evidence that these persons actually were supporting trustfulness within the group, and that their position as "trusting and trusted individuals" was vital for the development of trust within the group. This involved activities related to establishing and strengthening relations with colleagues at a distance, as well as exploitation of formerly established relations. Actions were also taken to moderate and solve conflicts within the group, and to deploy individual networks to help others to establish new relations: Even though most of the trust brokering was related to establishment of dyadic relations, indications of network building activities through third parties were evident.

THE EMERGENCE OF TRUST BROKERING IN MEDIATED ENVIRONMENTS

Trust brokering as such is not a new phenomenon. The existence of middlemen to enhance trust has been recognized as important and exploited actively for ages. In the beginning of the 20th century, the sociologist Georg Simmel wrote about the sociological significance of a "third element" in social relations. When a dyad was extended with a third person that acted as a neutral mediator, he argued that this tended to moderate conflicts and create a stronger focus on group based interest rather than individual needs (Wolff, 1950). The importance of using third parties to foster trust is also increasingly being recognized as important for trust development on cooperation and negotiations between companies (McEvily & Zaheer, 2004; Wall, Stark, et al., 2001). Yet this perspective is largely neglected in studies of trust in distributed groups.

As we have explained here, when such brokers succeed in lowering conflicts and establishing trust between two or more sub-groups, we can see this as trust brokering. There are reasons to believe that in current and emerging distributed organizations—as well as in temporal and time-limited organizations—trust brokering will become much more important. One reason for this is simply that distributed collaboration becomes more common. Often, however, this emerges in settings that challenge trust and trustfulness. As in the case of NOMO, the merger, or company acquisitions, initial conflicts and discomfort due to power differences and insecurity regarding future work tasks were created. Such settings call for an active approach to the development of trust, rather than a passive one expecting trust to emerge and develop over time as a result of regular interactions.

Another equally important issue is that changing competitive environments requires the rapid establishment of groups and teams, often with a limited time-frame. Despite the fact that groups may work over distance, collaboration—and trust—needs to be developed fast. Active trust brokering may here suggest a strategy for the development of trust in distributed groups and teams more efficiently than traditional approaches. Focusing on the network of relations opens for integration and trust building through a limited number of central connections rather than between all nodes in a network.

Finally, the issue of developing knowledge and common ideas in organizational environments is getting increasingly complex, as modern organizations tend to become more networked. In some cases this also represents a development of a "networked individualization" where the relations between individual employees are work tasks (Wellman, 2002; Wellman, Quan-Haase, Boase, Chen, Hampton, & Diaz, 2003). A high level of complexity makes it difficult for individuals to know or understand what others are doing. Trust brokers can in such organizations be central for connecting people with similar ideas and projects and make them work together. As such, trust brokering can be a key factor for transmission of tacit knowledge that usually depends on higher levels of trust (Hansen, 1999).

Implications for Further Research

Several contributions have recognized the challenge of developing trust in distributed groups, and different solutions have been suggested for remedying the difficulties. Research in this area tends to emphasize different facets of research as decisive for the trust building in the distributed groups. At least three central factors have been much studied: the timing of the interaction, the quality of the communication, and the duration of interaction in the group. The timing argument holds that face-to-face interaction should be regular during the lifetime of the group, or more intense in the beginning of the collaboration (Jarvenpaa & Leidner, 1999; Maznevski & Chudoba, 2000; Zolin & Hinds, 2002). The quality of interaction argument, on the other hand, emphasizes that changes in the communication content, in particular by the managers in the group, will support the trust development (Jarvenpaa & Leidner; Panteli, 2005). And finally, the duration argument argues that trust is enhanced by longer durations and time of interaction (Wilson, Straus, & McEvily,

2006). As an implication of these arguments, trust in distributed groups should develop in much the same way as in co-located groups, although it will take a longer time.

Within this chapter, distributed work groups are seen from a structural perspective. This approach helps us to see that trust development is largely established and sustained by a limited set of individual actors. The trust brokering argument holds that a closer focus on individual roles and their relations within a social network represents a supplementary and more detailed perspective on the development of trust in distributed groups. Rather than seeing the group as one closed unit, it provides a more fine-grained analysis of trust as a product of particular relational positions and patterns within a network of distributed workers. This is a novel approach to studies of distributed work group, and we believe that it should be further explored.

Although the concept of trust brokering has been explored through an inductive, and small scale study, both the identified phenomenon and the concept seem to refer to generic organizational processes. Therefore, we believe that it has value as a description of mechanisms of developing trust, in particular for distributed settings where trust processes are challenged and contested. As a theoretical concept it is rooted in social network theory, as well as in general theories about development of relational trust in organizations. Yet it reflects a wider stream of research over the last decades focusing on the value of doing "boundary work" to connect individual groups to larger units.[14]

Still, the concept needs to be further clarified and compared to other network related role descriptions such as hubs, central connectors, and boundary spanners, as well as gate-keepers. More empirically oriented studies focusing on trust brokering activities, as well as on the impact of such activities on trust within the groups would be of interest. Our study of Omega suggests that trust brokering activities seems to be highly dependent on multiple communication channels, as well as a deliberate use of face-to face interaction. It would, however, be of interest to know more about the use of media for support trust brokering activities. Variations in the use of communication channels, suggested that different communication media were used for different purposes and to support different kinds of relations and ties. In this chapter we have also suggested that trust brokering based on affective and cognitive bonds follow rather different tracks. It would be interesting to explore further the similarities and dissimilarities between these two dimensions of trust brokering.

A further exploration of the role of trust brokers on distributed groups can also be developed in a more methodological direction, utilizing more sophisticated techniques for detecting and analyzing trust brokers and brokering mechanisms. Within the area of social network studies several paths are optional, including the use of positional role analysis and traditional broker indicators (Borgatti & Foster, 2003; Breiger, 2004; Fernandez & Gould, 1994; Hanneman, 2001). The nature of trust brokering as described here, however, may in particular be to call for a combination of quantitative and qualitative network studies, focusing on both structural aspects as well as the individuals work to establish and sustain social relations in distributed networks.

Implications for Organizations

We have used this case study as a tool for developing the concept of trust brokering, based on the observation that individuals may facilitate collaboration and networking within a distributed organization, where trust is seen as something that, to a certain degree, can be actively addressed. The idea of trust in distributed groups as affected by brokerage allows for a more active approach to trust in organizations. This position is somewhat contrary to the view that trust is a by-product of other activities (Elster, 1983); instead, trust bro-

kering may be seen as "functional equivalent" to trust emerging over time (Giddens, 1994). The concept of trust brokering also shows an affinity to the concept of *active trust*, trust that has to be energetically treated and sustained (Beck & Beck-Gernsheim, 1994).

One practical implication is that organizations may actively assign individuals as trust brokers when setting up distributed work groups. This might include giving them particular and formal responsibilities and resources to develop relations, or one may take effort to enhance the development of social relations more indirectly through enhanced social interactions. Where the goal of traditional approaches would seek to develop trust on a broad scale, the trust-brokerage approach would emphasize the need for a few, but strong, relations across the boundaries. An alternative strategy is to develop the groups around existing relations where trust exists in advance. If there are pre-existing trusting relations spanning across the distant groups, this may kick-start the development of trust within the group.

A central issue for the development of trust in distributed networks is how to stimulate the development of trustful and stronger ties. For companies wanting to develop ties across boundaries and distances, the establishment of meeting places, communities, and fora where relations and networks can develop, becomes important strategy elements. Trust brokers can be central in the planning and development of such meeting places, and they can support them in the development of boundary-crossing relations and structures. Collaboration in projects might be one example of such fora, but more informal arrangements can also be introduced, such as professional interest groups.

Trust brokering should, however, not be seen as a highly fixed role description within a group. As emphasized by the definition suggested in this article, we see this as an ongoing activity. This implies that trust brokering activities may be performed by several persons in a group, shifting over time. Neither should this necessarily be seen as a formalized role; brokering activities will in most groups take place when there is a need to develop trust and someone feels obliged or called to support the development of a group.

Organizations should, however, be aware of the risks that may be ascribed to the trust brokers. Earlier studies of individuals located in boundary-crossing positions suggest that this can be a vulnerable position, where there are risks of being targets of cross pressure and role conflicts (Friedman & Podolny, 1982; Krackhardt, 1999). A higher awareness of the actions and processes involved in trust brokering might help to avoid negative consequences, such as overwork, stress, or burnout.

CONCLUDING REMARKS

A trusting relationship is usually characterized by having positive expectations about other parties' actions and doings, with few options of controlling this directly. We have argued that in settings where options for interaction, observation, and control diminish, like distributed work, and work in time-limited teams, trust becomes more vital. This is particularly critical for groups and organizations that are engaged in knowledge-based work, with high interdependencies in the tasks and high degrees of uncertainty. While regular interaction over time may enhance this, the particular setting of distributed work makes this difficult to achieve. This is what has been described as "the paradox of trust" in distributed work (Handy, 1995).

This chapter has argued that the development of trust in distributed groups can be strengthened by trust brokers who work actively to connect employees and build (or thereby building) trust across distributed groups. By studying a case of distributed product developers as a network of relations, we found that both cognitive and affective relational trust was facilitated by trust

brokers, centrally located between two national operations. Their active development of stronger relations within the group seemed to enhance the trust within the group, and helped to solve "the paradox of trust" in the distributed group of product developers. Thus, the answer to the difficulties of enhancing trust is not necessarily to develop more trust on a general basis among all the involved employees. Another option is to enhance the development of trust through a limited number of centrally located trust brokers.

REFERENCES

Adler, P. S., & Heckscher, C. (2006). Towards collaborative community. In P. Adler & C. Heckscher (Eds.), *The firm as a collaborative community* (pp. 11-105). New York: Oxford University Press.

Beck, U., & Beck-Gernsheim, E. (1994). *Individualization*. London: Sage.

Bernhardt, H. R., Killworth, P., et al. (1982). Informant accuracy in social network data V. *Social Science Research, 11*, 30-66.

Boissevain, J. (1974). *Friends of friends. Networks, manipulators and coalitions.* Oxford: Basil Blackwell.

Boon, S.D. & Holmes, J. G. (1991). The dynamics of interpersonal trust: resolving uncertainty in the face of risk. In R. A. Hinde & J. Groebel (Eds.), *Cooperation and Presocial Behavior.* (pp. 190-211). Cambridge, UK: Cambridge University Press.

Borgatti, S. P., Everett, M.G., et al. (2002). *Ucinet 6 for windows.* Harvard: Analytic Technologies.

Borgatti, S. P., & Foster, P. C. (2003). The network paradigm in organizational research: A review and typology. *Journal of Management, 6*(23), 991-1013.

Breiger, R. (2004). The analysis of social networks. In M. Hardy & A. Bryman (Eds.), *Handbook of data analysis.* London: Sage.

Brown, H. G., Poole, M. S., et al. (2004). Interpersonal traits, complementarity and trust in virtual collaboration. *Journal of Management of Information Systems, 20*(4), 115-137.

Burt, R. (2005). *Brokerage and closure. An introduction to social capital.* New York: Oxford University Press.

Burt, R., & Knez, M. (1996). Trust and third-party gossip. In R. M. Kramer & T. R. Tyler (Eds.), *Trust in organizations. Frontiers of theory and research* (pp. 68-89). Thousand Oaks: Sage.

Cohen, D. & Prusak, L. (2001). *In good company: How social capital makes organizations works.* Boston, MA: Harvard Business School Press.

Coleman, J. (1988). Social capital in the creation of human capital. *American Journal of Sociology, 94*, 95-120.

Cross, R., & Prusak, L. (2002). The people who make organizations go or stop. *Harvard Business Review, 80*(6), 104-112.

Cummings, J. N. (2004). Work groups, structural diversity, and knowledge sharing in a global organization. *Management Science, 50*(3), 352-364.

Cummings, J. N., & Cross, R. (2003). Structural properties of work groups and their consequences for performance. *Social Networks, 25*, 197-210.

Dirks, & Ferrin (2001). The role of trust in organizational settings. *Organizational Science, 4*(12), 450-467.

Eisenhardt, K. M. (1989). Building theories from case study research. *Academy of Management Review, 14*(4), 532-550.

Elster, J. (1983). *Sour grapes: Studies in the subversion of rationality.* Cambridge, MA: Cambridge University Press.

Fernandez, R. M., & Gould, R. V. (1994). A dilemma of state power: Brokerage and influence in the national health policy domain. *American Journal of Sociology, 99*, 1455-91.

Friedman, R. A., & Podolny, J. (1982). Differentiation of boundary spanning roles: Labor negotiations and implications for role conflict. *Administrative Science Quarterly, 1*(37), 28-47.

Giddens, A. (1994). Risk, trust, reflexivity. In *Reflexive Modernization.* U. Beck, A. Giddens, & S. Lash (Eds.), Cambridge, UK: Polity Press, 184-197.

Granovetter, M. S. (1973). The strength of weak ties. *American Journal of Sociology, 81*, 1287-1303.

Handy, C. (1995). Trust and the virtual organization. How do you manage people whom you not see? *Harvard Business Review, 73*(3), 40-50.

Hanneman, R. (2001). *Introduction to social network methods.* Department of Sociology, University of California, Riverside.

Hansen, M. (1999). The search-transfer problem: The role of weak ties in sharing knowledge across organization subunits. *Administrative Science Quarterly, 44*, 82-111.

Haragadon, A. B. (1998). Firms as knowledge brokers: Lessons in pursuing continuous innovation. *Management Review, 40*(3), 209-227.

Hartley, C., Brecht, M., et al. (1977). Subjective time estimates of work tasks by office workers. *Journal of Occupational Psychology, 50*, 23-36.

Hossain, L., & Wigand, R. (2004). ICT enabled virtual collaboration through trust. *Journal of Computer-Mediated Communication, 10*(1).

Jarvenpaa, S. L., & Leidner, D. E. (1999, Nov-Dec). Communication and trust in global virtual teams. *Organisational Science, 10*, 791-815.

Julsrud, T., Bakke, J. W., et al. (2004). *Status and*

strategies for the knowledge intensive Nordic workplace. Oslo, Norway: Nordic Innovation Center.

Julsrud, T., Schiefloe, P. M., et al. (2006). *Networks and trust in distributed groups.* University of Trondheim: NTNU.

Julsrud, T. E. & Schiefloe, P. M. (2007). Trust and stability in distributed work groups: A social network perspective. *International Journal of Networking and Virtual Organisations.* Forthcoming.

Kanawattanachai, P., & Yoo, Y. (2002). Dynamic nature of trust in virtual teams. *Strategic Information Systems, 11,* 187-213.

Kilduff, M., & Corley, K. G. (2000). Organizational culture from a network perspective. In N. Ashkanasy, C. P. M. Wilderom, & M. F. Peterson (Eds.), *Handbook of organizational culture and climate* (pp. 211-221). New York: Sage.

Kilduff, M., & Tsai, W. (2003). *Social networks and organizations.* London: Sage.

Krackhardt, D. (1992). The strength of strong ties: The importance of philos in organizations. In N. Nohria & R. Eccles (Eds.), *Network and organizations: Structure, form and action* (pp. 216-239). Boston: Harvard University Press.

Krackhardt, D. (1999). The ties that torture: Simmelian tie analysis in organizations. *Research in the Sociology of Organizations, 16,* 183-210.

Krackhardt, D., & Brass, D. (1994). Intraorganizational networks. In S. Wasserman & J. Galaskiewicz (Eds.), *The micro side. Advances in social network analysis* (pp. 207-229). Thousand Oaks, CA: Sage.

Krackhardt, D., & Kilduff, M. (2002). Structure, culture and simmelian ties in entrepreneurial firms. *Social Networks, 3*(24), 279-290.

Kramer, R. M., & Cook, K. S. (2004). Trust and distrust in organizations: Dilemmas and ap-proaches. In R. M. Kramer & K. S. Cook (Eds.), *Trust and distrust in organizations. Dilemmas and approaches* (pp. 1-17). New York: Russel Sage Foundation.

Kramer, R. M., & Tyler, T. R. (1996). Whither trust. In R. M. Kramer & T. R. Tyler (Eds.), *Trust in organizations: Frontiers of theory and research.* Thousand Oaks, CA: Sage Publications.

Lave, J., & Wenger, E. (1991). *Situated learning. Legitimate peripheral participation.* Cambridge, UK: Cambridge University Press.

Lewicki, R. L., & Bunker, B. B. (1996). Developing and maintaining trust in work relationships. In R. Kramer & T. Tyler (Eds.), *Trust in organizations. Frontiers of theory and research.* Thousand Oaks, CA: Sage.

Lewis, J. D., & Weigert, A. (1985). Trust as a social reality. *Social Forces 63*(4), 967-985.

Marchington, M., Grimshaw, D. (Eds.). (2005). *Fragmenting work. Blurring organizational boundaries and disordering hierarchies.* Oxford: Oxford University Press.

Mayer, R. C., Davis, J. H., et al. (1995). An integrated model of organizational trust. *Academy of Management Review, 3*(20), 709-734.

Maznevski, M. L., & Chudoba, K. M. (2000). Bridging space over time: Global virtual team dynamics and effectiveness. *Organizational Science, 11*(5), 473-492.

McAllister, D. J. (1995). Affect- and cognition-based trust as foundations for interpersonal cooperation in organizations. *Academy of Management Journal, 1*(38), 24-59.

McEvily, B., & Zaheer, A. (2004). Architects of trust: The role of network facilitators in geographical clusters. In R. M. Kramer & K. S. Cook (Eds.), *Trust and distrust in organizations. Dilemmas and approaches* (pp. 189-213). New York: Russel Sage Foundation.

McKnight, D. H., Cummings, L. L., et al. (1995). *Trust formations in new organizational relationships* (Information and Decision Sciences Workshop). University of Minnesota.

Meyerson, D., Weick, K., et al. (1996). Swift trust and temporary groups. In R. Kramer & R. Tyler (Eds.), *Trust in organizations. Frontiers of research and theory* (pp. 166-195). Thousand Oaks, CA: Sage.

Mishira, A. K. (1996). Organizational response to crisis: The centrality of trust. In R. M. Kramer & T. R. Tyler (Eds.), *Trust in organizations: Frontiers of theory and research* (pp. 261-287). Thousand Oaks, CA: Sage.

Nissenbaum, H. (2004). Will security enhance trust online, or supplant it? In R. M. Kramer & K. S. Cook (Eds.), *Trust and distrust in organizations. Dilemmas and approaches* (pp. 155-185). New York: Russel Sage Foundation.

O'Leary, M., Orlikowski, W., et al. (2002). Distributed work over the centuries. Trust and control in the Hudson Bay Company. In P. Hinds & S. Kiesler (Eds.), *Distributed work*. Cambridge, MA: MIT Press.

Panteli, N. (2005). Trust in global virtual teams. *Ariadne 43*. Retrieved April 2007 from http://www.ariadne.ac.uk/issue43/panteli/

Piccoli, G., & Ives, B. (2003). Trust and the unintended effects of behavior control in virtual teams. *MIS Quarterly, 27*(3), 365-395.

Podolny, J., & Baron, J. (1997). Resources and relationships, social networks and mobility in the workplace. *American Sociological Review, 62*, 673-693.

Ragin, C. A. (1994). *Constructing social research*. Thousand Oaks, CA: Sage.

Star, S. L., & Griesmer, J. (1989). Institutional ecology, "translation" and boundary objects: Amateurs and professionals in Berkley's Museum of Vertebrate Zoology. *Social Studies of Science, 19*, 387-420.

Wall, J. A., Stark, J. B., & Standifer, R. L. (2001). A current review and theory development. *Journal of Conflict Resolution, 45*(3), 370-391.

Wasserman, S., & Faust, K. (1994). *Social network analysis. Methods and applications*. New York: Cambridge University Press.

Wellman, B. (2002). Little boxes, globalization, and networked individualism? In M. Tanabe, P. v. d. Besselaar, & T. Ishida (Eds.), *Digital cities II: Computational and sociological approaches* (pp. 10-25). Berlin: Springer.

Wellman, B., Quan-Haase, A., Boase, J., Chen, W., Hampton, K., Diaz, I., et al. (2003) The social affordances of the internet for networked individualism. *Journal of Computer Mediated Communication, 8*(3). Retrieved April 2007 from http://www.ascusc.org/jcmc/vol8/issue3/wellman.html

Wenger, E. (1998). *Communities of praxis. Learning, meaning and identity*. Cambridge: Cambridge University Press.

Wilson, J. M., Straus, S. G., et al. (2006). All in due time: The development of trust in computer-mediated and face-to-face teams. *Organizational Behavior and Human Decision Processes, 99*, 16-33.

Wolff, K. H. (Ed.). (1950). *The sociology of Georg Simmel*. New York: The Free Press.

Zolin, R., & Hinds, P. (2002). Trust in context: The development of interpesonal trust in geographically distributed work teams (CIFE Working Paper). S. University. Stanford, Center for Integrated Facility Engineering.

ENDNOTES

[1] There is no single way to define distributed work groups. We will here follow Zolin and Hinds and define this in a general way, as group-based work where members are located in different cities or countries, supported by use of information and communication technology (2002).

[2] This discussion of networked environments has even wider implications, since the development of organizations and organizational units with more limited timeframes presents challenges quite similar to the "paradox of trust" in distributed work.

[3] The term has been used by former authors to coin individual actors work to integrate different units. For instance, Cohen and Prusak (2001) describe this as "someone who vouch for people and make introductions to help spread trust throughout an organization" (p. 35). The term "network facilitators" has been described by McEvily and Zaheer (2004) as organizations and institutions deliberately and intentional act to promote and sustain trust (p.208). The term "knowledge brokers" has in a similar way been applied to describe organizations that support innovation by connecting, recombining, and transferring to new contexts otherwise disconnected pools of ideas (Haragadon, 1998).

[4] On the concept of by-products of social activities, see Elster (1983).

[5] In addition, there are also factors related to the trustor (the person that are going to trust the other part) that affects the perceived trustworthiness of a person. The term "propensity to trust" is usually used to denote the general willingness of a party to trust others (Mayer et al., 1995; Brown, Poole, et al., 2004). Not only differences in personalities but also individual experiences and values can affect the willingness to trust others in general.

[6] The idea of structural holes has been criticized for not paying sufficient attention to content of the relations. Analyzing different types of relations in a high technological engineering company, Podolny and Baron found that structural holes were advantageous for strategic network content, but not for relations involving social support and trustfulness (1997).

[7] Please note that all names are pseudonyms, as well as the names of the group (Omega) and the organization (NOMO)

[8] In technical terms, the Norwegian unit acquired the Swedish and Danish units, but the term *merger* was commonly used, both by the interviewees and in internal publications; hence this term is used throughout the presentation of the case.

[9] Results from this study are reported elsewhere (Julsrud, Schiefloe, et al., 2006).

[10] Such self-reported frequency data are not expected to be objectively accurate, but are expected to allow comparison across relations, and to indicate relative strength of interactions within a group (Hartley, Brecht, et al., 1977; Bernhardt, Killworth, et al., 1982)

[11] Closer description of social network measures and techniques can be found in Wassemann and Faust (1994) and in the UCINET software manuals (Borgatti, Everett, et al., 2002).

[12] We will here prefer symmetrical rather than directional ties to reduce complexity in the presentation, even though this represents a reduction in the richness of the empirical material. A more thorough analysis of the mediation of the social relation should, however, analyze directional as well as symmetrical ties.

[13] This point is elaborated explicitly by O'Leary and his colleagues in an historical analysis of trust and control in the Hudson Bay Company (O'Leary, Orlikowski, et al., 2002)

[14] Related terms include boundary spanning agents in the field of intra organizational networks (Friedman & Podolny, 1982; Marchington & Grimshaw, 2005), legitimate peripheral participation in the field of communities of practice (Lave & Wenger, 1991; Wenger, 1998), and boundary objects related to actor network theory (Star & Griesmer, 1989).

Chapter X
Trust Building in E-Negotiation

Noam Ebner
Sabanci University, Turkey, and The United Nations' University for Peace, Costa Rica

ABSTRACT

As the use of e-communication proliferates, more and more types and subtypes of relationships are taking place online. Within the general framework of this book, this chapter focuses on one specific type of relationship: the relationship between people negotiating online via the communication channels offered by information technology. As the global market expands and business and personal relationships are increasingly taking place online, it is common to conduct negotiation processes in the online venue. This chapter focuses on the challenges to inter-party trust in e-negotiation, and on means for overcoming these challenges. It explains the critical role trust plays in negotiation and portrays the ways in which the communication medium through which a negotiation is conducted affects the dynamics of trust-building and trust-breaking. The author lists eight major obstacles to trust formation in e-negotiation and suggests methods not only for avoiding or defusing trust-breaking situations, but for engaging in proactive trust-building.

NEGOTIATION: IT IS EVERYWHERE AND ONLINE, TOO

The literature of negotiation grants a broad definition to the term negotiation, encompassing many different types of interpersonal interactions and relationships. Far from limiting the negotiation process to the activity of people in pin-striped suits sitting in a board room, negotiation is defined so as to include any interaction in which two or more people attempt to decide on the allocation of scarce resources (Thompson, 2004). A "scarce resource" might be stocks, oil, or territory, but it might just as easily be time, money, attention, affection, pleasure, or any other concept to which parties attach value. Any back and forth communication aimed at reaching an agreement is considered to be within the realm of negotiation activity (Fisher & Ury, 1991). In short, we all negotiate all the time—with our employers and employees, our colleagues and friends, our spouses and children, and with many, many others.

As e-communication becomes an increasingly natural medium through which to conduct business and personal relationships, we find that many of the varied interactions occurring in the e-world actually fall into the category of negotiation activity. In the organizational venue, a sales manager writes a memo to the Human Resources Department in an attempt to influence the hiring of a particularly promising field representative; two board members exchange e-mail messages with the goal of trying to form a voting coalition before a board meeting; and a manager sends out a group message to everyone on his staff list in an attempt to get them all to come in to work over the weekend. All these people are engaging in e-negotiation. On the interorganizational level, a purchaser in New York negotiates terms with a supplier in Singapore whom he is likely never to meet face to face. Increasingly, the same holds true for more "local" interactions. Attorneys representing rival corporations located in the same city might attempt to work out a settlement on a patent infringement dispute completely through e-mail exchanges. On the interpersonal level, one of these attorneys might negotiate with her travel agent through e-mail for a better price on a vacation deal she is considering or with her husband regarding her preferred travel destination. There is a great deal of experiential spill-over between the "personal" and "business" negotiations one takes part in, providing a wide range of settings for gaining experience, practicing and improving.

This chapter focuses on one major aspect of e-negotiation: the challenge of building trust in a negotiation relationship that is formed and maintained online. As interpersonal trust is both a relational and a contextual construct (Naquin & Paulson, 2003), we will be focusing on trust-building in negotiation processes which are conducted via any *text-based* channel that allows, to some extent, for both *contextual and relational communication*. This includes communication methods such as e-mail exchange, posting on a bulletin board or uploading text messages onto a negotiation support system (NSS).[1] Two other hallmarks of the negotiation process discussed in this chapter are that it takes place *entirely* online, through *asynchronous communication*. This focus makes our conclusions and suggestions particularly suited to negotiations conducted via e-mail, the most commonly used (and most widely researched) form of online communication.[2]

Our approach is a "theory-to-practice" one: after reviewing the literature on the ways that trust is developed and affected in e-negotiation, we will translate the theories proposed into practical, prescriptive suggestions for how to behave so as to generate trust for ourselves in our online negotiation opposite. The goal of this chapter is to enable readers, in their role as e-negotiators:

1. To understand the vital role that trust plays in these interactions
2. To identify negotiation process-moments which are pregnant with potential for trust building (or trust breaking)
3. To apply, at these critical junctures, tested methods for trust building which facilitate negotiation processes and improve their outcomes

TRUST AND THE COMMUNICATION CHANNEL

Literally, the term negotiation does not only connote two people exchanging knowledge and resources, it also conveys the meaning of successfully overcoming obstacles. One negotiates a river or a sharp turn or negotiates his way through a difficult period. When negotiating with another person, there are two distinct levels on which a negotiator needs to overcome obstacles. The first level involves achieving the goal, or solving the problem that brought him to the table in the first place; the "obstacle" is the scarceness of the resource he lacks and needs. The second level involves the process that the negotiator must go through in order to achieve that goal. On this level, the obstacles that need be overcome are process related obstacles such as communication difficulties or mutual lack of trust, which inhibit agreement and cause conflict to emerge and escalate.

In any negotiation, the medium through which the negotiation is conducted is also the medium through which these second level obstacles must be overcome. The medium itself, however, always poses obstacles that are inherent in its own nature. While the medium of e-communication certainly does posses certain characteristics potentially beneficial to the conduct of a negotiation, it is also rife with obstacles which need to be overcome in order for the process to succeed.

The most difficult challenge amongst all the second level, process-oriented obstacles to successful e-negotiation processes is that of trust. While creating, strengthening, and maintaining trust is certainly a challenge inherent in any negotiation framework, it is particularly challenging—and inestimably valuable—in the online environment. Before moving on to examine the particular challenges posed to trust-building by the e-communication channel, let us first define trust in the context of negotiation and examine the role trust plays in any negotiation process, whether face-to-face or online.

TRUST IN THE CONTEXT OF NEGOTIATION

While many attempts have been made to define trust, several authors have pointed out that there is no one way to define it and any definition that is offered has been affected by the particular perspective of the definer (Boyd, 2003; Koehn, 2003; Wang & Emurian, 2005). In the specific field of negotiation, there are also various definitions of trust. For the purposes of this chapter, we will

suggest combining three elements suggested in the literature:

- **Expectations:** One expects that his cooperation will be reciprocated by the other (Pruitt & Kimmel, 1977)
- **Risk:** Only when one is at risk, dependent or vulnerable, can his behavior or expectations demonstrate trust (Boyd, 2003)
- **Uncertainty:** Trust can manifest only when there is a degree of uncertainty regarding the other's future behavior; if his behavior is pre-programmed, trust becomes moot (Gambetta, 1988)

Building on these notions, we suggest that in the context of a negotiation process trust is an expectation that one's cooperation will be reciprocated, in a situation where one stands to lose if the other chooses not to cooperate. Or, practically speaking, we need our negotiation opposite to trust us every time we ask him to bet on our unguaranteed cooperation. We need our opposite to be willing to go out on a limb at various times during the process and at its culmination. Throughout the process, we need him to be willing to divulge information, despite the fact that he cannot be absolutely certain we will not use it to harm him. We need to have him be willing to invest time in discussing options for achieving our goals, even though he is not guaranteed we will reciprocate with matching efforts for searching for solutions to his own issues. At a negotiation's end, we often need the other to agree to a particular solution, in which he must make a concession without absolute certainty that we will stand by our own word and reciprocate. The leap of faith necessary for all these to happen is generated by trust.

The Role of Trust in Negotiation

Trust is essential for any success in negotiation. The professional and academic literature on negotiation devotes enormous effort to understanding the concept of trust as well as to exploring methods for building, maintaining, and restoring it.[3] Trust has been identified as an element playing a key role in enabling cooperation (Deutsch, 1962), problem solving (Pruitt, Rubin, & Kim, 1994) achieving integrative solutions (Lax & Sebenius, 1986; Lewicki & Litterer, 1985) and dispute resolution (Moore, 2003). Negotiators are trained and advised to seek out and create opportunities for trust building whenever possible, and as early as possible in the course of a negotiation process (Lewicki & Litterer). Trust is considered a vital precondition for sharing information, arousing generosity and empathy and reciprocating trust-building moves in a negotiation process. When trust in a negotiation opposite is lacking, negotiators fear that information imparted to the other might be used to one's own detriment (Nadler & Shestowsky, 2006). A trust-filled environment might enable negotiators to contemplate the worst outcome of the process as being a mutually agreed upon "no-deal," which holds promise of a continuing relationship and possible future interactions, dictating cooperative behavior patterns in the negotiation process. Distrust, on the other hand, causes parties to focus on how their cooperative behavior can be used against them by the other to cause them actual loss. This triggers defensive behavior—negotiators withhold information, attack the other's position and statements, threaten him, and lock themselves into positions from which they cannot easily withdraw (Lewicki & Litterer, 1985).

Routes to Establishing Trust in Negotiation

In negotiation, there are three primary routes for establishing trust:

- *Deterrence based trust*, also called *calculus based trust*, is trust premised on our perception that our negotiation opposite will act

as he committed himself to as a result of a subjective cost/benefit analysis we conduct estimating our opposite's own self interest. As our negotiation opposite will always be on the lookout to benefit by breaking trust, he will keep trust only if his payoff is greater that way. If he will gain more by breaking trust—for example, if we lack the ability to punish him for violating trust—he will do so (Paulson & Naquin, 2004). We will trust our opposite only as long as we think he considers trustworthy behavior to be in his own self interest. This does not need to be abstract. Negotiators can introduce what Axelrod (1984) called "changes in the payoff structure" (p. 134) —enforceability schemes, punitive measures for breaking trust, or positive rewards for cooperation— into their negotiation processes (Schelling, 1980) and into their final agreements (Lewicki, Saunders, Minton, & Barry, 2002) so as to manipulate that "self interest."

- *Knowledge-based trust* is grounded in the other's predictability. By knowing our negotiation opposite well enough to predict his responses and behaviors, we can estimate how far he can be trusted. The more information we have on his previous experiences and preferences, how he thinks, and what his value-system looks like, the better we can anticipate his behavior. This form of trust can result from in-depth study of our negotiation opposite; it can also develop over time, as a function of having a history of interactions through which our knowledge of him was obtained (Lewicki et al., 2002).

- *Identification based trust* is based on a perceived sharing of characteristics, traits, plights and backgrounds. People tend to trust negotiation opposites who seem to have elements in common with them. Even when other factors in the negotiation process may cause our negotiation opposite misgivings regarding our trustworthiness, what he perceives as our shared elements might tip his people-judgment scales in favor of trusting us. Of course, this is all a matter of degree; two negotiators discovering that they attended the same school breeds one level of connection, whereas two negotiators realizing they completely empathize or identify with the other party's needs and desires will experience wider and more robust trust (Lewicki et al., 2002). As we shall see, this route to trust is the one most severely affected by the e-communication medium.[4]

These three routes to establishing trust can also be viewed as pillars, each of which can support the negotiation process on its own or can be shored up by the others. At any given moment, the measure of faith needed by our negotiation opposite might be most suitably supported by a particular type of trust.[5]

Who Do Negotiators Trust?

Breaking the above down into practical terms, in seeking a normative answer to the practical question of how to encourage our negotiation opposite to trust us, we can posit that a negotiation opposite is more likely to trust us if:

1. We appear similar to the negotiation opposite in various ways
2. We show a positive attitude towards the negotiation opposite
3. We are dependent upon the negotiation opposite, who holds some power of reward or punishment over us
4. We ourselves initiate trusting, cooperative behavior, which invites reciprocation
5. We make concessions, thereby signifying our willingness to pay a price in order to find a joint solution (Lewicki & Litterer, 1985)
6. The negotiation opposite is likely to have a future interaction with us (Thompson, 2004)
7. We have been helped by the negotiation opposite in the past (therefore laying the groundwork for expectations that we will reciprocate)
8. We are known to have been helpful and cooperative in the past—towards others or, more particularly, towards our current negotiation opposite himself (Pruitt et al., 1994)
9. The negotiation opposite feels he knows and understands us to a degree granting him insight into our system of needs, norms, and values

This is not an abstract menu of conditions that might arise; this is a prescriptive guide to actions that can gain us our negotiation opposite's trust. Of course, some of these elements are counterintuitive and include putting ourselves at a certain degree of risk. That is the nature of trust-building in negotiation: it is a cyclic, dynamic of risk-taking, where small risks taken can result in great rewards.

TRUST IN THE ONLINE ENVIRONMENT

The Internet, in general, has developed into an environment fraught with distrust. No matter for what a person uses the Internet, he is likely to encounter a situation in which he must place his trust in a software platform, a Web site, an e-vendor or another individual. Since all of these interactions have been used exploitatively in the past, users approach the Internet with a large degree of distrust (Wallace, 1999). Future development of the Internet, from a financial perspective, depends to no small extent on the success of e-commerce, which is *absolutely* dependent—perhaps more than on any other element—on trust (Wang & Emurian, 2005). As Rule (2002) summarized the problem, "Transactions require trust, and the Internet is woefully lacking in trust" (p. 98).

Communication Media, Interpersonal Behavior, and Trust

Even before the advent of Internet-based e-communication, research showed that people using technology to communicate at a distance tend to experience low levels of interpersonal trust. Communication between physically distant parties is more susceptible to disruption and deterioration than face-to-face dialogue. Comparing telephone-based communication with face-to-face communication has shown that whereas face-to-face interactions foster rapport, trusting behavior, and cooperation, their absence leads to more distrusting, competitive, and contentious behavior (Drolet & Morris, 2000).

In the trust-devoid environment of the Internet, these findings do not only hold true, they are intensified. Research on e-communication has shown that the e-channel is conducive to people developing a sense of disinhibition; parties ignore the possible adverse consequences of negative online interactions because of physical distance, reduced accountability, and a sense of anonymity (Wallace, 1999). People attempting to work together utilizing e-communication tend to act more contentiously than face-to-face counterparts, which results in more frequent occurrences of swearing, name calling, insults, and hostile behavior (Kiesler & Sproull, 1992). Viewed through the perspective of trust, it would seem that online communicators find it difficult to build the trust necessary for their opposites to perceive them, and construe their intentions, positively. As a result, they often lose control of the process and of the relationship.

Trust Reduction in E-Negotiation

Unsurprisingly, research conducted from a negotiation perspective found that these findings on communication-at-a-distance also hold true when the relationship between the communicators is that of negotiators. Early research showed that negotiators are apt to act tough and choose contentious tactics when negotiating with people at a distance (Raiffa, 1982). As researchers began to focus on e-negotiation, it became apparent that e-negotiators feel less bound by normatively appropriate behavior than face-to-face negotiators. This results in an increased tendency to make threats and issue ultimata (Morris, Nadler, Kurtzberg, & Thompson, 2002), to adopt contentious, "squeaky wheel" behavior, to confront each other negatively and to engage in flaming (Thompson & Nadler, 2002).

Viewing these findings through the perspective of trust, it would seem that e-negotiators work through a communication channel which makes trust-building particularly challenging. This insight (which may ring true intuitively for many readers and, for others, tap into their own practical experience) is supported both by indirect and direct measurements of trust in e-negotiation processes.

Trust in negotiation is not only a difficult notion to grasp and define; it is also quite difficult to measure. One way to indirectly assess the degree to which an e-medium is conducive to trust is to measure the degree to which parties negotiating through it behave cooperatively throughout the negotiation process; cooperation is viewed as behavior manifesting *only* in trust-filled environments. Another indirect assessment method is to measure the degree to which parties are able to achieve integrative, win/win outcomes; such outcomes being viewed as possible only when parties trust each other enough to discuss their true needs, preferences, and priorities (Lax & Sebenius, 1986). The majority of experiments measuring these two indicators have shown that in e-negotiation, as opposed to face-to-face negotiation, one is less likely to encounter cooperation in the process, and less likely to achieve integrative outcomes. These results support the notion that the e-channel reduces trust between negotiators (Nadler & Shestowsky, 2006).[6]

These indirect measurements are reinforced by findings that e-negotiators, when questioned directly about the degree of trust they felt in negotiation processes, reported lower levels of trust than did face-to-face negotiators (Naquin & Paulson, 2003).

THE CHALLENGE OF TRUST BUILDING IN E-NEGOTIATION

Why is trust especially difficult to build in e-negotiation processes? A review of the research conducted so far points to eight major obstacles. While some of these obstacles also manifest in face-to-face negotiations, it is the way they reinforce each other in e-negotiation that makes trust-building in that process such a challenge. After introducing the obstacles briefly, we will elaborate on how each of them manifests in e-negotiation, and make suggestions for how negotiators can counter them in an effort to enable trust to emerge despite the channel-imposed obstacles. The obstacles are:

1. **Lack of contextual cues:** E-negotiators are denied many of the non-verbal cues that we rely on in interpersonal communication for assessing another person's trustworthiness.
2. **Sinister attribution effect:** The tendency to put the worst possible face on another's intentions and meanings increases in e-communication. As a result, e-negotiators will perceive the other's intentions through the most distrusting lens possible.
3. **Low expectations of trust:** E-negotiators have low expectations regarding the other's trustworthiness walking into the process, and this becomes a self-fulfilling prophesy.
4. **Anonymity and the faceless other:** The mutual invisibility inherent in e-negotiation facilitates trust-breaking behavior. It is

easier to cause damage to a faceless other, particularly when we feel protected by a shield of anonymity and physical distance.

5. **Confusing physical distance with interpersonal distance:** The feeling of distance and separation inherent to e-negotiation results in a sense of non-identification with the other, posing challenges to identity-based trust.
6. **The challenge of e-empathy:** Building trust by showing empathy for a negotiating opposite is a challenge even in a rich communication channel; in e-negotiation, it is an even greater challenge and therefore is often ignored.
7. **Pace problems:** The Internet incorporates two clashing characteristics: instant access to anything and anyone, and frequent asynchronous communication. This duality gives rise to pace-related expectations between negotiators that cannot be met, which breeds distrust.
8. **Negotiating in a new landscape:** The Internet itself, still a novelty to many, is viewed with distrust even by its most fervent advocates. Additionally, many lay and professional negotiators may be inexperienced at e-negotiation, not yet adept at trust-building through the e-channel.

We will proceed to examine these obstacles one by one:

1.   Lack of Contextual Cues
Human beings rely on contextual cues (such as another person's facial expressions, body language, tone of voice, etc.) to interpret messages. In fact, most of a message's meaning is perceived through these cues, rather than through the words actually spoken (Thompson, 2004). When communicating through any channel, we actively, if unconsciously, seek out such cues. For example, because we cannot see the face of the person on the telephone with us, we strain to infer meaning

from the tone of his voice. Text-based online communication is a very lean channel for contextual cues to pass through, rendering valueless most of the methods we instinctively use to transmit our trustworthiness to others, and neutralizing the senses we have developed to analyze cues from the other so as to asses his credibility. E-negotiators are denied many of the cues that have been found to inspire trust in face-to-face settings, including facial expressions, vocal inflections, physical proximity and touch.[7] Experiments comparing interactions through face-to-face, audio, video and text-based communication, found the last to be the *least* supportive of trust-building (Bos, Olson, Gergle, Olson, & Wright, 2002).

Alertness to this obstacle helps avoid many pitfalls, primarily the sinister attribution effect (See 2. The Sinister Attribution Effect). It encourages negotiators to strive for a friendly level of relationship in which emoticons as well as very open, direct and exact language can be used. When limited to text-only communication, we must bear in mind that sarcasm, cynicism, and humor can easily be misconstrued. If the meaning of a sentence we write is ambiguous, we need to provide the reader with the proper tone of voice to "hear" it in by adding in pointers such as "Forgive my sarcasm, but ..." or "Isn't it funny that ...," even though we would not do this in a face-to-face setting.

2. The Sinister Attribution Effect

The absence of contextual cues causes the e-negotiator to focus on the actual content of the communicator's message. While this is certainly useful, it may also have negative effects. For even if the message we communicated was not designed to insult or inflame our negotiation opposite, it can sometimes seem to convey negativity from his point of view. For instance, he might easily perceive any answer but "yes" to be threatening, an implication that "I will withhold from you that very thing you need." The combination of a negative predisposition towards anything but

"yes" and the lack of the contextual cues that help give messages their intended meaning, causes message-readers to remain uncertain regarding the writer's behavior and intentions. In such situations, the sinister attribution effect—the tendency to interpret another's behavior in the least positive way possible, and to infer his bad intentions from the negative way we perceive his behavior—may come to dominate the relationship (Kramer, 1995; Thompson, 2004). As a result, our opposite will view just about anything we say or do as being a negative, trust-breaking action arising from our negative, untrustworthy character or intentions. This is reinforced by the negotiation medium's characteristic of being recorded and reaccessible; our opposite can read and reread our e-mail message, ruminate on it over time and enter consecutive, spiraling, anger cycles—all before forming a devastating reply to our now-sinister message (Nadler & Shestowsky, 2006). A simple "Sorry, that's not enough for me" message can be perceived as treacherous and threatening, resulting in deep distrust.

The message-exchange dynamic of e-negotiation also contributes to the sinister attribution effect. In e-communication there are fewer opportunities to ask for a quick clarification or to make a snap correction. Research shows that e-negotiators ask fewer clarifying questions than face-to-face negotiators. The blanks get filled in by assumptions (Thompson & Nadler, 2002), never a good idea. The collapse of these assumptions later on will be perceived, through the filter of the sinister attribution effect, as a breaking of trust. The power of the sinister attribution effect in e-negotiation is clearly demonstrated by experiments showing that e-negotiators are more likely to suspect their opposite of lying than are face-to-face negotiators, even when no actual deception took place (Thompson & Nadler, 2002).

As negotiators, we need to constantly remember the power of the sinister attribution effect. If we fail to achieve something we need to, our ability to blame it on our negotiation opposite's misperception will be of little consolation. We need to actively help our opposite avoid misunderstanding or misinterpreting our meaning and intentions. The more ambiguous our messages are, the more opportunities we provide our opposite for sinister attribution. Analysis of failed e-negotiations shows that they tended to include unclear messages, long general statements and irrelevancies (Thompson, 2004). Each of these provide ample breeding ground for the sinister attribution effect; by avoiding them, we are assisting our opposite to understand our message as we intend him to. We need to add on "just to clarify" statements even when these would seem superfluous in face-to-face interactions. We might relate to the sinister attribution effect head on, by writing something such as "We both know how things get misunderstood in e-mail communication, so let me be as clear as possible on this point ... ." Finally, our basic message, offer or statement must be kept clear. Even if we elaborate in hopes of being understood better, of humanizing the conversation, or of forming rapport, every message should end with a very clear "to summarize" paragraph.

3.   Low Expectations of Trust

E-negotiators enter the process with a lower level of pre-negotiation trust in their opposite than do participants in face-to-face negotiations (Naquin & Paulson, 2003). This initial low expectation regarding interpersonal trust ties into the sinister attribution effect by reinforcing the tendency to seek out reasons to distrust rather than to recognize trustworthy actions. This becomes a self-fulfilling prophesy: expecting to find us untrustworthy, our negotiation opposite indeed finds us to be so. As a result, participants in e-negotiation also experience lower levels of *post*-negotiation trust than participants in face-to-face negotiations (Naquin & Paulson, 2003). This continues the cycle by lowering our trust expectations even further in anticipation of our next e-negotiation process. Participants in e-negotiation show less desire for future interactions with their negotia-

tion opposite than do participants in face-to-face negotiations, partially due to a negative perception of post-negotiation trust (Naquin & Paulson, 2003).[8] These findings are particularly important when one takes into account that they hold true even when there is no objective difference in the negotiation outcome. In other words, while a particular outcome, achieved through face to face negotiation, will build a degree of trust causing negotiators to look favorably on the possibility of future negotiation interaction with their opposite, the same outcome will achieve less in terms of inter-party trust-building and desire for future interaction if reached through e-negotiation (Naquin & Paulson, 2003). These findings demonstrate how difficult it is to build and maintain trust in e-negotiation: e-negotiators' initial trust levels are low, and they tend to remain low, relative to those of face-to-face negotiators. Of course, this is not to say that an individual e-negotiator, having concluded a successful negotiation in which any degree of trust was built, will not carry this trust over into a subsequent negotiation with the same opposite. However, the effects of this carry-over will be weak relatively to a comparable face-to-face negotiation process

It would seem that the best way to decrease the effects of diminished prenegotiation trust is, first and foremost, to recognize that the initial misgivings we may feel walking into an e-negotiation process are normal. They are a part of the playing field, and *not,* as we usually tell ourselves, an intuitive insight into our opposite's true nature or intentions. We should also pay attention to the other's behavior. If we sense that he seems to be affected by a case of prenegotiation distrust, we needn't be offended. We might even consider raising the issue head-on, empathizing with the way he feels and asking what we can do to dispel his doubts.

4.   Anonymity and the Faceless Other

A major challenge to the e-negotiator's wish to generate trust in his opposite arises from the nature of the encounter itself. Each party senses a degree of anonymity and distance, sitting behind his computer screen far away from his faceless opposite. This feeling of remote detachment leads both to assumptions that he can get away with trust-breaking behavior and to a lowering of moral inhibitions against doing so. While many e-negotiations take place between identifiable and accountable parties, it is as if something in the e-medium induces negotiators to forget this. Given a state of negotiating parties with no previous relationship or immediately perceived common ground, a no trust/no cooperation atmosphere is quick to evolve, with spirals of escalation likely as the sinister attribution effect kicks in (Nadler & Shestowsky, 2006).

However, one should keep in mind that small efforts can change this state of affairs substantially. The more one works at "unmasking" the other—or, more proactively, unmasking oneself towards the other—the more likely one is to find opportunities for trust-building. The more our opposite perceives us as an *identifiable other*, as opposed to an anonymous, faceless e-mail address, the more likely he is to share information, rely on us and trust in us (Nadler & Shestowsky, 2006). As a result, we need to take special care to incorporate an unmasking process into our e-negotiation processes.

The concept of using pre-negotiation social interaction to create a positive and unmasked environment for an upcoming negotiation process is widely discussed and advocated in negotiation literature. Negotiators are advised to create "instant relationships," absent a past relationship with one's negotiating partner. This process might be dubbed bonding (Shapiro & Jankowski, 1998) or building rapport (Drolet & Morris, 2000; Thompson & Nadler, 2002). Negotiators are advised to seek out and create opportunities to build a positive rhythm of interaction before reaching the table—indeed, even as they are walking to it. The more parties are "in sync" with each other before the negotiation process starts, and the more this improves throughout its course, the more likely they are to work together, coordinate and trust each other.

Holding preliminary face-to-face meetings has proven to be a highly effective means for building trust that carries over into e-negotiations (Rocco, 1998); even the most effective means (Zheng, Veinott, Bos, Olson, & Olson, 2002). It has also been suggested that adding face-to-face interactions as a support measure in the middle of an ongoing e-negotiation can have a positive effect (Cellich & Jain, 2003). Prescriptively speaking, this would suggest that when e-negotiating with a total stranger one should try to incorporate one face-to-face meeting into the negotiation dynamics. This might take the form of an informal, out-of-context encounter, such as stopping by your opposite's office to introduce yourself when you are in his physical location for other purposes. Notwithstanding the value of doing so, this will often be impossible to do with e-negotiation opposites; special effort must therefore be dedicated to an online unmasking process.

While in face-to-face encounters making introductions and light, social conversation comes naturally, in e-negotiation this tendency diminishes somewhat. This might be due to the semi-formal nature of written communication; to the asynchronous nature of e-mail exchanges; to the fact that writing is not as easy as talking; or to geographical and cultural distance between parties (someone in another hemisphere might not share our weather or our affinity for a particular sports team). As a result, e-negotiators need to consciously dedicate time and effort to the unmasking process. Experiments have indicated that even minimal prenegotiation contact, at the most basic level of "schmoozing" via preliminary e-mail introductory messages or brief telephone exchanges, has the potential for building trust, improving mutual impressions and encouraging the reaching of integrative outcomes (Nadler & Shestowsky, 2006). The easiest way to start this off is to send our negotiation opposite a short introductory letter before the negotiation process begins or at its initiation, letting him know, in brief, who we are, where we are and—depending on the context—a bit about ourselves. More likely than not, he will reciprocate with a note of his own, or incorporate his own introduction into the first substantive message he sends. By relating to this in the next message we send, we can create a cycle of unmasking. By looking for excuses to drop a small piece of personal information into the process, we give our opposite an opportunity to recognize the person behind the e-mail address and to reciprocate. By saying "I know what you mean" about something he shared and adding on something new about ourselves, we recognize his situation, disclose personal information of our own and invite reciprocal recognition. If our opposite mentioned children, we can mention our own. If our opposite talked about a pressure-period at work, we can mention our own 80-hour week. This allows the unmasking process to continue, and also moves the process towards empathy and identification-based trust.

Negotiators are constantly reminded of the value of turning a negotiation process into an ongoing relationship. This is particularly pertinent in protracted or recurring negotiation processes, such as between a purchaser and a supplier who negotiate price before each shipment. By dropping the other a line between negotiation rounds, we remind him that we share an ongoing relationship. By keeping a promise we made to get back to him with something, or sending him an article we think he will be interested in, we can reinforce this new ongoing relationship narrative. We can set up, in advance, opportunities we can use to 'touch base' later on—during pre-negotiation schmoozing, or during the course of the negotiation itself.

5. Confusing Physical Distance with Interpersonal Distance

Another challenge to trust-building in e-negotiation is that online communication causes a perception of difference, of otherness between the two parties. We feel anonymous and distant from the other, and that he is anonymous and distant from us. We know nothing about him, other than that we need something from him. This perception of otherness leads us to subconsciously assume that the other is nothing like us, and that his attitudes, personality and interests conflict with ours. While the unmasking process described above can potentially begin to dispel this, we suggest creating stronger identification-based trust and closing perceived distances between ourselves and the other by searching for shared group membership. When individuals perceive themselves as belonging to the same group as another, their perceptions of the other become more positive and their level of trust in him increases. While in-group members appreciate each other more and intuitively assume they share positive traits and attributes, they perceive out-group members, as "others", assuming that if they differ in one attribute, they surely must have inferior qualities and negative intentions. The power of in-group fraternity is impressive, even when the shared attribute of the in-group is something trivial or innocuous such as "coffee drinkers." The development of positive attitudes and identification-based trust towards in-group members plays out online much as it does in face-to-face encounters (Wallace, 1999). This results in a greater likelihood of agreement between in-group members (Moore, Kurtzberg, Thompson, & Morris, 1999).

In face-to-face encounters we can learn about our opposite from a variety of cues and then use the knowledge to form an in-group affiliation (a nonpolitically-correct example might be noticing our negotiation opposite patting his pockets absentmindedly, and seeing that as our cue to ask if he would like to join us for a cigarette break). However, in online exchanges this is a more challenging prospect. E-negotiators need to be carefully attuned to the other's messages in order to discover things to connect to. We must glean all the information we can from our opposite's introductory e-mail and any personal information he may have included in later messages. We should note issues he stresses as having particular importance to him. For example, if our opposite writes "I think sticking to schedule is very important" (as opposed to "let's get this done on time"), we should see that as our cue to state that we also see dedication to promptness as a positive attribute; this gives him the opportunity to perceive the two of us as belonging to the "punctual" group. If we do not have a clear enough picture of our opposite's traits or values to build on, we can proactively ask questions seeking out identity-based similarities (Nadler & Shestowsky, 2006).

We have mentioned the danger of being perceived by our negotiation opposite as a member of an out-group, and how online communication can cause this perception to form almost automatically. However, mindful use of online communication can actually help us protect ourselves against the forming of such perceptions. This is because e-communication renders invisible many of the stereotypes upon which people base instant group-affiliation judgment, such as age, gender, race and status (Wallace, 1999). Additionally, while carefully reading our opposite's messages to seek out shared in-group affiliation, we can also identify whom he might perceive as belonging to an out-group. This will allow us to avoid divulging details that may paint us as belonging to that group.

6.   The Challenge of E-Empathy

One of the basic directives of any book or course focusing on negotiation is that a negotiator should show empathy for his opposite (Mnookin, Peppet, & Tulumellow, 2000; Ury, 1991) Showing empathy is counterintuitive for most negotiators, who worry that showing *any* concern for the other's predicament or emotions will necessitate making a concession to him. They therefore prefer sweeping the other's predicament under the carpet and keeping a poker face. But in truth, empathizing with another does not require giving anything

up; it does, on the other hand, show the other our understanding and recognition. This can have the effect of eliciting reciprocation, increasing the other's willingness to listen, and lessening his tension and potential contentiousness. Moreover, it goes a long way towards building trust. In order to have these effects, empathy must be accurate (infer the specific contents of the other's experiences and emotions) and include a supportive response (some form of constructive or empowering input related to the other's needs).

The important role of empathy in negotiation has been shown to hold true in the online environment as well. E-negotiators showing empathy are trusted by their negotiation opposites more then those who do not (Feng, Lazar, & Preece, 2004). Nonetheless, showing empathy for another person via a communication channel with limited contextual cues is quite a challenge. We cannot nod understandingly, or smile and lay a supportive hand on our opposite's own. The e-channel necessitates special methods for showing e-empathy.

Many of the most basic communication tools negotiators are advised to employ are especially valuable for their facilitating the showing of empathy to one's negotiation opposite. Three good examples are:

- Active listening: Listening to our opposite carefully, in a manner demonstrating our absolute focus on him and our ability to contain everything he has to say
- Reflecting: Paraphrasing or repeating the content of our opposite's message, showing him that we understand his factual input, appreciate the emotional importance he attaches to what he said or to the situation and offering him the chance to correct misunderstandings on our part
- Asking pertinent, productive and to-the-point questions showing interest in our opposite, his needs and concerns (Ury, 1991)

While some aspects of these tools might be difficult to transfer to the online medium, this does not mean that showing empathy is impossible or inhibitively clumsy. Additionally, the fact that our opposite's message is recorded by the channel or indeed included in the message we send back to him (such as an original e-mail being automatically quoted when we answer a message by clicking reply) does not mean that applying the tools of listening or reflecting is superfluous. Showing empathy in e-negotiation is as necessary as it is in face-to-face encounters—we just need to find suitable methods for it.

A good way to demonstrate online listening is to stress, in replying to our opposite's messages, that we have read what he sent. For example, "I read your message carefully" or "Reading your letter last night, I realized ..." might seem like casual opening lines, but they convey a powerful message to your opposite: you have been heard. Similarly, other messages touching on the way we handle our opposite's message, such as "I showed your letter to my boss" or "I waited till I was in the office to read your e-mail, in order to give it my full attention" transmit an attentive intake process for his messages. Yet another method for online active "listening" is to relate to parts of our opposite's message specifically, in a manner that shows we read it carefully. Writing "I know you wrote that you'd be out of the office for a couple of days, but I wanted you to get this as soon as possible" transmits that in reading his message we took note even of non-central details. Alternatively, you might choose to insert your reply text into the original text of your opposite's message, signifying that you are relating to every point he made.

While demonstrating listening in e-negotiation may take some creativity, two characteristics of e-communication can actually make reflecting simple: messages are recorded, and word processing can be used to use the original message in a new way. If we don't want to invest time or physical effort paraphrasing or rewriting what our opposite wrote in your own words, we can easily go over his message and create a paraphrased summary to send back to him through copying, pasting, and editing.

When using questions to further understand the other's position and needs, we can utilize the benefits of recorded messages and of word processing to connect our question to specific parts of our opposite's message. This not only makes the question more to the point and part of the flow of conversation, it also incorporates elements of active listening and reflecting. For example, instead of writing "When do you need the computers delivered?," we might write "In your last e-mail you wrote 'I'm pressed for time and need these units ASAP.' I appreciate that you're in a real rush, however, I know it will take a while to get the computers organized for delivery. Can you tell me a bit more about our time frame?"

Word processing abilities also help us avoid perceived trust-breaking behavior based on misunderstandings. If a message is ambiguous, we can copy and paste it right back to our opposite, highlight what we don't understand and ask him to clear things up. This will not only help us avoid mistakes and breaking trust, it actually builds trust as our message conveys the subtext of "I'm listening to you, and it is very important for me to make sure I'm not misunderstanding you".

7. Pace Problems: The Challenge of Asynchronous Communication

The art of negotiating solely by exchanging written messages through postal mail is a long-forgotten one. We have become accustomed to exchanging opinions through synchronous communication, either face-to-face or over the telephone. Even in cases where there are time-gaps between actual offers—such as when lawyers exchange drafts of a contract back and forth over the course of a few months—much of the actual discussion of the issues takes place synchronously. E-negotiators need to relearn the art of asynchronous communication. This is not intuitive, for one of the Internet's promises which many have become used to and even reliant upon is instant access to anything and anyone. Our synchronous-communication upbringing, combined with our expectations of instant access, clash with the basic nature of asynchronous communication. As a result, e-mail communication often involves an anxiety that blends distrust of the channel with distrust of the other. When we send a message and do not receive a response promptly, not only do we question whether the other received the message, we begin to wonder why (if indeed he *has* received it) he is taking so long to respond? E-negotiators often forget, or at least disregard, the asynchronous nature of e-communication, and build expectations based on an assumption that they can control the rate of message exchange. When the other fails to live up to those expectations, frustration and sinister attribution are quick to follow (Thompson & Nadler, 2002).

Even when there is no sense or expectation of immediacy, the rule that frequent message exchanges, as opposed to communication broken by intervals, is conducive to trust-building within groups (Wallace, 1999; Walther & Bunz, 2005) holds true for the dyadic group of two people negotiating as well. Unresponsiveness or lengthy breaks between messages foster anxiety, a fertile breeding ground for distrust. As Billy Joel put it: "To insure yourself, you've got to provide communication constantly." A good rule of thumb to follow in order to avoid allowing time-gaps to develop is to always respond to an e-mail within 24 hours, even if only to say that we are working on, or considering, what our negotiation opposite has written, and will get back to him shortly (Katsh & Rifkin, 2001)

Once the trust-threatening elements of asynchronous communication are neutralized, it can actually be a very conducive channel for trust building. It can help control our response time—to our own advantage. Synchronous communication necessitates responding to our opposite's behavior on the spot, whereas asynchronous communication allows us to avoid knee-jerk reactions or escalatory cycles and to think proactively. The slower pace allows us to fashion and frame our response thoughtfully and productively. It enables us to verify details instead of giving off-the-cuff responses that may later turn out to be inaccurate. The ability to read over a message, or to ask a friend or colleague to take a look at it and tell us what he thinks, can help us avoid the pitfall of perceiving our opposite negatively due to the sinister attribution effect. These potentialities, unique to asynchronous e-communication, hold the promise of enabling trust building in e-negotiation in ways denied face-to-face negotiators.

8. Negotiating in a New Landscape

One of the primary problems for creating trust in online communication is, as Boyd (2003) put it, the medium's "novelty, and its attendant mystery" (p. 394). He suggests that a user's distrust of the medium due to inexperience can spill over to harboring suspicions towards a Web site's credibility or an e-negotiation opposite's trustworthiness. A successful experience with e-negotiation will therefore lead to a progressively lower trust threshold, and the spillover between trust towards the channel and trust towards a negotiation opposite might actually begin to work the other way around: a successful e-negotiation experience can cause a negotiator to feel optimism, which may facilitate the forming of trust during his next e-negotiation experience. This would suggest that trust-building might become easier as the medium's novelty declines over the course of the next generation.

Attempts have been made to provide structural solutions to the distrust caused by a disparity of negotiating skill and power, as well as by distance and anonymity, on the Internet. For example, eBay's rating system provides potential clients the opportunity to view the degree to which previous clients were satisfied with a particular vendor. While this system has been very successful in creating an environment of trust sufficient to hold the eBay community together, attempts to build a Web-wide rating system have not yet succeeded to the degree of being seen as an integral part of the Internet's commercial infrastructure (Rule, 2002). In the context of one-on-one e-negotiation, structural solutions are even harder to envisage.

E-negotiation is a growing and evolving phenomenon. We believe we are now suffering the birth pangs of the start-up phase, after which experience, adeptness, and trust in the medium will become factors positively affecting interpersonal trust. In the meantime, the best we can do is to choose our negotiation channel carefully. Our degree of comfort with the medium is something to consider seriously—we are more likely to take mis-steps and to arouse our negotiation opposite's suspicions if we are not adept at e-communication. However, we must also take into account our opposite's comfort and skill with the medium. The more the opposite distrusts the medium or feels insecure negotiating through it, the more this is likely to be transferred into distrust of us. If we sense this is affecting the negotiation process adversely, we might suggest utilizing an alternate means of communication.

FUTURE TRENDS: MORE AND MORE

Increasingly, negotiation relationships are taking place online. The growth of e-commerce from isolated transactions into a multitrillion dollar marketplace in which every inhabitant of the globe is a potential participant guarantees the future conduct of countless e-negotiation processes. The increasing use of virtual workplaces, as supplementary venues for group interaction or as the major venue for a business' internal activity, reinforces this trend. This is further supported by the rapidly rising degree to which people feel comfortable moving other types of relationships online, which, while non-transactional in the traditional sense, certainly incorporate elements of negotiation activity.

As the scope of e-negotiation itself widens, the need for trust-building tools and abilities will become more acute. Some tools will probably develop on their own, while others need to be initiated and directed. Richer language and contextual cues such as abbreviations, emoticons and other enhancements are constantly developing. Already colloquial, they will gradually spill over into more formal communication, providing a richer channel for e-negotiation, one which is able to convey meaning in a familiar, emotionally accurate, and trust building manner.

We have seen that one fundamental way to affect trust-building in e-negotiation is to thoroughly familiarize negotiators with the e-medium as a negotiation channel and to train them to be adept at its use. As negotiators improve in using the medium, they will achieve better results, and their increased trust in the medium will spill over to facilitate building interpersonal trust. We believe that this practice-based improvement of the field will be complemented by the teaching of e-negotiation as an important part of the professional skill-set imparted in business and law schools, as well as in the training programs of other disciplines.

CONCLUSION

Trust plays a crucial role in all negotiation processes. The degree to which parties trust each other often delineates the degree to which they will cooperate with each other in the process, share information, and search for integrative agreements. A negotiator's ability to inspire his opposite's trust in him can be considered one of his greatest assets. However, many factors at work in the negotiation process tend to affect trust negatively, causing trust-building and -maintenance to be an uphill battle in the best of cases.

As human interactions take place online with increasing frequency, negotiators find themselves dealing with each other in the online environment. Not only are most of the recognized challenges to trust development in face-to-face communication manifest in e-communication, these are often magnified due to particular characteristics of e-negotiation. In addition, negotiators find themselves dealing with new, unique challenges to trust development and maintenance particular to the communication medium used for these processes. Online communication, like any other communication channel, contains inherent channel-related obstacles to trust-building and trust-maintenance. Awareness of the role trust plays in negotiation and of the challenges posed to it by the e-communication channel will help the e-negotiator steer clear of most of these pitfalls. This awareness is also the starting point for wider adeptness in building trust in e-negotiation processes. By supplementing it with the increased familiarity, training, and experience in e-negotiation that the future promises, we anticipate that the e-negotiator will be provided with an enhanced skill-set, suitable for allowing him to actively create an atmosphere of trust between himself and his negotiation opposite, resulting in a more cooperative process as well as a more integrative and beneficial outcome.

ACKNOWLEDGMENT

The author wishes to thank Rochi Ebner for her assistance and insightful comments.

REFERENCES

Axelrod, R. (1984). *The evolution of cooperation.* New York: Basic Books.

Bos, N., Olson, J., Gergle, D., Olson, G., & Wright, Z. (2002). Effects of four computer-mediated communications channels on trust development. In *Proceedings of SIGCHI Conference on Human Factors in Computing Systems* (pp. 135-140). New York: ACM Press. Retrieved August 23, 2006, from http://portal.acm.org

Boyd, J. (2003). The rhetorical construction of trust online. *Communication Theory, 13*(4), 392-410.

Cellich, C., & Jain, S.C. (2003). *Global business negotiations: A practical guide.* Mason, OH: Thomson/South-Western.

Conley Tyler, M., & Raines, S. (2006). The human face of online dispute resolution. *Conflict Resolution Quarterly, 23*(3), 333-342.

Deutsch, M. (1962). Cooperation and trust: Some theoretical notes. In M.R. Jones (Ed.), *Nebraska symposium on motivation* (pp. 275-318). Lincoln: University of Nebraska Press.

Drolet, A.L., & Morris, M.W. (2000). Rapport in conflict resolution: Accounting for how face-to-face contact fosters mutual cooperation in mixed-motive conflicts. *Journal of Experimental Social Psychology, 36*(1), 26-50.

Feng, J., Lazar, J., & Preece, J. (2004). Empathy and online interpersonal trust: A fragile relationship. *Behavior & Information Technology, 23*(2), 97-106.

Fisher, R., Ury, W., & Patton, B. (1991). *Getting to yes* (2nd ed.). New York: Penguin Books.

Gambetta, D. (1988). Can we trust trust? In D. Gambetta (Ed.), *Trust: Making and breaking cooperative relations* (pp. 213-237). Oxford: Basil Blackford.

Katsh, E., & Rifkin, J. (2001). *Online dispute resolution.* San Francisco: Jossey Bass.

Kiesler, S., & Sproull, L. (1992). Group decision making and communication technology. *Organizational Behavior and Human Decision Processes, 52*(1), 96-123.

Koehn, D. (2003).The nature of and conditions for online trust. *Journal of Business Ethics, 43*(1), 3-19. Retrieved August 23, 2006, from http://proquest.umi.com

Koeszegi, S.T., Srnka, K.J., & Pesendorfer, E.-M. (in press). Electronic negotiations: A comparison of different support systems. *Die Betriebswirtschaft.*

Kramer, R.M. (1995). Power, paranoia and distrust in organizations: The distorted view from the top. In R.J. Bies, R.J Lewicki, & B.H. Sheppard (Eds.), *Research on negotiation in organizations* (pp. 119-154). Greenwich, CT: JAI Press.

Lax, D.A., & Sebenius, J.K. (1986). *The manager as negotiator.* New York: Free Press.

Lewicki, R.J., & Litterer, J. (1985). *Negotiation: Readings, exercises and cases.* Boston: Irwin.

Lewicki, R.J., Saunders, D.M., Minton, J.W., & Barry, B. (2002). *Negotiation: Readings, exercises and cases* (4th ed.). Boston: McGraw-Hill/Irwin.

Mnookin, R.H., Peppet, S.R., & Tulumello, A.S. (2000). *Beyond winning: Negotiating to create value in deals and disputes.* Cambridge, MA: The Belknap Press of Harvard University Press.

Moore, C. (2003). *The mediation process* (3rd ed.). San Francisco: Jossey Bass.

Moore, D., Kurtzberg, T., Thompson, L., & Morris, M. (1999). Long and short routes to success in electronically mediated negotiations: Group affiliations and good vibrations. *Organizational Behavior and Human Decision Processes, 77*(1), 22–43. Retrieved July 3, 2007, from http://www.cbdr.cmu.edu/mpapers/emn.pdf

Morris, M., Nadler, J., Kurtzberg, T., & Thompson, L. (2002). Schmooze or lose: Social friction and lubrication in e-mail negotiations. *Group Dynamics, 6*(1), 89-100. Retrieved August 23, 2006, from http://gateway.uk.ovid.com/gwl/ovidweb.cgi

Nadler, J., & Shestowsky, D. (2006). Negotiation, information technology and the problem of the faceless other. In A.W. Kruglanski & J.P. Forgas (Series Eds.) & L. Thompson (Ed.), *Negotiation theory and research. Frontiers of Social Psychology.* New York: Psychology Press.

Naquin, C.E., & Paulson, G.D. (2003). Online bargaining and interpersonal trust. *Journal of Applied Psychology, 88*(1), 113-120. Retrieved July 3, 2007, from http://gateway.uk.ovid.com/gwl/ovidweb.cgi

Paulson, G.D., & Naquin, C.E. (2004). Establishing trust via technology: Long distance practices and pitfalls. *International Negotiation, 9*(2), 229-244.

Pruitt, D.G., & Kimmel, M. (1977). Twenty years of experimental gaming: Critique, synthesis, and suggestions for the future. *Annual Review of Psychology, 28*, 363-392.

Pruitt, D., Rubin, J., & Kim, S. (1994). *Social conflict: Escalation, stalemate, and settlement* (3rd ed.). New York: McGraw Hill.

Raiffa, H. (1982). *The art and science of negotiation.* Cambridge, MA: Harvard University Press.

Rocco, E. (1998). Trust breaks down in electronic contexts but can be repaired by some initial face-to-face contact. In *Proceedings of the SIGCHI Conference on Human Factors in Computing Systems* (pp. 496-502). Los Angeles: ACM Press. Retrieved August 23, 2006, from http://portal.acm.org

Rule, C. (2002). *Online dispute resolution for business.* San Francisco: Jossey Bass.

Schelling, T.C. (1980). *The strategy of conflict* (2nd ed.). Cambridge, MA: Harvard University Press.

Shapiro, R.M., & Jankowski, M.A. (1998). *The power of nice.* New York: John Wiley.

Thompson, L. (2001). *The mind and heart of the negotiator* (2nd ed.). Upper Saddle River, NJ: Prentice Hall.

Thompson, L., & Nadler, J. (2002). Negotiating via information technology: Theory and application. *Journal of Social Issues, 58*(1), 109-124.

Ury, W. (1991). *Getting past no.* New York: Bantam Books.

Wallace, P. (1999). *The psychology of the Internet.* New York: Cambridge University Press.

Walther, J.B., & Bunz, U. (2005). The rules of virtual groups: Trust, liking and performance in computer-mediated communication. *Journal of Communication, 55*(4), 828-846. Retrieved July 3, 2007, from http://bunz.comm.fsu.edu/JoC2005_55_4_virtual.pdf

Wang, Y.D., & Emurian, H.H. (2005). An overview of online trust: Concepts, elements, and implications [Electronic version]. *Computers in Human Behavior, 21*(1), 105-125. Retrieved July 3, 2007, from http://nasa1.ifsm.umbc.edu/cv/TrustCHB.pdf

Zheng, J., Veinott, E., Bos, N., Olson, J.S., & Olson, G.M. (2002). Trust without touch: Jumpstarting long-distance trust with initial social activities [Electronic version]. In *Proceedings of the SIGCHI 2002 Conference on Human Factors in Computing Systems* (pp. 141-146).

ENDNOTES

[1] The past 20 years have seen the development of many different means for communication via information technology media, and negotiation processes utilize them all. Besides commonly used methods such as e-mail, instant messaging, and message posting, a variety of software platforms known as negotiation support systems (NSS) have been developed, designed to assist negotiators communication, exchange offers and consider them. NSS differ in nature, ranging from simple communication channels through which negotiators post e-mail-type messages to each other, to sophisticated optimization-algorithm-based software capable of analyzing input from two negotiators and suggesting solutions designed to meet their preferences in a manner they might not be able to come up with on their own (Koeszegi, Srnka, & Pesendorfer, in press; Rule, 2002). Some commonly used NSS cut out the need for communication altogether by engaging parties in a "blind bidding" process in which they only communicate with the software platform, which compares their offers and then lets them know whether they have reached a deal or not. It could be argued that through these NSS a negotiation process has been created in which interpersonal trust is circumvented—so long as parties trust the platform, they do not need to trust each other and do not need to engage in any trust-building communication or activities.

[2] Most of the conclusions reached in this chapter hold true for synchronous communication as well; with other conclusions, suggestions for adaptations will be provided. Similarly, they hold true for negotiation processes utilizing multiple mediums, where online communication, face-to-face meetings, instant messaging, or phone conversations complement each other to form hybrid processes. We will touch on the value of creating such interactions. Once aware of the potentialities and challenges inherent to "pure" e-negotiation, negotiators will be able to make informed and conscious choices regarding the medium suitable for any particular part of a negotiation process.

[3] Most of the negotiation literature focuses on face-to-face settings, even assuming that this is the default method in which negotia-

tion occurs, and that knowledge of all other communication channels used for negotiation can be extrapolated from face-to-face findings by making marginal adaptations. In short, as Nadler and Shestowsky (2006) comment, "Traditional approaches to research on negotiation do not typically consider the possibility that the type of communication media used by negotiators could be a factor affecting the negotiation itself" (p. 145). As a result, much of the research conducted showing the positive benefits of trust-building in negotiation is based on experiments conducted in face-to-face settings. Most of the writing about e-negotiation seems to take for granted that trust plays the same crucial role in the online environment as it does in face-to-face settings, although this may not have been decisively proven. Working under this assumption, the main thrust of research and writing goes on to explore the considerable challenges to building and maintaining trust over such a tenuous medium. Having provided this caveat, this chapter goes the same route.

[4] That is not to say that the other routes are not hampered by the medium. Distance from our negotiation opposite inhibits our ability to observe him and learn about him, challenging the development of knowledge-based trust. The difficulty to monitor and enforce agreements at a distance poses challenges to developing deterrence-based trust.

[5] Lewicki et al. (2002) suggest that these three paths to trust are often taken sequentially. Typically, deterrence-based trust forms first. As parties gain knowledge and experience of each other, knowledge-based trust forms. Identification-based trust, at least in its more extreme forms of complete empathy and identification, will typically be the last to form. Of course, in many relationships only one type of trust— usually deterrence-based trust—is manifest.

[6] As a result, most practitioners and researchers have adopted the assumption that e-negotiation, as a rule, involves less inter-party trust and results in fewer integrative agreements. Others have noted experiments challenging these findings (Conley Tyler, & Raines, 2006; Nadler & Shestowsky, 2006), indicating that more careful examination needs to be done, which might differentiate between different e-communication platforms or examine e-negotiation's suitability to specific types of disputes (Conley et al., 2006).

[7] These cues are at best replaced with a limited range of emoticons. It is unclear whether this may help or interfere, much as with "real" terms, people use and interpret emoticons differently, paving the way for misunderstanding. In formal communication, the use of emoticons has not yet become the norm, leaving communicators with only the text-based channel.

[8] This trust-downgrading cycle manifests not only vis-à-vis one's negotiation opposite, it also ties into one's own negotiation satisfaction and self-trusting: online negotiators tend to feel less satisfied with their outcomes and less confident in the quality of their performance than face-to-face negotiators (Naquin & Paulson, 2003).

Chapter XI
Antecedents of Consumer Trust in B2C Electronic Commerce and Mobile Commerce

Dan J. Kim
University of Houston Clear Lake, USA

ABSTRACT

Despite the importance of trust in electronic commerce including mobile commerce, there is insufficient theory and model concerning the determinants of consumer trust in business-to-consumer electronic commerce. Thus, the purpose of this chapter is to (1) identify the major antecedents of a consumer's trust in electronic commerce and mobile commerce contexts through a large-scale literature review, (2) develop an integrative trust antecedent reference model summarizing the antecedents of consumer trust, and (3) discuss six categories of mobile applications as future trends of technologies and key issues related to consumer trust area in electronic commerce. In addition, to provide the validity of the proposed reference model, this chapter also proposes a research model derived from the reference model and discusses the constructs of the proposed model in detail. The chapter concludes that building trust is not simply an issue related to consumer-technology-buyer, but it is a complex issue that involves the interactions of key elements (buyer, seller, third-party, technology, and market environment) at least.

INTRODUCTION

Trust is important in exchange relations because it is a key element of social capital (Mayer, Davis, & Schoorman, 1995), and is related to firm performance, satisfaction, competitive advantage, and other favorable economic outcomes. Trust is identified as an important factor in several literatures, including marketing, behavioral science, and electronic commerce (Beatty, Mayer, Coleman, Reynolds, & Lee, 1996; Czepiel, 1990; Dirks & Ferrin, 2001, 2002; Hoffman, Novak, & Peralta, 1999; Jarvenpaa, Knoll, & Leidner, 1998; Kramer, 1999). According to the study conducted by Urban, Sultan, and Qualls (2000), consumers make electronic commerce (e-commerce) transaction decisions based on trust. Therefore, lack of trust is one of the most frequently cited reasons for online consumers not engaging in exchange relationships with Internet vendors in e-commerce (Lee & Turban, 2001).

Mobile commerce (m-commerce) extends current e-commerce channels into more convenient "anytime, anyplace, and personalized" environment. As an emerging subset of e-commerce, m-commerce faces the same problems troubling e-commerce plus a few of its own due to the limitations of mobile technology (Siau & Shen, 2003). The limitations include restricted computation powers, memory, small screens, low-resolution displays, tiny multifunction keypads, battery life, unfriendly user interface for mobile devices, low bandwidth, unstable network connection, relatively high usage cost, and vulnerability of wireless data transmission. Therefore, building consumer trust in m-commerce is a particularly intimidating task due to the unique limitations of mobile technology.

Since consumer trust plays an essential role in online transactions, it is important to identify antecedents that affect a consumer's trust in e-commerce and m-commerce areas. Several researchers and professionals (Ba, Whinston, & Zhang, 1999; Beatty et al., 1996; Brynjolfsson & Smith, 2000; Czepiel, 1990; Hoffman et al., 1999; Jarvenpaa et al., 1998; Ratnasingham, 1998; Urban et al., 2000) have focused on various issues of trust in e-commerce. Even so, some scholars (Ratnasingham, 1998) have argued that the study of trust has been problematic for several reasons. These include problems with the definition of trust, confusion between trust and its antecedents, difficulties of observing and measuring trust, the tendency of particular disciplines to provide only partial descriptions of trust antecedents, and a lack of specificity about who the parties are (e.g., trustor and trustee) in research contexts in which trust is relevant (Mayer et al., 1995).

This chapter attempts to consider some of the above issues. First, we identify the major antecedents of a consumer's trust in electronic commerce and mobile commerce contexts through a large-scale literature review, second, develop an integrative trust antecedent reference model summarizing the antecedents of consumer trust, and finally discuss six categories of mobile applications as future trends of technologies and key issues related to consumer trust area in electronic commerce. In addition, this study also proposes a theoretical research model derived from the integrative trust antecedent reference model and discusses the constructs of the proposed model in detail to provide the validity of the reference model.

BACKGROUND: ANTECEDENTS OF TRUST

Trust Antecedents in E-Commerce Studies

Several researchers have tried to categorize antecedents or factors of a consumer trust (Barney & Hansen, 1994; Doney & Cannon, 1997; McKnight, Choudhury, & Kacmar, 2002b; Walczuch, Seelen, & Lundgren, 2001; Zucker, 1986). Zucker (1986) proposed three major ways to build trust: (1) process-based (e.g., reputation, experience), (2) characteristic-based (e.g., disposition), and (3) institutional-based (e.g., third-party certification). Mayer et al. (1995) defined trust as a behavioral intention based upon the expectations of another person. Based on this definition, they proposed a model of dyadic trust in organizational relationships that includes the characteristics of both the trustor and trustee that influence the formation of trust. The three characteristics included in the model, representing the perceived trustworthiness of the trustee, are benevolence, integrity, and ability. Doney and Cannon (1997) developed five distinct trust building processes in business relationships: (1) calculative process (trustor calculates the costs and/or rewards of a target acting), (2) prediction process (trustor develops confidence that target's behavior can be predicted), (3) capability process (trustor assesses the target's ability to fulfill its promises), (4) intentionality process (trustor evaluates the target's motivations), and (5) transference process (trustor draws on proof sources from which trust is transferred to the target). They also categorized characteristics of supplier firm, salesperson, and the relationship into four types. Barney and Hansen (1994) and Lewis and Weigett (1985) defined the three levels of customer trust: (1) strong trust, (2) semistrong trust, (3) weak trust. Bhattacherjee (2002) proposed three key dimensions of trust: (1) trustee's ability, (2) benevolence, and (3) integrity, based on cross-disciplinary literature review on dimensions of trust. Recently, Kim, et al. (2005) identified four different entities of e-commerce market structure: consumer, seller, third party, and technology. Based on the four entities, they investigated the determinants of online trust and divide the determinants into six dimensions: consumer-behavioral, institutional, information content, product, transaction, and technology dimension.

Trust and National Culture

National culture also influences individual and organizational trust development processes (Doney, Cannon, & Mullen, 1998). Hofstede (1991, 1994) revealed the five cultural dimensions: individualism/collectivism, uncertainty avoidance, power distance, masculinity/femininity, and long/short term orientation on life. *Individualism* refers to the degree the society reinforces individual or collective achievement and interpersonal relationships; *uncertainty avoidance* refers to the degree of tolerance for uncertainty and ambiguity within the society—that is, unstructured situations; *power distance* refers to the degree of equality, or inequality, between people in the country's society; *masculinity* refers to the degree the society reinforces, or does not reinforce, the traditional masculine work role model of male achievement, control, and power; and *long/short*

term orientation of life refers to the degree the society embraces, or does not embrace, long-term devotion to traditional, and forward thinking values (Hofstede, 1980, 1991, 1994).

Based on Hofetede's framework and using individualism/collectivism and power distance as independent variables, Strong and Weber (1998) examined the theory that trust is culturally determined in organization's contexts. They concluded that differences in trust exist globally between cultures. Griffith, Hu, and Ryans (2000) designated the United States and Canada as *Type I culture* with an "individualistic-small power distance-weak uncertainty avoidance" type of culture to contrast with *Type II culture* countries (Chile and Mexico) with "collectivistic-large power distance-strong uncertainty avoidance" characteristics. Although no significant difference in the strength of the trust-commitment relationship was found between Type I and Type II cultures, the study discovered that Type I cultures have a higher possibility of forming a trusting relationship with other Type I cultures, rather than with Type II cultures.

Several cultural studies (Mayer & Tan, 2002; Park & Jun, 2003; Png, Tan, & Wee, 2001; Soh, Kien, & Tay-Yap, 2000; Tan, Wei, Watson, Clapper, & McLean, 1998; Tan, Wei, Watson, & Walczuch, 1998) have shown that the dimensions of national culture affect the development, adoption, and impact of information communication technology (ICT) infrastructure and its applications in the field of information systems. However, only a handful of studies (Gefen & Heart, 2006; Jarvenpaa, Tractinsky, Saarinen, & Vitale, 1999; Lim, Leung, Sia, & Lee, 2004; Pavlou & Chai, 2002) to date have aimed at the effect of national culture on trust in computer-mediated electronic commerce transactions.

Jarvenpaa et al. (1999) used Hofstede's dimensions to compare Internet trust in individualistic and collectivistic cultures to conduct a study on a cross-cultural validation of an Internet consumer trust model. They found that consumers in different cultures may have differing expectations of what makes a Web merchant trustworthy. Although no strong cultural effects were found regarding the antecedents of trust, their study ignited examinations of cultural differences in the antecedents of trust and the levels of trust in the context of e-commerce. Incorporating Hofstede's three cultural dimensions (i.e., individualism/collectivism, power distance, and long-term orientation) along with the theory of planned behavior, Pavlou and Chai (2002) conducted an empirical study to explain e-commerce adoption across cultures using data from consumers in the United States and China. The results of the study support the theory that cultural differences play a significant role in consumers' e-commerce adoption. Lim et al. (2004) identified two national culture dimensions (i.e., individualism-collectivism and uncertainty avoidance) and their interaction that influences Internet shopping rates across countries. They also found that trust mediates the relationship between cultural differences and Internet shopping adoption decisions. Cross-validating the scale of trust and its antecedents in both the U.S. and Israel, a cross cultural study by Gefen and Heart (2006) found that trust beliefs may be a relatively unvarying aspect of e-commerce but the effects of predictability and familiarity on trust beliefs may differ across national cultures.

Trust Antecedent in M-Commerce Studies

Mobile commerce is defined as business activities and processes related to an e-commerce transaction conducted through wireless communications networks that interface with mobile devices (Tarasewich, Nickerson, & Warkentin, 2002). Several studies (Anckar & D'Incau, 2002; Booz, 2000; Kannan, Chang, & Whinston, 2001; Malhotra & Segars, 2005; Siau, Lim, & Shen, 2001) identified the following distinctive mobile capabilities or values which drive one of the most promising innovative application services in near future:

ubiquity, time-criticality, spontaneity/immediacy, constancy, convenience, personalization, location discovery, and so forth.

Ubiquity is the ability to allow mobile users to obtain information and conduct mobile transactions any place through Internet-enabled mobile devices. *Time-criticality* refers to the ability to access time-sensitive information immediately (Malhotra & Segars, 2005; Sadeh, 2002). A similar value to time-criticality, *spontaneity/immediacy* refers to the mobile capability for mobile users to get information and complete transactions in real-time. *Constancy* refers to the accessibility to network applications anytime and anywhere (Baldi & Thaung, 2002; Clarke, 2001; Malhotra & Segars, 2005). The constancy feature of mobile service provides the mobile value related to *convenience*. Since mobile devices are personal devices, they contain individual information as well as personal preferences. Thus, *personalization* refers to the ability to customize content and uses of mobile devices (Sadeh, 2002). Another mobile value is *location discovery* which allows mobile service providers to do location-based marketing and to deliver promotional offerings based on a user's current geographic position (Clarke, 2001). Since mobile devices are always on and carry user identity, the location of the mobile user can be tracked (Baldi & Thaung, 2002; Kannan et al., 2001; Malhotra & Segars, 2005).

Studies on trust in m-commerce are scarce due to the novelty of mobile commerce area. Siau and Shen (2003) developed a framework for building customer trust in mobile commerce. They identified two components of customer trust in mobile commerce: (1) mobile technology and (2) mobile vendor. Another study of trust in m-commerce conducted by Siau, Sheng, and Nah (2003) proposed a framework for trust in mobile commerce which outlines the variables influencing trust building in mobile commerce. Table 1 provides a summary of selected studies of antecedents/processes of trust in e-commerce.

Table 1. Selected studies of antecedents/processes of trust in e-commerce and m-commerce

Study Topic and Author(s)	Category of Antecedents	Subcategories or Set of Antecedents
Three levels of customer trust (Barney & Hansen, 1994; Lewis & Weigert, 1985)	Strong trust	Interactions, cognitive trust (e.g., the similarity), emotional trust
	Semistrong trust	Rational-calculation-based trust (e.g., a company's reputation, the threat of punishment)
	Weak trust	Transferred trust (e.g., a well developed market or word-of-mouth)
Three central modes of trust production (Zucker, 1986)	Process-based	Reputation, brands, gift-giving
	Characteristic-based	Family background, ethnicity, sex
	Institutional-based	Professional, firm associations, bureaucracy, banks, regulation
Three dimensional generic typology of trust (Mayer et al., 1995)	Ability	Competency, experience, institutional endorsements, knowledge-ability
	Integrity	Fairness, fulfillment, loyalty, honestly, dependability, reliability,
	Benevolence	Concern, empathy, faith, receptivity
Five distinct trust building processes (Doney & Cannon, 1997)	Calculative process	Firm's reputation, size, willingness to customize, confidential information sharing, length of relationship with firm, length of relationship with salesperson
	Prediction process	Length of relationship with firm, salesperson likeability, salesperson similarity, frequent social contact with salesperson, frequent business contact with salesperson, length of relationship with salesperson
	Capability	Salesperson expertise, salesperson power
	Intentionality	Firm's willingness to customize, firm's confidential information sharing, salesperson likeability, salesperson similarity, frequent social contact with salesperson
	Transference	Firm's reputation, supplier firm size, trust of supplier firm, trust of salesperson
Trust of a supplier firm and salesperson (Doney & Cannon, 1997)	Characteristics of the supplier firm and firm relationship	Reputation, size, willingness to customize, confidential information sharing, length of relationship
	Characteristics of the salesperson and salesperson relationship	Expertise, power, likeability, similarity, frequent business contact, frequent social contact, length of relationship
A trust model for consumer Internet shopping (Lee & Turban, 2001)	Trustworthiness of Internet merchant	Ability, integrity, benevolence
	Trustworthiness of Internet shopping medium	Technical competence, reliability, medium understanding
	Context factors	Effectiveness of third party certification, effectiveness of security infrastructure
	Other factors	Individual trust propensity, etc
An integrative typology of trust (McKnight, Choudhury, & Kacmar, 2002a)	Disposition to trust	Faith in humanity, trusting stance
	Institution-based trust	Situational normality, general competence, integrity, benevolence, structural assurance
	Trusting beliefs	Competence beliefs, benevolence beliefs, and integrity beliefs
	Trusting intentions	Willingness to depend, subjective probability of depending
Online trust: a stakeholder perspective (Shankar, Urban, & Sultan, 2002)	Web site characteristics	Navigation, user friendliness, advice, error free
	User characteristics	Internet savvy, past Internet shopping behavior, feeling or control
	Other characteristics	Online medium, trustworthiness of firm, perceived size of firm
Psychological antecedents of consumer trust (Walczuch & Lundgren, 2004)	Personality-based	Extraversion, neuroticism, agreeableness, conscientiousness, openness to experience, propensity to trust
	Perception-based factors	Perceived reputation (e.g., word-of-mouth), perceived investment, perceived similarity, perceived normality, perceived control, perceived familiarity
	Experience-based	Experience over time, satisfaction, communication
	Knowledge-based factors	Information practices, security technology
	Attitude	Computers & the Internet, Shopping

Table 1. continued

	Consumer-Behavioral Dimension	Demographic factors, experience, familiarity, individual culture, traditions, privacy, etc.
Process-oriented Multidimensional Trust Formation (Kim, Song, Braynov & Rao, et al., 2005)	Institutional Dimension	Reputation, accreditation, authentication, approvals (e.g., advisors and guarantors), customer communities (e.g., eBay's feedback forum), legal requirements, authorities, and so forth
	Information Content Dimension	Accuracy, currency, completeness, non-bias, credibility, Web site brand royalty, entertainment, usefulness, etc.
	Product Dimension	Durability, reliability, brand equity, quality, variety, customization, competitiveness and availability, etc.
	Transaction Dimension	Transparency, pricing and payment options, financial planning (complexity), sales-related service (refund policy, after-sales, etc.), promotions, delivery fulfillment, etc.
	Technology Dimension	Quality of media transmission, interface design and contents, security, reversibility, digital certificate, public-key cryptography (infrastructure), authenticity, integrity, confidentiality, non-repudiation, attributes of the system (benevolence, competency, predictability), and so forth
A Trust-based Consumer Decision Making (Kim, Ferrin, & Rao, in press)	Cognition (observation)-based	Privacy protection, security protection, system reliability, information quality, and so forth
	Affect-based	Reputation, presence of third-party seals, referral, recommendation, buyers' feedback, word-of-mouth, and so forth
	Experience-based	Familiarity, Internet experience, e-commerce experience, and so forth
	Personality-oriented	Disposition to trust, shopping style, and so forth
Framework of trust-inducing features (Wang & Emurian, 2005)	Graphic design	Use of three-dimensional, dynamic, and half-screen size clipart, symmetric use of moderate pastel color of low brightness and cool tone, use of well-chosen, good-shot photographs
	Structure design	Easy-to-use navigation, accessible information, navigation reinforcement, application of page design techniques
	Content design	Brand-promoting information, disclosure of all aspects of the customer relationship, seals of approval or third-party certificate, use of comprehensive, correct, and current product information, use of a relevant domain name
	Social-cue design	Inclusion of representative photograph or video clip, use of synchronous communication media
Customer Trust in Mobile Commerce (Siau & Shen, 2003)	Mobile Technology	Initiate trust formation (feasibility) Continuous trust development (reliability, consistency)
	Mobile Vendor	Initiate trust formation (familiarity, reputation, information quality, third-party recognition, attractive reward, Continuous trust development (site quality, competence, integrity, privacy policy, security controls, open communication, community building, external auditing)
Trust in mobile commerce (Siau et al., 2003)	Vendor Characteristics	Reputation, brand reputation, availability, privacy policy
	Web site Characteristics	Web site design, ease of input and navigation, readability, accuracy, richness
	Technology of wireless services	Connection speed, coverage area, transaction data, authentication
	Technology of mobile services	User interface, ease of input and navigation, readability
	Other factors	Third-party regulation, word-of-mouth

AN INTEGRATIVE TRUST ANTECEDENT REFERENCE MODEL

The literature review depicts that various factors and entities influence the complex process of engendering customer trust in e-commerce. A process- oriented, multi-dimensional trust formation model recently proposed by Kim et al. (2005) is well reflected in the actual online exchange process. The model consists of six dimensions of trust formation process and four different entities representing three ingredients of e-commerce transactions: trustor (buyer), trustee (seller), and environment (third party and technology). Although the model describes a holistic, multi-dimensional trust formation processes in a succinct manner the phenomena of trust formation in e-commerce transaction, it does not capture some environmental factors which influence trust formation process such as cultural factors, national industry characteristics, market regulations, ethics, social context, and so forth. Therefore, along with the four entities of e-commerce markets suggested by Kim et al. (2005), I suggest five entities of e-commerce markets, to include buyer, seller, third-party and social context, technology, and market environment factors. Finally, after reclassifying and reorganizing determinants of trust in e-commerce and m-commerce areas, an integrative trust antecedent reference model (see Figure 1) is proposed in an effort to synthesize existing literature on enhancing consumer trust in e-commerce and m-commerce.

The integrative trust antecedent reference model shows that cultivating consumer trust involves the interactions of five entities at least. A *buyer* (i.e., trustor) has several subdimensional factors influencing his or her trust belief such as personal characteristics (e.g., propensity to trust, individual culture, demographic elements, and so on), individual experiences (e.g., familiarity, Web experience, self-efficacy, and so on), and individual perceptions (e.g., perceived privacy, perceived security, perceived normality, and so on). As a trustee, a seller also possesses several sub-dimensional factors.

Plank, Reid, and Pullins (1999) suggested a definition of trust toward multiple objects: salesperson, product, and company. According to their definition of trust, trust is a global belief on the part of the buyer that the salesperson, product, and company will fulfill their obligations as understood by the buyer. In e-commerce context, a seller could be multiple objects: Web site, product, and company. Thus, three subdimensional factors of an e-commerce *seller* (i.e., trustee) are vendor (company) characteristics (e.g., size, reputation, ability, integrity, and benevolence), Web site elements (e.g., information quality, usefulness, usability, system reliability, and so on), and product service factors (e.g., product quality, product reliability, product variety, after-sales service, delivery fulfillment, and so on). *Third-party and social context* are important entities in e-commerce transactions. Third parties are impartial organizations which include individual mechanisms delivering business confidence through an electronic transaction (Kim et al., 2005). Social contexts are about how the trustee is viewed by the people around. Third-party services include assurance seals and business certification services, escrow service, and so on. Examples of social context are buyers' reviews and feedbacks, referral, word-of-mouth, and so forth.

Technology is the major entity which makes a difference between e-commerce and traditional brick-and-mortar transactions because all e-commerce transactions take place primarily through wired and/or wireless network infrastructure. Network infrastructure and end-unit devices for electronic transactions are identified as subdimensional factors. Network reliability, connection quality, speed, and coverage area for wireless networks and user interface, easy to use, and reliability for mobile units are specifically important. Although Web site characteristics could be classified as technology subdimensions, they are arranged as a seller side component because a Web site is a seller's storefront. Finally, electronic market (e-market) environmental factors are another important entities influencing consumer trust in e-commerce. *E-market environment* has several subdimensional factors that include regulations and structural assurances (e.g., standardization policies, market regulations, structural assurances, and so forth), ethics (e.g., fair information practices, information and property rights, and so forth), and national culture and industry characteristics (e.g., nationality, economical structure, government support, and so forth).

FUTURE TRENDS AND KEY ISSUES RELATED TO CONSUMER TRUST

The exponential growth of wired broadband and wireless mobile networks will be expected to drive the future development of e-commerce and provide new opportunities in m-commerce beyond e-commerce (Maamar, 2003). Enhancing the current e-commerce applications and business models in the market, there are six categories of mobile applications which utilizing the major unique features of mobile technology (i.e., anytime, anywhere, and personalized service).

The six categories of mobile applications are: (1) commerce transaction applications (e.g., mobile-shopping, micro-payments, bill payment, mobile banking, mobile trading, hotel reservation, and so forth), (2) communication applications (e.g., e-mail, char/SMS, multi-media SMS, mobile conferencing, broadcast, news flash, and so forth), (3) content delivery applications (e.g., information browsing, and directory service, interactive online gaming, music/video/game downloading, off-line games, flight schedules, weather information,

Figure 1. An integrative trust antecedent reference model

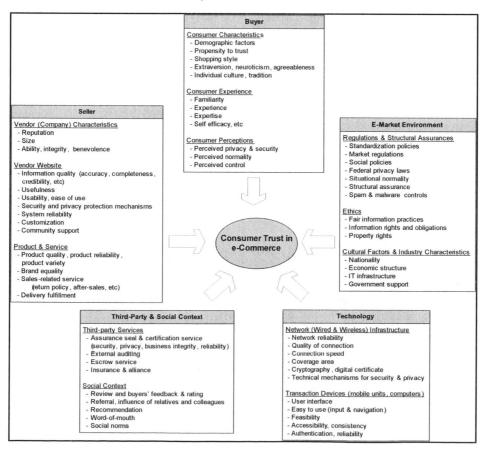

and so forth),(4) community applications (finding buddies, mobile blog, dating, mobile community for referral and recommendation, and so forth), (5) customization (e.g., scheduling, location based services, personal dieting, information filtering, and so forth), (6) connection (e.g., mobile tracking, mobile inventory management, geographic positioning systems, and so forth).

While there are many potential advantages of the new "niche" technology, there are many problems and issues as well. Using the five entities of the integrative trust antecedent reference model, some key challenges are identified in e-commerce and m-commerce areas (Cavoukian & Gurski, 2002; Maamar, 2003; Yeo & Huang, 2003).

1. Issues related to trustors (buyers)

- User comfort level of e-commerce transaction
- Privacy and security issues because of tracking and location based service
- Restricted data collection and control of personal information
- Individual culture
- Experiences
- Self-efficacy
- Different perceptions

2. Issues related to trustees (sellers)

- Pricing issue
- Marketing issue
- Consumer retention issue
- Fulfillment issue
- Customization and advertising issues
- Web interface development issue for mobile devices
- Information quality
- Application development issue

3. Issues related to third-party and social context

- Effectiveness of third-party assurance services
- Fair feedback and rating systems
- Open community in e-commerce and m-commerce areas
- Social influence

4. Issues related to technology

- Wired and wireless technology infrastructure
  - Global standardizations of new technologies
  - The lack of network security
  - Slow bandwidth and efficient use of limited bandwidth
  - Strong encryption technology
  - Open source technology
  - Mobile payment issues
  - Virus and malware (spyware, adware, phishing, and hacking) control issues
- Transaction device technology
  - Small display screen
  - Comfortable user interface
  - Open platform for wireless devices
  - Computational power—hardware and software

5. Issues related e-market environment

- Cultural issues
- Market regulations and social polices
- International and inter-states taxation
- Information and property rights
- Digital dividend
- Government regulation and support issues

Supplemental Study

In order to provide the validity of the proposed integrative trust antecedent reference model, a research model titled "Antecedents of Consumer Trust in B2C E-Commerce" is developed. The research constructs of the model are discussed in detail below.

A RESEARCH MODEL: ANTECEDENTS OF CONSUMER TRUST IN B2C E-COMMERCE

In traditional commerce, trust is affected by the characteristics of customers and the selling party (salespersons and company) and interactions between the two parties involved (Burt & Knez, 1996; Doney & Cannon, 1997; Shapiro, Sheppard, & Cheraskin, 1992; Swan, Bowers, & Richardson, 1999). It is also true in electronic commerce. Therefore, drawing from a part of the integrative trust antecedent model, three categories of antecedents influencing a consumer's trust toward an electronic commerce vendor are selected. The three categories and some trust antecedents from previous studies are summarized as follow:

Consumer side antecedents

1. **Consumer personality-oriented:** Disposition to trust, shopping style, culture, and so forth
2. **Consumer experience-oriented:** Familiarity, ease of use, Internet experience, e-commerce experience, satisfaction, and so forth
3. **Consumer perception (observation) toward e-commerce vendor Web site:** Presence of third party assurance services, privacy protection, security protection, information quality, system reliability, and so forth.

The personality-oriented and experience-oriented antecedents are related to the characteristics of consumers, which are not easy to improve and manage by selling party perspectives. The perception-oriented antecedents are associated with salespersons (Web sites), company (brand image), and interactions (interface) between the two parties. In light of the difficulty of controlling all antecedents at the same time, this study proposes a research model mainly focusing on the perception-oriented antecedents with some personality and experience-oriented antecedents. Consumer disposition to trust, culture, familiarity with a selling party, ease-of-use, and Internet experience are included in the research model because some studies have shown evidence that they are strong antecedents of consumer trust (Gefen, 2000; Luhmann, 1979; Mayer et al., 1995; Rotter, 1971).

Even though we are interested in the antecedents of trust, there is concern that some antecedents of trust may have a direct effect on purchase intention (McKnight & Chervany, 2002; McKnight, Cummings, & Chervany, 1998). Therefore it is necessary at least to propose the direct effects from antecedents to a consumer's purchase intention. Figure 2 shows the research model including direct paths from antecedents to trust and intention, and the description of each construct and their relationships with trust are following.

An online *consumer trust (TRUST)* is defined as a consumer's subjective belief that the selling party or entity will fulfill its transactional obligations as the consumer understands them and as such transactions are enabled by electronic processes. Trust plays a vital role in almost any commerce involving monetary transactions (Gefen, 2002; Jarvenpaa et al., 1999; Urban et al., 2000). Internet business is much more based on the consumer's trust in the processes, in contrast to that of traditional business involving brick and mortar stores, where trust is based on face-to-face personal relationships. Peter Grabosky, in *The*

Figure 2. Research model: Antecedents of consumer trust in B2C e-commerce

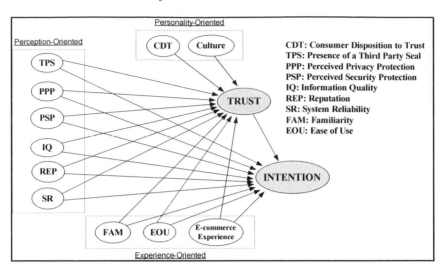

Nature of Trust Online, supports the idea that the key to success in Internet business is the establishment of trusted processes (Grabosky, 2001). This fact mandates that Internet sellers create an environment in which a prospective consumer can be relaxed and confident about any prospective transactions. Thus we propose that a consumer trust positively influences a consumer's purchase intention of electronic transaction.

Intention to purchase (INTENTION) refers to the degree to which a consumer intends to purchase from a certain vendor through the Web. The theory of reasoned action (TRA) presumes that volitional behavior is determined by intentions to act. Ajzen and Fishbein (1980) point out that behavior intention (intention to purchase, in this study) is a predictor of actual behavior (purchase), and there is a strong correlation between behavioral intentions and actual behavior (Sheppard, Hartwick, & Warshaw, 1988; Venkatesh & Davis, 2000). Consumer's purchase intention is one of the interesting variables for most e-shopping vendors.

Consumer disposition to trust (CDT) refers to a customer's personality traits that lead to generalized expectations about trustworthiness, which is a consumer-specific antecedent of trust. Since consumers have different developmental

experiences, personality types, and cultural backgrounds, they differ in their inherent propensity to trust (Gefen, 2000). This tendency is not based upon experience with or knowledge of a specific trusted party, but it is the result of ongoing lifelong experience and socialization (Kahneman, 2003; McKnight et al., 1998; Rotter, 1971). If a consumer has a high tendency to trust others in general, this disposition is especially influential when customers have not had an extensive personal interaction with the selling parties (McKnight et al., 1998; Rotter, 1971). Consumer disposition to trust is an antecedent of trust, but it is not directly related to a consumer behavior intention.

Culture is defined by Hofstede (1994) as "the collective programming of mind which distinguishes one national group or category of people from another" (p. 5). Several studies (Mayer & Tan, 2002; Png et al., 2001; Soh et al., 2000; Tan, Wei, Watson, Clapper, et al., 1998; Tan, Wei, Watson, & Walczuch, 1998) have shown that the dimensions of national culture affect development, adoption, and impact of information communication technology (ICT) infrastructure and its applications in the field of information systems. Even though culture is a crucial aspect of trust, it has been overlooked by previous e-commerce studies. Only a handful of studies (Gefen & Heart, 2006;

Jarvenpaa et al., 1999; Lim et al., 2004; Pavlou & Chai, 2002) to date have aimed at the effect of culture on trust in computer-mediated electronic commerce transactions. Since e-commerce transactions are sometimes required international interactions, understanding the cross-national aspects (i.e., culture) of trust building is essential (Gefen & Heart, 2006).

Familiarity with the online selling party (FAM) is a consumer experience-oriented antecedent of trust, which refers to the degree of consumer's acquaintance with the selling party. Familiarity would include enough knowledge to search for products and information and to order through the Web site's purchasing interface. Familiarity is a "precondition or prerequisite of trust" (Luhmann, 1979), which is an antecedent of trust because familiarity leads to an understanding of the current actions while trust deals with beliefs about the future actions of other entities (Gefen, 2000). For example, a consumer's familiarity based on previous good experience with salesperson (i.e., Web site), their services (i.e., searching products and information, and so forth) let the consumer create concrete ideas of what to expect for the future. As in electronic commerce in general, the more customers are familiar with such a selling party, the more their favorable expectations (trust) are likely to have been confirmed. It is thus hypothesized that more familiarity with a selling party should affect customer's trust on the selling entity.

Ease of use (EOU) of a Web site primarily deals with ease of navigation, ease of searching for products and information, and ease of understanding content. These trappings, along with the user's movement throughout the site, are as integral to the overall user experience as the transaction the user wants to execute. Like the importance of user interface design for software development, the Internet Web site interface design has received enormous research attention, since poorly designed sites have an adverse influence on consumer's shopping behavior (Lohse &

Spiller, 1998). We posit that ease of use increases a consumer's trust toward the selling party.

The relationship between *e-commerce experience* and trust is found to be strongly associated (Gefen, 2000). In the traditional "brick-and-mortar" business environment, trust is mainly build through repeated successful transaction experiences (Lunn & Suman, 2002). It could be true at the "brick-and-click" or "pure-click" business environments. Thus, a positive e-commerce transaction experience is an antecedent of consumer trust, which is also directly related to a consumer purchase intention.

The presence of a third party seal (TPS) refers to the assurance of Internet vendors by third party certifying bodies (e.g. banks, accountants, consumer unions, and computer companies). Recently, a wide variety of third party seals were introduced to help create trust in electronic commerce. The purpose of seals is to provide assurance to consumers that a Web site discloses and follows its operating practices, that it handles payments in a secure and reliable way, that it has certain return policies, or that it complies with a privacy policy that says what it can and cannot do with the collected personal data (Castelfranchi & Tan, 2001; Koreto, 1997; Shapiro, 1987). An example of the third party involved in the trust of online transactions is TRUSTe, a non-profit, privacy seal program. The TRUSTe trust mark on Web sites informs buyers that the owners have openly agreed to disclose their information gathering and dissemination practices, and that their disclosure is backed by credible third-party assurance (Benassi, 1999). The basic argument of the presence of a seal and consumer trust is that the seals on a vendor's site issued from certificate authorities may assure consumers that the site is a reliable and credible place to do business. Therefore, when Internet customers see the seal on a given site, it creates extra trust in that selling site.

Perceived privacy protection (PPP) refers to a consumer's perception of the likelihood or intention of Internet vendors to protect consumers' personal information, which is collected during electronic transactions, from unauthorized use or the disclosure of confidential information. At the time of a transaction, the online seller collects the names, e-mail addresses, phone numbers, and home addresses of buyers. Some sellers pass the information on to telemarketers. For many online consumers, loss of privacy is a main concern. In a recent survey, 92% of survey respondents indicated that they do not have confidence that companies will keep their information private, even when the companies promise to do so (Light, 2001). These increasing consumer concerns are forcing sellers to take privacy protection measures to increase their trustworthiness and thereby to encourage online transactions. Consumers often perceive that one of the obligations of a seller is that the seller should not share or distribute the buyer's private information. Since this is a perceived obligation of the seller under the contract, buyers will be more likely to trust a seller who they believe will protect personal privacy.

Perceived security protection (PSP) refers to a consumer's perception that the Internet vendor will fulfill security requirements, such as authentication, integrity, encryption, and nonrepudiation. How a consumer perceives security protection when making online transactions depends on how clearly she or he understands the level of security measures implemented by the seller (Friedman, 2000). When an ordinary consumer finds security features (e.g., a security policy, a security disclaim, encryption, a safe shopping guarantee, SSL technology, and so forth) in the seller's Web site, he or she can recognize the seller's intention to fulfill the security requirements during the online transactions. This positively affects the trustworthiness of the seller as far as security is concerned, and, thus the consumer feels comfortable completing the transaction.

Even the definition of information is a complex concept and quality of information may be interpreted in multiple ways (e.g., accuracy, relevance, timelines, reliability, sufficiency, and

so forth), *information quality (IQ)* refers to a consumer's general perception of the accuracy and completeness of Web site information as it relates to products and transactions. It is well recognized that information on the Internet varies a great deal in quality, ranging from highly accurate and reliable, to inaccurate and unreliable, to intentionally misleading. As well, it is often very difficult to tell how frequently the information in Web sites is updated and whether the facts have been checked or not (Pack, 1999). Thus, potential purchasers on the Internet are likely to be particularly attentive to the quality of information on a Web site because the quality of information should help them make good purchasing decisions. To the extent that consumers perceive that a Web site presents quality information, they are more likely to have confidence that the vendor is reliable, and therefore will perceive the vendor as trustworthy. As buyers perceive that the Web site presents quality information, they will perceive that the seller is interested in maintaining the accuracy and currency of information, and, therefore, will be more inclined to fulfill its obligations and be in a better position to fulfill its obligations.

Reputation of selling party (REP) refers to the degree of esteem in which public consumers hold a selling party. Positive reputation has been considered a key factor for creating trust in organizations by marketing (Doney & Cannon, 1997; Ganesan, 1994) and electronic commerce (Jarvenpaa et al., 1999). Reputation building is a social process dependent on the past interactions (e.g., whether that business partner was honest before) between consumers and selling party (Zacharia & Maes, 2000).

A positive reputation provides information that the selling party has honored or met its obligations toward consumers in the past, or, in the case of a negative reputation, that it has failed to honor or meet its obligations. Based on this reputation information, a consumer may infer that the selling party is likely to continue in its behavior. In the case of a positive reputation, one is likely to infer that the company will honor its specific obligations to oneself, and therefore conclude that the selling party is trustworthy. By the same reasoning, an individual may conclude that the selling party will not honor its specific obligations, and hence conclude that it is untrustworthy. A positive reputation generates a feeling of trust and willingness to engage in the transaction.

System reliability (SR) refers to the consumer's perception that a Web vendor system is always available and fast and makes few errors at all levels, that the transaction record is correct, and that services will not fail during a transaction. As a technical dimension to support electronic commerce, system reliability considers key factors such as the following: access is always fast and available, very few errors are allowed at all levels, the transaction record is correct and remains correct, and services do not fail during a transaction. For example, a site may not totally fail but site access may become so slow that sales may be lost. This is not a hard failure, but may be classified as a soft failure. Even under soft failure, consumer's trust regarding that site may be negatively affected.

CONCLUSION

Wired and wireless technologies bring together a broad range of evolution or revolution influencing today's business life. Many studies have indicated that trust is critical for the growth and success of e-commerce. Since we already have observed the negative consequences of a lack of confidence and trust on the growth of e-commerce, trust issues including security and privacy concerns must be addressed in the early stage of mobile commerce development. In the electronic business world, building trust is not simply an issue related to consumer-technology-buyer, but it is a complex issue that involves the key interactions of five elements (i.e., buyer, seller, third-party, technology, and market environment) at least.

ACKNOWLEDGMENT

This study is supported in part by the Faculty Research and Support Fund (FRSF) (Award #908) of the University of Houston Clear Lake.

REFERENCES

Ajzen, I., & Fishbein, M. (1980). *Understanding attitude and predicting social behavior.* Englewood Cliffs, NJ: Prentice-Hall.

Anckar, B., & D'Incau, D. (2002). Value creation in mobile commerce: Findings from a consumer survey. *The Journal of Information Technology Theory and Application (JITTA), 4*(1), 43-64.

Ba, S., Whinston, A. B., & Zhang, H. (1999). *Building trust in the electronic market through an economic incentive mechanism.* Paper presented at the 1999 International Conference on Information Systems.

Baldi, S., & Thaung, H. (2002). The Entertaining Way to M-Commerce: Japan's Approach to the Mobile Internet—A Model for Europe? *Electronic Markets, 12*(1).

Barney, J. B., & Hansen, M. H. (1994). Trustworthiness as a source of competitive advantage. *Strategic Management Journal, 15,* 175-190.

Beatty, S. E., Mayer, M., Coleman, J. E., Reynolds, K. E., & Lee, J. (1996). Customer-sales associate retail relationships. *Journal of Retailing, 72*(3), 223-247.

Benassi, P. (1999). TRUSTe: An online privacy seal program. *Communications of the ACM, 42*(2), 56-59.

Bhattacherjee, A. (2002). Individual trust in online firms: Scale development and initial test. *Journal of Management Information Systems, 19*(1), 213-243.

Booz, A. H. (2000). The wireless internet revolution [Electronic Version]. *Insights: Communications, Media & Technology Group, 6*(1). Retrieved from http://www.boozallen.com/media/file/34103.pdf

Brynjolfsson, E., & Smith, M. (2000). Frictionless commerce? A comparison of Internet and conventional retailers. *Management Science, 46*(4), 563-585.

Burt, R., & Knez, M. (1996). Trust and third-party gossip. In R. M. Kramer & T. R. T. (Eds.), *Trust in organizations: Frontiers of theory and research* (pp. 68-89). Thousand Oaks, CA: Sage Publications.

Castelfranchi, C., & Tan, Y.-H. (2001). *Trust and deception in virtual societies.* Norwell, MA: Kluwer Academic Publishers.

Cavoukian, A., & Gurski, M. (2002). Privacy in a wireless world. *Business Briefing: Wireless Technology.* Retrieved from http://www.ipc.on.ca

Clarke, I. (2001). Emerging value propositions for m-commerce. *Journal of Business Strategies, 18*(2), 133-148.

Czepiel, J. A. (1990). Service encounters and service relationships: Implications for research. *Journal of Business Research, 20*(1), 13-21.

Dirks, K. T., & Ferrin, D. L. (2002). Trust in leadership: Meta-analytic findings and implications for research and practice. *Journal of Applied Psychology, 87*(4), 611-628.

Dirks, K. T., & Ferrin, D. L. (2001). The role of trust in organizational settings. *Organization Science, 12*(4), 450-467.

Doney, P. M., & Cannon, J. P. (1997). An examination of the nature of trust in buyer-seller relationships. *Journal of Marketing, 61*(2), 35-51.

Doney, P. M., Cannon, J. P., & Mullen, M. R. (1998). Understanding the influence of national culture on the development of trust. *Academy of Management Journal, 23*(3), 601-620.

Friedman, B. (2000). Trust online. *Communications of the ACM, 43*(12), 34-40.

Ganesan, S. (1994). Determinants of long-term orientation in buyer-seller relationships. *Journal of Marketing, 58*, 1-19.

Gefen, D. (2000). E-commerce: The role of familiarity and trust. *Omega: The International Journal of Management Science, 28*(5), 725-737.

Gefen, D. (2002). Reflections on the dimensions of trust and trustworthiness among online consumers. *ACM SIGMIS Database, 33*(3), 38-53.

Gefen, D., & Heart, T. (2006). On the need to include national culture as a central issue in e-commerce trust beliefs. *Journal of Global Information Management, 14*(4), 1-30.

Grabosky, P. (2001, April 23). The nature of trust online [Electronic Version]. *The Age*, pp. 1-12. Retrieved from http://www.aic.gov.au/publications/other/online_trust.html

Griffith, D. A., Hu, M. Y., & Ryans, J. K. (2000). Process standardization across intra- and inter-cultural relationships. *Journal of International Business Studies, 31*(2), 303-325.

Hoffman, D. L., Novak, T. P., & Peralta, M. (1999). Building consumer trust online. *Communications of the ACM, 42*(4), 80-85. Association for Computing Machinery.

Hofstede, G. (1980). Motivation, leadership, and organization: Do American theories apply abroad? *Organizational Dynamics, 9*(1), 42-63.

Hofstede, G. (1991). *Cultures and organizations: Software of the mind.* London: McGraw-Hill.

Hofstede, G. (1994). *Cultures and organizations: Software of the mind: Intercultural.* London: HarperCollins.

Jarvenpaa, S. L., Knoll, K., & Leidner, D. E. (1998). Is anybody out there? Antecedents of trust in global virtual teams. *Journal of Management Information Systems, 14*(4), 29-64.

Jarvenpaa, S. L., Tractinsky, N., Saarinen, L., & Vitale, M. (1999). Consumer trust in an Internet store: A cross-cultural validation [Electronic Version]. *Journal of Computer Mediated Communications, 5*(2). Retrieved from http://jcmc. indiana.edu/vol5/issue2/jarvenpaa.html

Kahneman, D. (2003). Maps of Bounded Rationality, Psychology for Behavioral Economics. *American Economic Review, 93*(5), 1449-1475.

Kannan, P., Chang, A., & Whinston, A. (2001, January 3-6). *Wireless commerce: Marketing issues and possibilities.* Paper presented at the the 34th Annual Hawaii International Conference on System Sciences (HICSS-34).

Kim, D. J., Ferrin, D. L., & Rao, H. R. (Forthcoming). A trust-based consumer decision making model in electronic commerce: The role of trust, risk, and their antecedents. *Decision Support Systems.*

Kim, D. J., Song, Y. I., Braynov, S. B., & Rao, H. R. (2005). A multi-dimensional trust formation model in B-to-C e-commerce: A conceptual framework and content analyses of academia/practitioner perspective. *Decision Support Systems, 40*(2), 143-165.

Koreto, R. (1997). In CPAs we trust. *Journal of Accountancy, 184*(6), 62-64.

Kramer, R. M. (1999). Trust and distrust in organizations: Emerging perspectives, enduring questions. *Annual Review of Psychology, 50,* 569-598.

Lee, M. K. O., & Turban, E. (2001). A trust model for consumer Internet shopping. *International Journal of Electronic Commerce, 6*(1), 75-91.

Lewis, J. D., & Weigert, A. (1985). Trust as social reality. *Social Forces, 63,* 967-985.

Light, D. A. (2001). Sure, you can trust us. *MIT Sloan Management Review, 43*(1), 17.

Lim, K. H., Leung, K., Sia, C. L., & Lee, M. K. (2004). Is eCommerce boundary-less? Effects of individualism-collectivism and uncertainty avoidance on internet shopping. *Journal of International Business Studies, 35,* 545-559.

Lohse, G. L., & Spiller, P. (1998). Electronic shopping. *Communications of the ACM, 41*(7), 81-87.

Luhmann, N. (1979). *Trust and power.* Chichester, UK: Wiley.

Lunn, R. J., & Suman, M. W. (2002). Experience and trust in online shopping In B. Wellman & C. A. Haythornthwaite (Eds.), *The Internet in everyday life* (pp. 549-577). Blackwell Publishing.

Maamar, Z. (2003). Commerce, e-commerce, and m-commerce: What comes next? *Communications of the ACM, 46*(12), 251-257.

Malhotra, A., & Segars, A. H. (2005). Investigating wireless Web adoption patterns in the U.S. *Communications of the ACM, 48*(10), 105-110.

Mayer, R. C., Davis, J. H., & Schoorman, F. D. (1995). An integrative model of organizational trust. *Academy of Management Review, 20*(3), 709-734.

Mayer, M. D., & Tan, F. B. (2002). Beyond models of national culture in information systems research. *Journal of Global Information Management, 10*(1), 24-32.

McKnight, D. H., & Chervany, N. L. (2002). What trust means in e-commerce customer relationships: An interdisciplinary conceptual typology. *International Journal of Electronic Commerce, 6*(2), 35-60.

McKnight, D. H., Choudhury, V., & Kacmar, C. (2002a). Developing and validating trust measures for e-commerce: An integrative typology. *Information Systems Research, 13*(4), 334-359.

McKnight, D. H., Choudhury, V., & Kacmar, C. (2002b). The impact of initial consumer trust on intentions to transact with a Web site: A trust building model. *Journal of Strategic Information Systems, 11*(3-4), 297-323.

McKnight, D. H., Cummings, L. L., & Chervany, N. L. (1998). Initial trust formation in new organizational relationships. *Academy of Management Review, 23*(3), 473-490.

Pack, T. (1999). Can you trust Internet information? *Link - up, 16*(6), 24.

Park, C., & Jun, J.-K. (2003). A cross-cultural comparison of Internet buying behavior. *International Marketing Review, 20*(5), 534-553.

Pavlou, P. A., & Chai, L. (2002). What drives electronic commerce across cultures? A cross-cultural investigation of the theory of planned behavior. *Journal of Electronic Commerce Research, 3*(4), 240-253.

Plank, R. E., Reid, D. A., & Pullins, E. B. (1999). Perceived trust in business-to-business sales: A new measure. *The Journal of Personal Selling & Sales Management, 19*(3), 61-71.

Png, I. P. L., Tan, B. C. Y., & Wee, K.-L. (2001). Dimensions of national culture and corporate adoption of IT infrastructure. *IEEE Transactions on Engineering Management, 48*(1), 36-45.

Ratnasingham, P. (1998). The importance of trust in electronic commerce. *Internet Research: Electronic Networking Applications and Policy, 8*(4), 313-321.

Rotter, J. B. (1971). Generalized expectancies for interpersonal trust. *American Psychologist, 26*, 443-450.

Sadeh, N. (2002). *M-Commerce: Technologies, services, and business models.* Boston: Wiley.

Shankar, V., Urban, G. L., & Sultan, F. (2002). Online trust: A stakeholder perspective, concepts, impleications, and future directions. *Journal of Strategic Information Systems, 11*, 325-344.

Shapiro, S. P. (1987). The social control of impersonal trust. *American Journal of Sociology, 93*(3), 623-658.

Shapiro, D., Sheppard, B., & Cheraskin, L. (1992). Business on a handshake. *The Negotiations Journal, 8*, 365-377.

Sheppard, B. H., Hartwick, J., & Warshaw, P. R. (1988). The theory of reasoned action: A meta analysis of past research with recommendations for modifications in future research. *Journal of Consumer Research, 15*(3), 325-343.

Siau, K., Lim, E., & Shen, Z. (2001). Mobil commerce: Promises, challenges, and research agenda. *Journal of Database Management, 12*(3), 4-13.

Siau, S., & Shen, Z. (2003). Building customer trust in mobile commerce. *Communication of ACM, 46*(4), 91-94.

Siau, K., Sheng, H., & Nah, F. (2003). *Development of a framework for trust in mobile commerce.* Paper presented at the Workshop on HCI Research in MIS.

Soh, C., Kien, S. S., & Tay-Yap, J. (2000). Cultural fits and misfits: Is ERP a universal solution? *Communication of ACM, 43*(4), 47-51.

Strong, K., & Weber, J. (1998). The myth of the trusting culture. *Business & Society, 37*(2), 157-183.

Swan, J. E., Bowers, M. R., & Richardson, L. D. (1999). Customer trust in the salesperson: An integrative review and meta-analysis of the empirical literature. *Journal of Business Research, 44*(2), 93-107.

Tan, B. C. Y., Wei, K.-K., Watson, R. T., Clapper, D. L., & McLean, E. R. (1998). Computer-mediated communication and majority influence: Assessing the impact in an individualistic and a collectivistic culture. *Management Science, 44*(9), 1263-1278.

Tan, B. C. Y., Wei, K.-K., Watson, R. T., & Walczuch, R. M. (1998). Reducing status effects with computer-mediated communication: Evidence from two distinct national cultures. *Journal of Management Information Systems, 15*(1), 119-141.

Tarasewich, P., Nickerson, R. C., & Warkentin, M. (2002). Issues in mobile e-commerce. *Communications of the Association for Information Systems, 8*, 41-64.

Urban, G. L., Sultan, F., & Qualls, W. J. (2000). Placing trust at the center of your Internet strategy. *Sloan Management Review, 42*(1), 39-48.

Venkatesh, V., & Davis, F. D. (2000). A theoretical extension of the technology acceptance model: Four longitudinal field studies. *Management Science, 46*(2), 186-204.

Walczuch, R., & Lundgren, H. (2004). Psychological antecedents of institution-based consumer trust in e-retailing. *Information & Management, 42*, 159-177.

Walczuch, R., Seelen, J., & Lundgren, H. (2001). Psychological determinants for consumer trust in e-retailing. *Eighth Research Symposium on Emerging Electronic Markets* (*RSEEM 01*). Retrieved March 8, 2007, from http://www-i5.informatik.rwth-aachen.de/conf/rseem2001/

Wang, Y. D., & Emurian, H. H. (2005). An overview of online trust: Concepts, elements, and implications. *Computer in Human Behavior, 21*, 105-125.

Yeo, J., & Huang, W. (2003). Mobile E-commerce outlook. *International Journal of Information Technology & Decision Making, 2*(2), 313-332.

Zacharia, G., & Maes, P. (2000). Trust management through reputation mechanisms. *Applied Artificial Intelligence, 14*(9), 881-907.

Zucker, L. (1986). Production of trust: Institutional sources of economic structure (1840-1920). *Research in Organizational Behavior, 8*, 53-111.

Chapter XII
Trust in E-Commerce:
Risk and Trust Building

Loong Wong
University of Canberra, Australia

ABSTRACT

This chapter examines the importance of trust in business-to-consumer e-commerce. The author explores the issue of trust in the development and implementation of e-commerce and focuses on the context and role of users and consumers in transactions. The author contends that trust is more than a technical consideration and emphasizes the non-technical components such as community, identity, and experiences and their relevance to e-commerce. Despite the growing ubiquity of e-commerce, analysts and commentators continue to draw our attention to the issue of trust in e-commerce transactions. In particular, stories of "hacking," "phishing," and illegitimate online transactions have been an on-going public and private concern. These breaches are seen as cyber crimes and detrimental to the development of an efficient and effective business practice. Resolving these breaches are costly; businesses have to outlay financial resources not only to fix the breaches but, in the eyes of their clients, such breaches call into question the efficacy, integrity, and security of these businesses, creating both disquiet and a potential shift to alternative providers. For individuals, it boils down to an invasion of privacy and a lack of trust in the integrity of business systems and practices. This chapter examines the critical import of trust in business-to-consumer e-commerce. The chapter begins by exploring the issue of trust in the development and implementation of e-commerce; in particular, it focuses on the context and the central role of users and consumers in the transaction process. I argue that this development is an evolutionary one congruent with increasing complexities and the shift towards a risk society. The author argues that there is a growing virtualization of social life and that this virtualization plays an important role in our everyday lives. In particular, it transforms our views of agency, interactionism and community, generating both new identities and new possible spheres of autonomous action. Businesses have cashed in on these developments and sought to provide users with choices and ease of use, contributing to a pervasive and critical reception to e-commerce business practices. Via their Web sites and information conveyed, we learn to trust the information we receive. As such, we tend to equate trust with information. Trust becomes no more than a technical consideration. However, trust is and cannot simply be reducible to information. Its nontechnical components—the issues of community, identity and experiences—are critically important. As such, I seek to examine these issues in this chapter and their relationships to the building of trust and consequently, their relevance to e-commerce.

INTRODUCTION

E-commerce has become ubiquitous and according to some, will be a high trust community (Davidson & Rees-Mogg, 1997, p. 371). Yet, numerous studies point out that obstacles remain in the uptake of e-commerce for many consumers. One of the reasons was identified by Hoffman, Novak, and Paralta (1999): the fundamental lack of faith between most businesses and consumers remain a key consideration for many. Frauds, on-line scams, hacking and phishing are common occurrences and the everyday consumer is increasingly concerned over breaches of privacy and security. In their study of "Consumer Reactions to Electronic Shopping," Jarvenpaa and Todd (1997) found concerns with risk, both personal and performance, were recorded by over 50% of Web shoppers. On the other hand, Cheskin Research (1999) found that only 10% of participants in their survey on e-commerce usage considered little or no risk when purchasing on the Web. Clearly, there are significant differences in views. In e-commerce, critical and vital information essential for effecting transactions is carried from site to site.

Increasingly, there are concerns over security breaches and the misuse of data. For the consumer, companies that profess to be reliable and dependable can appear and disappear in an instant, jeopardizing many of their personal and economic details. Industry sources have, however, claimed that the rapid technological evolution of the Internet as a medium for social intercourse and commerce will in itself deliver new solutions and in the process offer new possibilities and context for trust creation and maintenance mechanisms (Bhimani, 1996). Such a technologically determinist viewpoint is indeed common and suggests that through the use of, and exposure to, these new technologies, users will adopt new forms of behavior explicitly linked to the technology itself. Further, it suggests that these forms of behavior will be novel—they neither grow out of, nor bear any relation to, users' everyday actions, experiences, or routine practices. It implies that there is a special and new category of human behavior which will come into being and is substantially different from the everyday systems of trust that we use to routinely order our behavior. However, it is posited that such a viewpoint is unsustainable and patently inaccurate. Instead, the chapter suggests that a more considered approach to the understanding of trust and the ways in which it affects people's e-commerce practices (and also their decision not to practise) is needed if we are to understand and further develop e-commerce.

First, the chapter examines how the notion of trust can be applied to consumer e-commerce, exploring the ways in which trust is relevant and applied by users engaging in shopping transactions. The chapter then draws on previous sociological research on trust, interaction, and everyday experiences, particularly in trying to show that trust is best understood as a non-technical or deterministic process. The chapter demonstrates how users approaching e-commerce bring with them previous experiences of trust and apply them to the new computer-mediated situations rather than being merely acted upon in e-commerce systems that affect their preferred actions and responses. The chapter then examines five areas at which interactions and e-commerce systems intersect and argue that these areas are critical for those building, managing and maintaining e-commerce projects and strategies.

WHAT IS TRUST?

Everyday, we place our trust in people, even strangers, and in the services these people provide. We trust that our friends, our accountants, and our lawyers will not betray our confidences, that the food we consume will not poison us, that the car we travel in will not break down, that people will listen to us when we talk to them, that our parents and children will tell us the truth and, indeed, the list goes on. If we do not place our trust so routinely in others, life would be practically unbearable, and we would be enveloped by all sorts of fearful possibilities and risks. Our life would rapidly descend into chaos or helplessness and we would rapidly be tagged as neurotic, schizophrenic, obsessive-compulsive. For most of us, this scenario is nowhere near our day to day reality—instead we learn to interact and trust people and strangers.

However, there are different levels of trust. Blanket trust is seldom applied toward another party. As Baier points out, there needs to be "an answer not just to the question, Whom do you trust? but to the question, What do you trust to them?" (Baier, 1986, p. 236). Indeed, to say I trust you seems almost always to be elliptical, as though we can assume some other phrase as "to do X" or "in matters Y" (Hardin, 1993, p. 506). It follows that trust is situational where A trusts B to do X and that X is often narrowly defined and that A distrusts B with regard to Y and that A has no conscious view of B's trustworthiness with regard to all other matters.

For the most part, much of the literature on trust written by marketing researchers implicitly embraces this view of blanket trust even though they often stress the multidimensionality of trust (Anderson & Naurus, 1990; Butler, 1991). Other disciplines have each approached the notion of trust differently. However, the majority of writers have sought to locate the discourse of trust within related discussions of security, confidence, vulnerability, uncertainty, and risk

(Abdul-Rahman & Hailes, 2000; Barber & Kim, 2000; Blois, 1999; Giddens, 1990; Lane & Bachman, 1998; Lewicki & Bunker, 1995; Lewis & Weigert, 1985b; Luhmann, 1988; Tan & Thoen, 2002; Yuan & Sung, 2004). Several factors, such as shared norms, repeated interactions, and shared experiences have been suggested to facilitate the development of trust (Baier, 1986; Bradach & Eccles, 1998; Lewis & Weigert, 1985; Mayer et Al., 1995). Another factor seen as critical in promoting trust and cooperation is the anticipation of future association (Anderson & Naurus, 1990; Powell, 1990); others see face-to-face encounters as irreplaceable for building trust and repairing shattered trust (Nohria & Eccles, 1992; O'Hara-Devereaux & Johansen, 1994).

Sociologically, the idea of trust has had a long intellectual career (Gambetta, 1988; Seligman, 1997; Silver, 1985; Sztompka, 1999) dating back to Garfinkel's works in the early sixties (Garfinkel, 1963). From his work, a number of crucial observations around trust develop:

- Trust is manifest in the actions of individuals
- We judge how to act based upon the trust we have in others
- Trust is used to the benefit of both parties involved in an action
- Where trust is offered is it generally expected in return
- Trust is offered based upon "expectancies" of other's behavior
- Trust is used to define one's relationship to others

For Garfinkel, trust is active, interactive, symbolic, and a transaction involving interactions and negotiations between individuals. There will be rules governing these interactions but they are context sensitive as can be seen from the differences in interactions between family members, friends in a social environment, business acquaintances. These rules and the routines, scripts, or "practical consciousness" that organize our everyday behavior are "tacit or taken-for-granted qualities [which] form the essential condition which allows actors to concentrate on tasks in hand" (Giddens, 1991, p. 36). Further, these organizational techniques must be mutual to all those involved in an encounter, and indeed have been incorporated into the business literature, for example, the concepts of embeddedness and networks, among others.

In the business discipline, trust is, however, often viewed from a rational-calculative and social perspective—the effecting of self-interest (Coleman, 1990; Gulati, 1995; Lane, 1998). Increases in trust decrease transaction costs and the converse applies (Barber & Kim, 2000; Casson, 1995). Trust, as such, mediates and manages risks (Sztompka, 1999, pp. 29-32). Socially, trust enables relationships and high-levels of cooperation and inter-organizational relationship (Kramer & Tyler, 1996; Zand, 1972), particularly where a contract does not or cannot fully specify the nature of a relationship between two parties. Thus, for example, business networks, close supplier relationships and economic clusters are key examples of such trust-engendered mechanisms, allowing companies to be cooperative, more effective, efficient and productive (Axelrod, 1990). They are also more predictable (Rose & Miler, 1992).

E-COMMERCE, TRUST, AND THE SOCIOLOGICAL IMAGINATION

The world of e-commerce is often seen as one in which teenagers are making their dotcom fortunes before they are old enough to have to pay taxes. Consumer computer technology has developed at such a speed that people who were using e-mail 10 years ago remember the advent of Mosaic at the end of 1993,[1] or still feel a little uncomfortable using a mouse. The industry moves so fast that it is very seductive to start thinking in the compressed perspective of "Internet time" in which things move faster, change quicker, and become

outdated almost immediately, or are blurred. For most people, the only access they have to computing (if at all) is at work and using computers is still a chore. Despite recent growth in PC ownership in homes, the technology is still not seen as a domestic one. Many homes do not have it and it does not have the same place as the television does within the majority of people's daily routines (see, for example, Lull, 1990; Moores, 1993). Because of this, manufacturers and e-commerce strategists have sought to harness the television to further increase their penetration, for example, via the integrated media player and digital television. Clearly, this strategy of co-opting the television is significant for it suggests that far from technology alone, familiarity with the technologies and its place in our daily lives and routines may actually be more important and may actually act as the prime basis for trust in e-commerce.[2]

In this chapter, the Internet can be seen as a place in which the rules and knowledge that have informed our everyday experiences are not seen to apply and as such it is a place of potentially high risk for those that venture into it. There have indeed been numerous reports of consumers' concerns over security on the Internet (Hancock, 1999). Given these concerns and its seemingly lawlessness, the Internet can be best seen as an exemplar of Giddens' vision of post-traditional society in which rather than going about our everyday interactions offering trust to others without thought or reflection, the winning of trust is constantly necessary (Beck, Giddens, & Lash, 1994). This, he argues, is due to living an urban existence in which we do not know most of the people or institutions we have contact with and are forced to inhabit situations that are beyond our control. In this "risk society" both our perceptions of risk and our exposure to actual risk is higher than before. In such an environment, our approach to using e-commerce can hardly be greatly different.

This chapter puts at the centre of its proposition that trust is a human quality that is observable through interaction (Golembiewski & MConkie, 1975) and that this interaction shapes interactions among transacting members. This chapter looks at e-commerce not as a mere technical development or even a new type of online interaction but as a technology which mediates already established and long-practiced routines of human behavior. As such, while not disregarding the importance of online security and the evolving systems that support it, this chapter believes that security has little to do with general consumer trust. This is the case not only because of consumers' lack of the expertise needed to make informed choices but, more importantly, because of their general lack of interest in the technology of security. This, among other reasons, is seen as a little protection against "disreputable or careless people" who will take credit card details and use them for their own gain and the owner's loss (Fukuyama, 1999; Ratnasingham, 1998). As such, it is of little surprise that significant numbers of Internet users have never bought anything online or have taken part in e-commerce. Clearly central here is the issue of trust, and numerous researchers have suggested that trust is a key actor in e-commerce adoption (Bons, Lee, & Wagenaar, 1998; Castelfranchi & Tan, 2002; Keen, 1999; Tan & Thoen, 2000).

However, while trust is important in understanding why consumers choose to use or refrain from using e-commerce and understanding how they make choices about which B2C retailer they use, the opposite side of this coin is that for the firms' consumers, trust offers initial and repeat purchases, strong brand loyalty and encourages word-of-mouth recommendation. Trust is a valuable (if intangible) asset which, as Fukuyama (1995) has persuasively argued, is firmly linked with economic success. Indeed, although the link between commerce and culture (Casson, 1995) has long been recognized, the nature of these links still remains under-explored and under-developed in many areas.

Garfinkel's (1963) views offer us an opportunity to further examine these links to better understand trust drawn from the observation of participants' action rather than through modeling or technological systems. In accepting Garfinkel's (1963) work as a basis for approaching consumer trust in e-commerce, it becomes clear that trust has very little relationship to technical security. In fact, it is suggested that people do not mistrust technology because they are inherently Luddite and have a view of technology as wrong, bad, or evil. But, because following Garfinkel's insights, trust is a basically a transaction involving individuals, rather than people and mediating technologies. It is apparent that phrases such as "I trust my PC to help my do my home accounts" or "I trust this network with my credit card details," are somewhat strange personifications. As neither of these objects can accept my trust it is not possible for me to bestow it.[3]

Trust, indeed, as many have pointed out, is something which one individual offers to another and is prepared to be vulnerable to the actions of another party, predicated on the premise that the other party will perform an appropriate action important to the trustor (see also Hosmer, 1995; Sako, 1994; Sztompka, 1999). In this formulation, to trust somebody means that you do not believe that it is necessary to safeguard against possible harmful actions by them. Indeed our actions might be unconscious and even habitual, narrowing down the set of possible acceptable actions by the other actor (Sako, 1994, p. 4). As Luhmann suggests, trust "reduces social complexity by going beyond available information and generalizing expectations of behavior in that it replace missing information with an internally guaranteed certainty" (Luhmann, 1979, p. 93).

This mutuality of interaction is as important in business to consumer e-commerce transactions as it is elsewhere (Gefen, 2000). The parties involved in the transaction must have a shared perspective of what is going on and the routines that generally (and acceptably) govern that interaction. Now the idea that both client and merchant know what is going on and behave accordingly may not appear to be a ground-breaking observation. If we, however, unpack the process a little, its profound relevance to successful e-commerce becomes apparent. Questions arise about who is doing what, what preconceptions are they bringing to the interaction and how responsive (or alive) are they to the situation.

Take, for example, a user new to e-commerce. When approaching this new experience, what people try to do is apply rules that have governed previous similar experiences. For the new e-commerce users this will most likely be that of shopping. However, even the most sophisticated and well developed e-commerce sites remove the situated experience usually associated with shopping: gone are geographical, temporal, tactile, and social experiences of shopping. Now, in routine forms of shopping such as regular food purchases (which forms the majority of household shopping expenditure along with petrol and mortgages) this may be a good thing but for leisure and recreational purchases (which make up the majority of B2C e-commerce) it means that the e-commerce experience stops being shopping and is reduced to buying things. What happens in such e-commerce interactions is that mutuality can begin to fall apart. Site designers and merchants believe, for their part, they are entering into a selling interaction while consumers are not necessarily involved in a buying interaction and therefore can often be uncertain as to what is required of them as their previous shopping routines fail to work. As the rules that usually apply to our social relationships cease to remain valid, people become uneasy. Such an uncertainty heightens the importance of trust for the consumer and increases the need for the merchant to respond to it. The interaction that online retailers often believe themselves to be involved in is one of collecting customer data first and selling the goods, second. Customers, especially those new to the Internet and e-commerce, will be looking for a shopping interaction. The two are bound to be irreconcilable. This is not to say that either is wrong or that

the future of e-commerce will not continue to be data collection orientated. Placed in interactional situations in which the customer is uncertain as to what is going on, entering fully into the interaction becomes difficult and potentially risky In such a situation, issues of trust are paramount especially when the retailer makes demands on the consumer without offering anything either upfront or in return. Further, if trust in e-commerce is simply reduced to a matter of consumers "learning" to understand e-commerce systems or putting into place increasingly sophisticated security and validation systems, the lack of real interactions means that it is highly likely to be less successful.

Despite this, most e-commerce writers have ignored sociological aspects of users' interactions and aims to reify trust when exploring network communications. For example, Gerck (1998) defines the goal of his research on trust as producing a practical definition of trust is one which allows considerations to "be viewed non-antromorphically [sic] when dealing with the concept of trust in communication engineering and security design." Such a view claims to solve the problem of trust by removing trust from the equation; trust has somehow been technologically integrated and therefore not problematic. In such an argument, it is however unclear who is trusting or what it is that trust is being placed in. This is a vital question when addressing trust's implication for business-to business (B2B) c-commerce.

According to Gerck (1998), "trust is that which is essential to a communication channel but cannot be transferred from a source to a destination using that channel." Clearly, here, Gerck is suggesting that we trust the information we receive. This is an over technological and deterministic view as we cannot trust information per se but rather, based on our previous experiences, the provider of that information or our own informed evaluation of it. Trust is not reducible to information; it "does not reside in integrated circuits or fibre optic cables" and cannot be digitized and transmitted (Fukuyama, 1995, p. 25). If these propositions have some validity, how can e-commerce solution providers and online merchants then promote and exploit trust? It is simply not enough to demonstrate that consumer trust is not related in any significant way to technical security without offering at least some suggestions as to what it is related to. The next section of this chapter discusses a few possibilities that derive from the framework above and examines their practical attendant effects.

HOW CAN E-COMMERCE FOSTER TRUST?

Keeping up-to-date in any broad manner with developments in the e-commerce industry has become difficult as its fortunes and directions fluctuate wildly. A plethora of organizations, market research, academic, industrial and journalistic, often with irreconcilable perspectives and always in need of revision further down the line map out their gyrations and, indeed, the new economy's speed makes it very easy to ignore the relatively stable aspects of business which would make strategies, decisions, and plans firmer (Porter, 2001).

Drawing on sociological insights, I have attempted to highlight a few key areas of concern relevant to e-commerce and consumer trust. These are community, flow, brand identity, personal experience, and the idea of institutions. In developing these five areas, I am not claiming that other areas are inconsequential; indeed, the sociological interest in other areas, for example, virtuality (Carrier & Miller, 1998) and mobility (Urry, 2000), among others, are equally significant. In concentrating on the five areas I have nominated, I have sought to proximate and relate some of these sociological concepts with the broader business discipline.

Community

The idea of online communities has become an attractive concept, and despite their notorious fluidity, they offer the online retailer a valuable resource for promoting user trust. They also promote site "stickiness" and add value to the site and the products being offered by providing (at little cost to the retailer beyond initial development) reviews, overviews, hints and tips, buying advice, and so forth. Although a richly discussed term, the following covers some of the main features of community as far as sociological research stands. *Community* stands as a convenient shorthand term for the broad realm of local social arrangements beyond the private sphere of the home and family but more familiar to us than the impersonal institutions of the wider society (Crow & Allan, 1994, p. 1). In such a context we can see that what is happening in communities which pivot around online retail sites is the development of relationships.[4]

The virtual basis of these communities is largely irrelevant to the interaction involved and does not mean that the relationships involved are in any way less real as members increasingly feel part of the community and begin to align themselves with the community and, by extension, the community host (Jones, 1995; Rheingold, 1993; Smith & Kollock, 1999). Being part of the community—especially if one is seen as a core or long-term member—begins to carry with it its own kudos or cultural capital as community membership begins to carry with it its own value. Therefore, as a community evolves, members will begin to actively demonstrate membership of the community, for example, through techniques such as displaying specialist knowledge of the community history and its members and outside of it through recommendation or favourable comparison with other, similar communities. This contributes not only to the sense of community found within the group but also the development of boundaries around it that marks it as separate from other online communities. In such communities, trust is central and is the glue binding members together—it fosters, maintains and helps develop community relationships.

By developing an imagined community (Anderson, 1991), for example, eBay, retailers become seen as a trusted part of the community rather than merely an institution. Further, the development of a community and its relationship to trust is a cyclic process: members of communities demonstrate trust in other members they know and also in the other members by virtue of their membership of the community. The longer people remain a member of the community the more they are likely to offer greater amounts of trust to the community. As Fukuyama observes, "community depends on trust, and trust in turn is culturally determined" (Fukuyama 1995, p. 25). Indeed, this is discernable in numerous examples of Web practices, for example, the peer-to-peer facilitating networks (Kaaza, YouTube and Napster) where members of the community develop trust with each other through thick interactive information transactions and exchanges and in e-commerce retailing, for example, eBay.

Since its founding in 1995, eBay hosted person-to-person online auctions for the members of its community. That community was composed of anonymous and remote individuals who were unlikely to have repeat dealings. Trade was impersonal with, for example, a seller knowing only the user name of bidders until the winner of the auction provided a shipping address. Buyers did not have an opportunity to inspect the goods on which they bid, and the winning bidder paid for the item prior to shipment (Livingston, 2005). Trades were neither supported by contracts nor in most cases by public enforcement of implicit contracts but rather by relied on trust engendered within the eBay community and its institutional practices. The rules and regulations on membership practices and its online reputation mechanism were based on feedback provided by the transacting parties (Dellacros & Resnick, 2003; Li & Lin, 2004). Members established informal norms, standards, provided feedback on other members' performance and policed the site. Standards were thus maintained and reputations managed. eBay's strategy, while far from perfect, illustrates the types of responses that can strengthen a reputation, reduce the cost of trust, lower transaction costs and amplify the value of community practices. In fact, there is anecdotal evidence suggesting that customers are relying more and more on online opinions when making their online purchasing decisions (Guernsey, 2000). The Web community, it appears, can enable businesses to grow and develop when properly harnessed.

Flow

Although Csikszentmihalyi's (1975, 1988) notion of *flow*,[1] the way computer users become absorbed in their activity to the exclusion of other things, is usually applied to athletes, it has also been used in understanding e-commerce (Hel, van Niekerk, Berthon, & Davies, 1999) and, more generally, to the online experience (Novak & Hoffman, 1997). According to Mihaly Csikszentmihalyi, flow is the "holistic sensation that people feel when they act with total involvement" (Csikszentmihalyi, 1975, p. 36). Csikszentmihalyi wanted to understand the experience of enjoyment which we do for the sheer joy of it (Csikszentmihalyi, 1975, p. 4). Flow is a positive, highly enjoyable state of consciousness that occurs when our perceived skills match the perceived challenges we are undertaking. When our goals are clear, our skills are up to the challenge, and feedback is immediate, we become involved in the activity. In the process, we lose our sense of self and time is distorted. The experience becomes autotelic or intrinsically rewarding (Csikszentmihalyi, 1990, p. 34).

This mode is characterized by a narrowing of the focus of awareness so that irrelevant perceptions and thoughts are filtered out by loss of self-consciousness, by a responsiveness to clear goals and unambiguous feedback, and by a sense of control over the environment. It is this common flow experience, an intense, immersive and emotional involvement, that people adduce as the main reason for performing the activity (Csikszentmihalyi, 1975). Thus, for flow to exist, there must be a level of challenge involved in the activity but that level of challenge must not be so great as to make those involved feel out-maneuvered so that they lose interest. Such challenge is often designed into computer games such as motor racing simulations. In such games when a player is performing badly and slips to the back of the race the cars at the front will start to gently slow down in order for there to be more of a possibility of the less skilled player to catch up. Conversely, the computer-driven cars will increase their speed and driving accuracy to maintain challenge and interest for the experienced gamers (Poole, 2000). Successful Web sites are not about navigating content, but rather about staging and managing experiences where participants, when in the flow state, shift into a common mode of experience as they become absorbed in their activity. Such flow, as Novak and her collaborators found, "is determined by: (1) high levels of skill and control; (2) high levels of challenge and arousal; (3) focused attention; and (4) enhanced interactivity and telepresence" (Novak, Hoffman & Yung, p. 24). They found that speed had the greatest effect on the amount of time spent online and on frequency of visits for Web applications. For repeat visits, the most important factors were skill/control, length of time on the Web, importance, and speed.

For the Web developer or e-commerce solution provider, this in itself presents a substantial challenge and in many ways runs contrary to common knowledge of site design in which everything must be simple and transparent for the users as they navigate through the site. To facilitate flow, designers have to ensure that visitors to a site are given clear information and receive feedback but it also needs to include a variable element of challenge to the interaction

users have with the site. The combination of goal orientated challenge, feedback, and interaction with other users in auctions is one such strategy. When the idea of flow is applied to e-commerce, this engagement encourages users' involvement with a site, increases the amount of time they will spend on it, and makes the likelihood of their returning often greater. To enter into a flow state, many of the same conditions that are necessary for trust are required. For example, there must be an established and recognizable set of rules that govern the interaction and which people can expect others to adhere and by which to judge their actions. Given this situation, despite the challenge involved which will itself fulfil users' expectancies, the interaction will avoid situations in which the site user is faced with the unexpected or situations in which mutual interaction fails to operate. Given such a relationship, it is highly likely that trust will develop. Again, here the example of eBay is instructive. On logging onto eBay, users are directed to their interests through a process of interactive flows, learn to absorb the rules and practices of the site, and are socialized into its institutional practices. These experiences can be seen in other sites which seek to engage users and to buy more products, making the site experiences more compelling and engaging.

Brand Identity

Brands have been critical in instances of information asymmetries and where consumers rationally depend on brand names in making their purchase decisions. This has not changed that much online. KPMG (1999) found that more than 50% of Internet users claimed that they would shop online more if major financial institutions or vendors guaranteed their transactions. The emphasis here of placing trust in organizations rather than systems is clear (even if the organization stands as a metonym for the individuals who make it). Theoretically, as trust is fostered through relationships, it would appear that the trust that e-commerce users demonstrate would favour familiar brands. If this was so it is likely that there will be growing number of users visiting a select number of the largest e-commerce providers.

Branding is also linked strongly with trust services such as Verisign, TrustUK, PayPal and *Which? WebTrader*. In situations in which a retailer is a member of such a professional or regulatory organization it is easy to simply assume that trust can be produced by a regulatory agency. However, the problem arises about why we should trust e-Trust or *Which?* in the first place. These services rely on a previous trustworthy relationship between the consumer and the trust service. Trust in a firm is not about doing business with them because PayPal will refund me the $50 that my credit card company may not if the company disappears with my money and fails to deliver the new television set I ordered from them. Such safety nets (like the emphasis Fukuyama places on hierarchy) are the product of distrust and act to minimize the effects of wrongful behavior. In a trusting relationship, the display of a trust service's logo or banner on the retailer's site assures the consumer that the business done there will conform to a set of norms which are already established, available for review, and familiar to the parties involved. It restores a sense of mutuality as discussed and highlighted above. If I have trust in the behavior sanctioned by the trust service then such legal assurance become less important (Fukuyama, 1995, p. 27). It is like being introduced to a friend of a friend. We assume, because of the trust relationship we have with our friend, that our new acquaintance will demonstrate similar points of view to ourselves, that they will not be untrustworthy, offensive, abusive, and so forth. In such close relationships, the recognition of the power of this trust can make it embarrassing to point out that we do not like our friend's friends and, conversely, to have a friend we have introduced act in an inappropriate manner. This ensures that trust is therefore not breached.

As consumers become better educated and seek greater protection and privacy online (Homburg & Furst, 2005), branding coupled with strategies to develop more efficacious privacy practices, for example, the appointment of chief privacy officers and the development of a more rigorous privacy guideline, offers protection to online consumers and, unwittingly, a distinct competitive advantage vis-à-vis their competitors (Andrews & Shen, 2000; Frombrun, 1996; Tadelis, 1999). Banks, finance companies, and credit providers have, in particular, crafted their strategies accordingly, and through their Web protection strategies, enhanced their reputation (Barr, Knowles, & Moore, 2003) and makes them more trustworthy in the eyes of their customers (Dellacros, 2005; Melnik & Alm, 2002; Windley, Tew, & Daley, 2006; Yu & Singh, 2000). Conversely, if consumer expectations (and their complaints) are not handled appropriately, this may lead to the magnification of prevailing negative perceptions and the brand is consequently affected. Custom and business invariably suffers.

Personal Experiences

All of the above three aspects of trust inform our personal experiences. That is, trust is a quality which grows out of, and informs, our local interactional experiences. This is because personal experience and narratives have profound importance in the creation of trust (see also Jones & Vijayasarathy, 1998; Lane & Bachmann, 1998). We make decisions to trust through our own (often limited or misrepresented) satisfied experiences and the anecdotes offered by those we trust more readily (and pragmatically) than by any rational evaluation of available facts. Trust, as such, may have components of rationality (especially when it is institutionally processed) but this does not mean it is a rational system.

As the Cheskin Research (1999) points out, "Trust is understood by most customers to be a dynamic process. Trust deepens or retreats based on experience." As such, there is a pattern to the development of trust through personal experience which applies as readily to e-commerce as it does to friendships. In the early stages of a relationship, the level of trust is low for both parties involved. As neither party knows much about each other, mutuality appears lows and the risk of having an offering of trust abused is potentially high. It is at this stage that trust services, consumer law, well formulated and displayed return policies play their major role in the trust process. This is when the new customer seeks reassurance that the level of damage they open themselves up to, the risks they take, and the amount of misappropriation that can be done by the company is limited through law and officially sanctioned regulation. It is only as the trust relationship builds through successive and successful interactions that more informal transactions can be comfortably entered into. As the level of satisfactory service the customer receives grows so does the level of trust they have in the retailer. Indeed, successful e-tailers often remarked on their customers' satisfaction.

Linked to this search for certainty is the growth in tolerance towards variability of service, for example, the occasional mix-up in order or slightly delayed delivery so long as recompense is made and apology is offered. Such repairs demonstrate to the consumers (as does an apology to a friend) that the relationship is valued and that there is a desire to maintain the relationship and its development. Such an observation, which effectively boils down to a commitment to provide a service or regular quality with a focus of customer recognition and satisfaction, is not new but as online retailing continues to be more about the service offered than the product sold the quality of that service becomes increasingly important. Again, here, I refer to the earlier brief discussion of eBay and its practices, where its feedback mechanism enables the development of customers' feedback, satisfaction, and the refinement of its base of customers' knowledge (Morgan, Anderson, & Mittal, 2005) as it seeks to enhance its appeal.

Institutions

Research has clearly shown that individuals involved in human-computer interaction have tended to rely on social attitudes and rules in vesting trust in machines. Nass and Moon (2000), for example, found that many people mindlessly and readily concede their trust to computers. Somehow, computers are seen as part of the institution of knowledge, science, and technology; hence, they are reliable and can only but induce trust. According to Zucker (1986), there has been a shift to certification institutions derived and supported by governments because local personal-trust networks are and have been disintegrating. Through their "power" and authority, these practices, protocols, standards, and regulations induce compliance and trust in the control procedures (see also Rea, 2001; Benassi, 1999; Keen, 1999; Lane & Bachman, 1997). Indeed, without this support, understanding, and the ability to exploit this social background, neither security nor trust will be effected and/or effective.

Norbert Elias (1994) has argued that the civilizing process is synonymous with reducing the unpredictability of encounters with strangers. Long distance trade and financial exchange promoted new forms of discourse and practice—written documents, orders, promissory notes and bills of exchange—during the Middle Ages (Braudel, 1981; Kerridge, 1988), allowing "strangers with no basis for trust to work with one another" (Fukuyama, 1995, p. 150). These new practices enabled and ensured promises were rendered more stable, mobile and containable. In contemporary virtual trading environments, while risks can be amplified, sociotechnical solutions have been advanced to fix the ensuing problems of trust and distrust/control (Kyas, 1997). Stability, predictability and normalcy is restored and maintained, and clearly the new ICT solutions civilize and induct us into the process of a new sociotechnical age. For example, users of the Internet have structural assurance that legal and technological safeguards protect them from privacy infractions, identity loss or online fraud. This institution-based trust provides assurances that things will go well, normalizes our roles and expectations, and increases our dispositions to trust (Baier, 1986; Benassi, 1999; Gefen, 2000; Lewis & Weigart, 1985; McKnight & Chervany, 2002; Shapiro, 1987; Zucker, 1986). Trust, as such, is constructed for and by people to enact some form of predictability and reliability. In e-commerce transactions, trust can be seen at different levels of interactions:

- Trust in the environment and infrastructure
- Trust in the computing agent and in mediating agents
- Trust in potential partners
- Trust in the authorities to enforce compliance, for example, protocols and procedures and laws

In the Internet context, beliefs that there are legal and regulatory protections for consumers clearly influence and effect trust to be built and developed. Trust and confidence is thus based on abstract systems resting on the validity of commonly acceptable and accepted technical and social norms and standards of business behavior and practice, and the power these technologies of trust invokes and maintains, as a new social contract emerges. Perhaps, the clearest manifestation of this institutional trust is best seen in discussions of communities of practice, global networks of innovation, and supply chain relations (Bachman, 2003; Lane & Bachman, 1998). In these communities and networks, members are concerned with both practical outcomes for customers and learning, combining an agency's focus on personal development with traditional community's foundation of shared purpose.

CONCLUSION

Trust is a complex and slippery subject; it "is a cultural norm which can rarely be created intentionally because attempts to create trust in a calculative manner would destroy the affective basis of trust" (Sako, 1994, p. 6). Because of this, Baier warned, trust "is a fragile plant, which may not endure inspection of its roots, even when they were, before the inspection quite healthy" (Baier, 1986, p. 260). In the case of trust, particularly in virtual economic environments, both fragility and complexity are preset, and sensitive handling of these issues is required if e-commerce is to be properly understood and effected. By looking at Web-based business-to-consumer e-commerce this chapter has placed trust within a context of everyday routines, interactions and local experiences. It is therefore suggested that regardless of who is involved or how business is conducted, e-commerce will continue to change our routine behavior and our approaches to trust interactions. As such, there will remain a need to build upon social science research in general, and interactional sociology in particular, in order to develop the initial observations that have been offered above. The task that this chapter leaves us with is not only to refine our understanding of trust for the online consumer but to see how this understanding corresponds to research in other fields and explore how it can inform the development of e-commerce solutions.

In this chapter, I have sought to show trust as a social process through which control is affected in the sense that people, actions, and events can be rendered relatively predictable. I have also argued that a sociologically-informed view of trust will readily reveal that e-commerce solutions predicated on technological solutions are therefore flawed and unable to deliver expected outcomes as they failed to understand the different logics induced by trust. Trust is central to predictability but is not rule-bound but rather is invoked by power relationships through which relationships are created, maintained, enacted and negotiated. Via standardization and a set of *communal* values, individuals are potentially controlled and controllable, predictable, and familiar. Individuals may thus seize upon, enact, and thereby reproduce mechanisms of trust governing their conduct and behavior.

From a practical standpoint, it is clear that successful e-commerce sites and practices need to integrate these complementary notions of community, flow, branding, personalization, and systemic practices into their Web business strategies. Businesses need to be cognizant and respond to the larger desire from consumers for voices and attention to those voices. Customers not only want to be heard but also want their personal experiences to be taken seriously through feedback mechanisms. In addition, they want businesses to respond actively to their negative comments and to devise appropriate strategies to respond to their concerns. In so doing, businesses also invariably manage their customers and socialize them into acceptable institutional arrangements.

While the five areas discussed in this chapter may help to consolidate and refine Web business strategies, it is also clear that culture may affect the notion of community participation, the perception of flows and, accordingly, color personal experiences. These cultural variables are often under-emphasized in much of the e-commerce literature but they can be particularly important. For example, the response and reception of mobile commerce in East Asia has been attributed to cultural (and institutional) practices. Clearly, prevailing institutional arrangements in different countries, such as the lack of bandwidth, censorship practices, access, flow, and personal experiences, also impact e-commerce experiences and these arrangements and issues need further consideration and research.

REFERENCES

Abdul-Rahman, A., & Hailes, S. (2000). Supporting trust in virtual communities. In *Proceedings of the Hawaii International Conference on System Sciences* (pp. 55-63).

Anderson, B. (1991). *Imagined communities: Reflections on the origin and spread of nationalism*. London: Verso.

Anderson, J.C., & Naurus, J.A. (1990). A model of distributor firm and manufacturing firm working partnerships. *Journal of Marketing, 54*, 42-58.

Andrews, S., & Shen, A. (2000). *Laws or regulations posing barriers to electronic commerce*. Washington, DC: Electronic Privacy Information Center.

Axelrod, R. (1990). *The evolution of co-operation*. Harmondsworth, UK: Penguin.

Bachman, T. (2003). Trust and Power as a Means of Coordinating the Internal Relations of the Organisation: A Conceptual Framework. In B Nooteboom & F. Six (Eds.) *The Trust Process in Organisations: Empirical Studies of the Determinants and the Process of Trust Development*. Cheltenham, UK: Edward Elgar.

Bachmann, R. (1998). Trust: Conceptual aspects of a complex phenomenon. In C. Lane & R. Bachmann (Eds.), *Trust within and between organisations: Conceptual issues and empirical applications* (pp. 298-322). Oxford: Oxford University Press.

Baier, A. (1986). Trust and antitrust. *Ethics, 96*, 231-260.

Balachander, S. (2001). Warranty signaling and reputation. *Management Science, 47*(9), 1282-1289.

Barber, K.S., & Kim, J. (2000). *Belief revision process based on trust: Agents evaluating reputation of information sources*. Retrieved March 9, 2007, from http://www.istc.cnr.it/T3/download/aamas2000/Barber-Kim.pdf

Barr, T., Knowles, A., & Moore, S. (2003). Trust in transactions: Australian Internet research. Paper presented at the Communications Research Forum 2003. Retrieved March 9, 2007, from http://www.dcita.gov.au/crf/papers03/barr3final.pdf

Beck, U., Giddens, A., & Lash. S. (1994). *Reflexive modernisation.* Cambridge, UK: Polity.

Benassi, P. (1999). TRUSTe: An online privacy seal program. *Communications of the ACM, 42*(2), 56-59.

Bhimani, A. (1996). Securing the commercial Internet. *Communications of the ACM, 39*(6), 29-35.

Blois, K.J. (1999). Trust in business to business relationships: An evaluation of its status. *Journal of Management Studies, 36*(2), 197-215.

Bons, R.W.H., Lee, R.M., & Wagenaar (1998). Obstacles for the development of open electronic commerce. *International Journal of Electronic Commerce, 2*(3), 61-83.

Bradach, J.L., & Eccles, R.G. (1989). Price, authority and trust: From ideal types to plural forms. *Annual Review of Sociology, 15*, 97-118.

Braudel, F. (1981). *Civilisation and capitalism, 15th-18th century.* London: Collins/Fontana.

Butler, J.K. (1991). Toward understanding and measuring conditions of trust. *Journal of Management, 17*, 643-663.

Carrier, J., & Miller, D. (1998) *Virtuality: A new political economy.* Oxford, UK: Berg Publishers.

Casson, M. (1995). *The organisation of international business: Studies in the economics of trust.* Aldershot, UK: Edward Elgar.

Castelfranchi, C. (2000). Why computers will (necessarily) deceive us and each other. *Ethics and Information Technology, 2*, 113-119.

Castelfranchi, C., & Falcone, R. (2001). *Social trust: A cognitive approach.* In C. Castelfranchi & Y.H. Tan (Eds.), *Deception, fraud and trust in virtual societies* (pp. 55-90). Dodrecht, The Netherlands: Kluwer.

Castelfranchi, C. & Tan, Y.H. (2002). The Role of Trust and Deception in Virtual Societies. *International Journal of Electronic Commerce 6*(3), 55-70.

Cheskin Research & Studio Archetype/Sapient (1999). *eCommerce trust study.* Retrieved March 9, 2007, from http://www.sapient.com/cheskin

Clayman, S.E. (1993). Booing: The anatomy of a disaffiliative response. *American Sociological Review, 58*, 110-130.

Coleman, J. (1990). *Foundations of social theory.* Boston: Harvard University Press.

Cranor, L.F. (1999). Internet privacy. *Communications of the ACM, 42*(2), 28-31.

Crow, G., & Allan, G. (1994). *Community life: An introduction to local social relations.* Hemel Hempstead, UK: Harvester Wheatsheaf.

Csikszentmihalyi, M. (1975). *Beyond boredom and anxiety.* San Francisco: Jossey-Bass.

Csikszentmihalyi, M. (1988) *Optimal experience.* New York: Cambridge University Press.

Csikszentmihalyi, M. (1990). *Flow: The psychology of optimal experience.* New York: Harper and Row.

Davidson, J., & Rees-Mogg, W. (1997). *The sovereign individual.* London: Pan.

Dellarocas, C. (2003). The digitization of the word of mouth: Promise and challenges of online feedback mechanisms. *Management Science, 49*(10), 1407-1424.

Dellarocas, C. (2005). Reputation mechanism design in online trading environments with pure moral hazard. *Information Systems Research, 16*(2), 209-230

Dellarocas, C., & Resnick, P. (2003). *Online reputation mechanisms: A roadmap for future research.* Summary Report of the First Interdisciplinary Symposium on Online Reputation

Mechanisms. Retrieved March 9, 2007, from http://www2.sims.berkeley.edu/research/conferences/p2pecon/papers/s8-dellarocas

Elias, N. (1994). *The civilising process.* Oxford: Blackwell.

Fombrun, C.J. (1996). *Reputation: Realizing value from the corporate image.* Boston: Harvard Business School Press.

Fukuyama, F. (1995) Trust: The social virtues and the creation of prosperity. New York, NY: Free Press.

Gambetta, D.G. (Ed.). (1988). *Trust, making and breaking of cooperative relations.* New York: Basil Blackwell.

Garfinkel, H. (1963). A conception of, and experiments with, "trust" as a condition of stable concerted action. In O.J. Harvey (Ed.), *Motivation and social interaction* (pp. 187-238). New York: Ronald Press.

Gefen, D. (2000). E-commerce: The role of familiarity and trust. *Omega, 28*(6), 725-737.

Gerck, E. (1998). *Towards real-world models of trust: Reliance on received information.* Retrieved March 9, 2007, from http://www.mcg.org.br/trustdef.com

Giddens, A. (1984). *The constitution of society.* Cambridge, UK: Polity.

Giddens, A. (1990). *The consequences of modernity.* Cambridge, UK: Polity.

Giddens, A. (1991). *Modernity and self-identity.* Cambridge, UK: Polity.

Giddens, A. (1994). Living in a post-traditional society. In U. Beck, A. Giddens, & S. Lash (Eds.), *Reflexive modernisation* (pp. 56-109). Cambridge, UK: Polity.

Golembiewski, R.T., & McConkie, M. (1975). The centrality of interpersonal trust in group processes. In G.L. Cooper (Ed.), *Theories of group processes* (pp. 131-85). London: John Wiley.

Guernsey, L. (2000). Suddenly, everybody's an expert on everything. Retrieved March 9, 2007, from http://www.nytimes.com/library/tech/00/02/circuits/articles/03info.html

Gulta, R. (1995). Does Familiarity Breed Trust? The Implications of Repeated Ties for Contractual Choice in Alliances. *Academy of Management Journal 38,* 85-112.

Hancock, B. (1999). Security Views. *Computers and Security 19*(7), 553-64.

Hardin, R. (1993). The street-level epistemology of trust. *Politics and Society, 21*(4), 505-529.

Hel, D., van Niekerk, R., Berthon, J.P., & Davies, T. (1999). Going with the flow: Web sites and customer involvement. *Internet Research, 9*(2), 109-116.

Hoffman, D., Novak, T.P., & Peralta (1999). Building consumer trust online. *Communications of the ACM, 42*(4), 80-85.

Hoffman, D.L., & Novak, T.P. (1996). Marketing in hypermedia computer-mediated environments: conceptual foundations. *Journal of Marketing 60,* 50-68.

Homburg, C., & Furst, A. (2005). How organizational complaint handling drives customer loyalty: An analysis of the mechanistic and the organic approach. *Journal of Marketing, 69*(3), 95-114.

Hosmer, L.T. (1995). Trust: The connecting link between organisational theory and philosophical ethics. *Academy of Management Review, 20*(2), 379-403.

Jarvenpaa, S.L., & Todd, P. (1997). Consumer reactions to electronic shopping. *International Journal of Electronic Commerce, 1*(2), 59-88.

Jones, S.G. (1995). Understanding community in the information age. In S.G. Jones (Ed.), *Cybersociety: Computer-mediated communication and community.* Thousand Oaks; London: Sage.

Jones, J., & Vijayasarathy, L.R. (1998). Internet consumer catalog shopping: Findings from an exploratory study and directions for future research. *Internet Research, 8*(4), 322-333.

Keen, P.G.W. (Ed.). (1999). *Electronic commerce relationships: Trust by design*. Englewood Cliffs, NJ: Prentice-Hall.

Kerridge, E. (1988). *Trade and banking in early modern England*. Manchester, UK: Manchester University Press.

KPMG (1999). *The new mass medium*. USA: Ziff-Davis/Dell/Intel.

Kramer, R.M., & Tyler, T.R. (Eds.). (1996). *Trust in organisations: Frontiers of theory and research*. Thousand Oaks, CA: Sage.

Kyas, O. (1997). *Internet security: Risk analysis, strategies and firewalls*. New York: International Thomson Publishing.

Lane, C., & Bachman, R. (1996). The social constitution of trust: Supplier relations in Britain and Germany. *Organisation Studies, 17*(3), 365-395.

Lane, C., & Bachman, R. (1997). Co-operation in Inter-Firm Relations in Britain and Germany: The Role of Social Institutions. *British Journal of Sociology 48*(2), 226-54.

Lane, C., & Bachman, R. (Eds.). (1998). *Trust within and between organisations: Conceptual issues and empirical applications*. Oxford, NY: Oxford University Press.

Latane, B., Liu, J.H., Nowak, A., Bonevento, M., & Zheng, L. (1995). Distance matters: Physical space and social impact. *Personality and Social Psychology Bulletin, 21*(8), 795-805.

Lewicki, R.J., & Bunker, B. (1995). Trust in relationships: A model of trust development and decline. In B.B. Bunker & J.Z. Rubin (Eds.), *Conflict, cooperation and justice* (pp. 133-73). San Francisco: Jossey-Bass.

Lewicki, R.J., & Bunker, B. (1996). Developing and maintaining trust in work relationship. In R. Kramer & T. Tyler (Eds.), *Trust in organisations: Frontiers of theory and research* (pp. 114-139). Thousand Oaks, CA: Sage.

Lewis, J.D., & Weigert, A. (1985a). Trust as a social reality. *Social Forces, 63*(4), 967-985.

Lewis, J.D., & Weigert, A. (1985b). Social atomism, holism and trust. *Sociological Quarterly, 26*(4), 455-471.

Li, D., & Lin, Z. (2004, December 5-8). Negative reputation rate as the signal of risk in online consumer-to-consumer transactions. In *Proceedings of ICEB 2004*.

Livingston, J. (2005). How valuable is a good reputation? A sample selection model of Internet auctions. *The Review of Economics and Statistics, 87*(3), 453-465.

Luhmann, N. (1988). Familiarity, confidence, trust: Problems and alternatives. In D. Gambetta (Ed.), *Trust: Making and breaking co-operative relations*. Oxford, UK: Basil Blackwell.

Luhmann, N. (1979). *Trust and power*. New York, NY: John Wiley.

Lull, J. (1990). *Inside family viewing: Ethnographic research on television's audience*. London: Routledge.

Mayer, R.C., Davis, J.H., & Shoorman, F.D. (1995). An Integrative Model of Organizational Trust. *Academy of Management Review, 20*(3), 709-34.

McKnight, D.H., & Chervany, N.L. (2002). What trust means in e-commerce customer relationships: An interdisciplinary conceptual typology. *International Journal of Electronic Commerce, 6*(2), 35-59.

Melnik, M.I., & Alm, J. (2002). Does a seller's e-commerce reputation matter? Evidence from eBay auctions. *Journal of Industrial Economics, 50*(3), 337-349.

Moores, S. (1993). *Interpreting audiences: The ethnography of media consumption*. Thousand Oaks, CA: Sage.

Morgan, N.A., Anderson, E.W., & Mittal, V. (2005). Understanding firms' customer satisfaction information usage. *Journal of Marketing, 69*(3), 131-151.

Morgan, R.M., & Hunt, S.D. (1994). The commitment-trust theory of relationship marketing. *Journal of Marketing, 58,* 20-38.

Nass, C., & Moon, Y. (2000). Machines and mindlessness: Social responses to computers. *Journal of Social Issues, 56*(1), 81-103.

Nohria, N., & Eccles, R.G. (1992). Face-to-face: Making network organisations work. In N. Nohria & R.G. Eccles (Eds.), *Networks and organisations* (pp. 288-308). Boston: Harvard Business School Press.

Novak, T.P., & Hoffman, D.L. (1997). A new marketing paradigm for electronic commerce. *The Information Society: An International Journal 13*(1), 43-54.

Novak, T.P., & Hoffman, D.L. (1997). *Measuring the flow experience among Web users* (Working Paper). Vanderbilt University. Retrieved March 7, 2007, from http://www.2O0.osgm.vanderbilt.edu/novak/flow.julv.1997/flow.htm

Novak, T.P., Hoffman, D.L., & Yung, Y.F. (2000). Measuring the customer experience in online environments: A structural modeling approach. *Marketing Science, 19*(1), 22-42.

Oakes, G. (1990). The sales process and the paradox of trust. *Journal of Business Ethics, 9,* 671-679.

O'Hara Devereaux, M., & Johansen, R. (1994). *Global work: Bridging distance, culture and time.* San Francisco: Jossey-Bass Publishers.

Poole, S. (2000). *Trigger happy: The inner life of video games.* London: Fourth Estate.

Porter, M. (2001). Strategy and the Internet. *Harvard Business Review (March),* 63-78.

Powell, W.W. (1990). Neither market nor hierarchy: Network forms of organisation. *Research in Organisational Behavior, 12,* 295-336.

Ratnasingham, P. (1998). The importance of trust in electronic commerce. *Internet Research, 8*(4), 313-321.

Rea, T. (2001). Engendering trust in electronic environments: Roles for a trusted third party. In C. Castelfranchi & Y.H. Tan (Eds.), *Deception, fraud and trust in virtual societies* (pp. 221-234). Dodrecht, The Netherlands: Kluwer.

Rheingold, H. (1993). *The virtual community: Homesteading on the electronic frontier.* New York: Addison-Wesley.

Rose, N., & Miller, P. (1992). Political power beyond the state: Problematics of government. *British Journal of Sociology, 43*(2), 173-205.

Sako, M. (1994). Price, quality and trust: Inter-firm relations in Britain and Japan. Cambridge, UK: Cambridge.

Schoorman, D.F., Mayer, R.C., & Davis, J.H. (1996). Including versus excluding ability from the definition of trust. *Academy of Management Review, 21*(2), 339-340.

Seligman, A. (1997). *The problem of trust.* Princeton, USA: Princeton University Press.

Shapiro, S.P. (1987). The Social Control of Impersonal Trust. *American Journal of Sociology, 93,* 623-658.

Silver, A. (1985). Trust in social and political theory. In G.D. Suttles & M.N. Zald (Eds.), *The challenge of social control* (pp. 52-70). Greenwich, CT, USA: Ablex.

Silver, A. (1989). Trust as a moral ideal: An historical approach. *Archives Europeenes de Sociologie, 30*(2), 69-87.

Silver, A. (1998). Two different sorts of commerce: Friendship and strangership in civil society. In J.Weintraub & K. Kumar (Eds.), *Private and public in thought and practice* (pp. 43-74). Chicago: University of Chicago Press.

Sitkin, S.B., & Roth, N.L. (1993). Explaining the limited effectiveness of legalistic "remedies" for trust/distrust. *Organisation Science, 4*(3), 367-392.

Smith, M.A., & Kollock, P. (Ed.) (1999). *Communities in cyberspace.* London: Routledge.

Sztompka, P. (1999). *Trust: A sociological theory.* Cambridge, UK: Cambridge University Press.

Tadelis, S. (1999). What's in a name? Reputation as a tradeable asset. *American Economic Review, 89*(3), 548-563.

Tan, Y.H., & Thoen, W. (2000). A generic model of trust in electronic commerce. *International Journal of Electronic Commerce, 5*(2), 61-74.

Tan, Y.H., & Thoen, W. (2002). Formal aspects of a generic model of trust for electronic commerce. *Decision Support Systems, 33*(3), 233-246.

Urry, J. (2000). *Sociology beyond societies.* London: Routledge.

Wang, H., Lee, M.K.O., & Wang, C. (1998). Consumer privacy concerns about Internet marketing. *Communications of the ACM, 41*(3), 63-70.

Williams, R. (1974). Television, technology and cultural form. London, UK: Fontana.

Williamson, O. (1993). Calculativeness, trust and economic organisation. *Journal of Law and Economics, 30*, 131-145.

Windley, P.J., Tew, K., & Daley, D (2006). *A framework for building reputation systems.* Retrieved March 9, 2007, from http://www.windley.com/essays/2006/dim2006/framework_for_building_reputation_systems

Yamagishi, T., & Yamagishi, M. (1994). Trust and commitment in the United States and Japan. *Motivation and Emotion, 18*(2), 129-166.

Yu, B., & Singh, M.P. (2000). A social mechanism of reputation management in electronic communities. In M. Klusch & L. Kerschberg (Eds.), *Proceedings of the 4th International Workshop on Cooperative Information Agents.*

Yuan, S.T., & Sung, H. (2004). A learning-enabled integrative trust model for e-markets. *Applied Artificial Intelligence, 18*, 69-95.

Zand, D.E. (1972). Trust and managerial problem solving. *Administrative Science Quarterly, 17*(2), 229-239.

Zucker, L.G. (1986). Production of trust: Institutional sources of economic structure 1840-1920. In B.M. Staw & L.L. Cummings (Eds.), *Research in organisational behavior* (pp. 53-111). JAI Press, CT, USA.

ENDNOTES

[1] The exact release dates of the GUI browser, Mosaic, are hard to pin down as versions for different operating systems were often released at different times. While the first official release is dated November 1993, x-mosaic dates back to December 1992. The release of Mosaic 3 in January 1997 marked the end of the browsers development by NCSA. See http://www.ncsa.uiuc.edu/SDG/Software?XMosaic.

[2] This focuses primarily on Web-based e-commerce throughout and assumes access via a desktop computer. This is not in any way a rejection of the importance of other platforms—from interactive mobile telephony to games consoles—will have in future access to e-commerce but a recognition of their current marginal ownership and use.

[3] Giddens' notion of "ontological security" (Giddens, 1984, 1990, 1991) and the work of Fukuyama (1995, 1998) demonstrates better than the brief remit of this paper can why this is not merely linguistic play.

[4] Elsewhere, Giddens (1991, p. 88) has suggested that quest for intimacy is a central feature of contemporary social life. He has suggested that relationships are possible and develop where mutual trust exists and intimacy formed through working at the relationship.

[5] This is not to be confused with Raymond Williams concept of flow—that is, the way items run into each other without marked separation—which is as applicable to Web sites as it is to television.

Chapter XIII
When Is a Duck Not a Duck? When It Is a Euro! Trust–Based Marketing Communications in Virtual Communities

Gianluigi Guido
University of Salento (Lecce), Italy

M. Irene Prete
University of Salento (Lecce), Italy

Rosa D'Ettorre
University of Salento (Lecce), Italy

ABSTRACT

This chapter tries to evaluate the effects of the propagation of a trust-based marketing message through selected below-the-Web technologies, which are those particular types of information technologies different from Web sites—such as e-mails, discussion lists, BBSs, newsgroups, forums, peer-to-peer, IRCs, MUDs and MOOs—that allow for the creation of virtual communities. A preliminary experiment on informal marketing communications, carried out over 12,000 accesses to below-the-Web communities and regarding the proposal to use the term "Ducks" for "Euros" in view of its similarity with the term "Bucks" for dollars, showed that below-the-Web technologies can be an appropriate tool for building trust among participants when four conditions for the existence of virtual communities are met: (a) a minimum level of interactivity; (b) a variety of communicators; (c) a virtual-common-public space; and (d) a minimum level of sustained membership.

INTRODUCTION

In the present economic environment, characterized by global competition, an increasing level of complexity, and a growing interconnection and interdependence, companies must manage new technological requirements for achieving success on the marketplace (Morgan & Hunt, 1994). On one hand, the Internet and digital technologies provide a powerful means for information searching and propagation without limits of time, place, and costs, as well as an effective tool for the development of computer-mediated communications (CMCs)—that is, those task-related and interpersonal exchanges of messages through electronic media that involve the use of computers (Hoffman & Novak, 1996) and which encourage the spread of new forms of relationships and social networks.

CMC technologies have a positive impact on two fundamental dimensions of communication: content and relation (Pastore & Vernuccio, 2004). As regards to content, new technologies make available multimedia differentiated mailings of a large mass of users, reducing the trade-off between reach (the dimension of the potential market) and richness (the product differentiation), and promoting online strategies for customized product positioning. As regards to relation, new technologies allow various types of communications—such as one-to-one, one-to-many, many-to-one, and many-to-many—encouraging users to play an active role in contents' generation.

In this way, the Internet offers the opportunity to accomplish socialization processes of content production and consumption activities, thus allowing companies to establish trusting computer-mediated relationships with their customers and allowing consumers to spontaneously express their expectations and desires. On the other hand, today's global marketplace gives firms no option but to face the growing level of competition through the modification of unilateral relationships in long-lasting trust-based multilat-eral relationships with markets (Castaldo, 2002; Urban, 2003). In the newly connected economy, the environmental complexity changes itself in its relational-based articulation, which needs *trust* as a fundamental resource to govern and regulate market relationships: companies are induced to develop partnerships and strategic networks with those economic parties which contribute to the generation of a corporate value—suppliers, customers, governments, and even competitors. In order to manage competition, individuals and corporations need to cooperate and work together (Morgan & Hunt, 1994); consequently, the creation, development, and maintaining of trust is a requirement for building durable and collaborative relationships (Sultan et al., 2002).

CMC technologies offer companies new opportunities to establish and nurture trust-based communications, allowing them the development of a multichannel strategy on the Internet. In particular, below-the-Web CMC technologies, that is those particular types of information technologies different from Web sites, such as e-mails, discussion lists, and so forth, and their cyberspace, that is the electronic place created by a computer system or by a computer network in which they are, represent the means by which companies and consumers can develop below-the-Web communities. By connecting a large number of computers worldwide, the cyberspace eliminates distances and creates a new place rich in information resources made available through computer networks.

The present chapter tries to evaluate the effects of the propagation of a trust-based marketing message through selected below-the-Web technologies, pursuing both a theoretical and an operational purpose. From a theoretical point of view, it tries to prove that such technologies—as a result of perceived competence and goodwill—are better able than Web sites to develop trust among members, since they generate *virtual communities* (Jones, 1997; Ridings, Gefen & Arinze, 2002). From an operative perspective, this study tries to verify, in a preliminary experiment concerning over 12,000 accesses to below-the-Web communities, their suitability for communicating trust-building messages to different users all over the world. Results of the experiment show that below-the-Web communities are indeed appropriate tools, using a minimum amount of resources, to build trust among their participants.

Below-the-Web Technologies and Communities

The taxonomy of below-the-Web technologies, which consider alternative CMCs to Web sites, is discussed, discriminating among them by considering the timing of messages. Specifically, in "asynchronous communications," members of the community exchange information without the contemporaneous presence of communicators, reading their messages and replying in different times (Baym, 1995), using technologies such as e-mail, discussion lists, BBS, newsgroups, forum, and peer-to-peer (Adams, Toomey, & Churchill, 1999). Whereas, in "synchronous communications," members of the community are online at the same time, interacting in real time, reading messages, and replying immediately (Baym, 1995), using technologies such as Internet Relay Chats (IRCs), Multi-User Dungeons (MUDs), and Multi-Object Oriented (MOOs).

E-mails consist of text-based electronic messages which are sent out over the Internet, generally from one single individual to another, or to small groups, allowing the establishment of a simple one-to-one interaction. E-mail, which can enclose photos, sound clips, video clips, computer files, or computer programs, is the most widespread form of Internet communication (Coon, 1996; Kollock & Smith, 1999).

Discussion lists are delivery lists where users discuss a particular topic: every single message transmitted to a group address is automatically copied and sent to all the e-mail addresses on a list, generating a flux of contents from and to each user

and allowing the formation of discussion groups when users transmit a series of messages and responses to the list (Kollock & Smith, 1999).

BBSs (also known as conferencing systems) are "private" computers (hosts) accessible only to a specific category of users, who exchange opinions and ask for information on technical matters (Rafaeli, 1986). They allow participants to perform functions such as downloading software and data, uploading data, playing games, reading news, and exchanging messages with other users. While e-mail and discussion lists are "push" media, in which messages are transmitted to individuals without them necessarily doing anything. On the contrary, BBSs are "pull" media, in which individuals must choose groups and messages they intend to read and actively demand them (Kollock & Smith, 1999).

Newsgroups consist of virtual discussion groups on specific interests in which messages are stored in a central location. Users can access by going to a particular newsgroup site through the Internet (e.g., Usenet, a world-wide distributed discussion system which consists of a set of newsgroups with names classified hierarchically by subject), or a specific Internet Service Provider (ISP), that is a commercial service that sells access to the Internet to individuals (e.g., AOL, America OnLine) (Blanchard, 2004).

Forums are information interchanges containing messages on specific or general themes allowing individuals to post messages and comment on other messages. They are hosted on a newsgroup, online service, or BBS (Anderson & Kanuka, 1997). Forums differ from Newsgroups in the fact that additional software is generally necessary to participate in a newsgroup or a newsreader, while visiting and participating in a forum normally does not require additional software beyond the Web browser.

Peer-to-peer (P-to-P or P2P) networks are distributed network architectures in which clients share a part of their own hardware resources, such as processing power, storage capacity, or network-linked capacity. Shared resources are directly accessible by other participants, without passing intermediary entities; they make accessible content offered by the network, such as file sharing containing audio, video, data ,or anything in digital format, and real time data, for instance telephony traffic (Schollmeier, 2002).

IRCs (Internet relay chats), which were first formed in 1988 by a researcher in Finland as a text-based way of chatting, are multi-user communication systems through which individuals can hold real-time online conversations (Menichella, 2000). Discussions of various topics are structured into different channels, each of them hosts a discussion on a particular subject with different participants which take part in real-time discussions. Users can join numerous channels at once by selecting a nickname, which allows them to communicate with each another (Coon, 1996).

MUDs (multi-user dungeon) and MOOs (multi-object oriented) are text-based virtual realities which allow several users a contemporaneous navigation in a large hypertext aimed at playing, communicating, socializing. They are basically online interactive role-playing games, which users can access through Telnet, a computer program/ protocol, which allows different types of Internet connected computers to communicate with each another (Coon, 1996; Kollock & Smith, 1999).

These below-the-Web technologies allow personal communications through the computer, creating new opportunities for real-time "chatting" among geographically dispersed individuals, and supporting social relations and below-the-Web communities. The construct of below-the-Web community is strictly connected to the notion of the "virtual community." Rheingold (1993), who first coined the term *virtual communities*, defined them as "social aggregations that emerge from the Net when enough people carry on those public discussions long enough, with sufficient human feeling, to form webs of personal relationships in

cyberspace" (p. 5). Below-the-Web communities differ from virtual communities simply for the reason that the former ones include aggregations that occur through digital technologies different from traditional Web sites. Below-the-Web communities offer several advantages beyond traditional Web sites both to companies and consumers. Companies can propagate their brand images, provide information on product and service characteristics, obtain feedback, and gather useful information about the development and the improvement of existing products, and the launch of new ones. Consumers can find information they need and get in touch with companies using a simple e-mail, reducing the fear of being identified or being obliged to reply to potentially embarrassing questions (Antognazza & Moeder, 1999). Moreover, below-the-Web communities can exist and provide information about products and services even if companies do not intend to propagate any news, do not have a website, or are not informed about them. They are valid tools for creating value for consumers and companies should consider them in their communication strategic planning.

THE CONSTRUCT OF TRUST IN BELOW-THE-WEB COMMUNITIES

Trust has been studied in many research fields (Sultan et al., 2002), leading to different definitions, devoid of a universally accepted conceptualization. By considering common grounds in management and marketing literature, trust can be treated as a cognitive construct which denotes the ability of a counterpart to maintain assumed obligations towards a particular trustor; it indicates the conviction that the trustee—characterized by distinctive elements, such as motivations, competences, and values—will behave in conformity with the trustor's expectancies (Castaldo, 2002). Trust, therefore, is a construct similar to "attitude"

(Fishben & Ajzen, 1975). That is, essential elements of trust and attitude, such as overall beliefs, feelings, values, and personal competences, have an influence on the intention (or willingness) to act, and, consequently on the behavior which, in turn, is coherent with the decision to trust.

The importance of trust is relevant in the Internet context (Reichheld & Shefter, 2000; Urban, Sultan, & Qualls, 2000), and in virtual organizations (Handy, 1995). Consequently, it is crucial to examine processes by which trust can be stimulated and encouraged both within companies and between companies and individuals. The nature and certain features of online contacts, for instance, the lack of face-to-face communications and visual cues and the ease with which members are able to hide personal traits (i.e., gender, age, occupation), may obstruct trust development. Conversely, the opportunity to share common interests and create intimacy and a cooperative interaction may encourage the development of trust in online communications (Ridings et al., 2002; Shankar, Sultan, & Urban, 2002). Considering the existing literature concerning online trust (Fukuyama, 1995; Handy, 1995; Ridings et al.; Urban et al., 2002; Zuboff, 1988), it is possible to assert that two of the main conditions for the creation of trust in below-the-Web communities are: (a) perceived competence (Butler, 1991; Gabarro, 1978; Jarvenpaa, Tractinsky, Saarinen, & Vitale, 1999; Mayer, Davis, & Schoorman, 1995; Sheppard & Sherman, 1998), and (b) perceived goodwill (Bhattacherjee, 2002; Mayer et al., 1995).

Perceived competence is the first condition for the creation of trust. It is the trustor's perception that the trustee possesses skills, ability, and knowledge needed to accomplish specified actions in order to achieve the expected performance or behavior (Castaldo, 2002; Mayer et al., 1995). This condition is appropriate in the context of below-the-Web virtual communities since they are usually means of linking people with common specialized reciprocal interest, life experience, professional occupation, or resource-sharing habits and are related to members' abilities concerning their mutual questions, encouraging information exchange, and knowledge sharing (Bhattacherjee, 2002; Ridings et al., 2002). Professional below-the-Web communities allow participants to exchange their qualified competences and knowledge such as working cultures, problem solving techniques, professional values and behaviors.

Perceived goodwill (or benevolence) is the expectation that a trustee will intend to do good to the trustor, beyond its individual intention, and even though the trustee is not obliged to cooperate and is not compensated for it, the trustee generally responds with a collaborative behavior with the purpose of giving support and care (Bhattacherjee, 2002; Mayer et al., 1995). Benevolence establishes faith and altruism, diminishing uncertainty and the tendency to defend opportunistic behaviors (Bhattacherjee, 2002). Benevolence is important in below-the-Web virtual communities for the reason that positive reciprocation is a fundamental element of a community. While for many Web sites and e-commerce environments, forecasting user expectations for conceiving or delivering benevolent services is problematical or expensive (Bhattacherjee, 2002), in below-the-Web communities contributing to prosocial motives and generous duties has been observed empirically (Wasko & Faraj, 2000). In the online context, perceived goodwill is also associated with integrity, which can be defined as the expectation that the trustee will behave accordingly with socially accepted principles of honesty or a series of values accepted by the trustor, for instance, telling the truth and giving realistically demonstrated information (Ridings et al., 2002). Integrity is relevant in the context of below-the-Web virtual communities since it is the acceptance of norms of reciprocity, strictly related with benevolence, which let a community develop.

THEORETICAL FOUNDATION

We hypothesize that trust can be nurtured and supported particularly through the use of below-the-Web technologies, through the creation of virtual communities. From a theoretical point of view, such information technologies—thanks to perceived competence and goodwill—are better able than Web sites to develop trust among their participants, because they allow for the creation of below-the-Web virtual communities. They are defined by Jones' (1997) theory of the virtual settlement, that is a cyberspace, with associated information technologies, which verifies the following four conditions: (a) a minimum level of interactivity, (b) a variety of communicators, (c) a virtual-common-public space, and (d) a minimum level of sustained membership. This goal is reached by evaluating how the various below-the-Web technologies verify the four requirements for the development of a virtual settlement and of its related community.

Jones' (1997) Virtual Communities

Jones (1997) stated that virtual communities are more than just a series of CMC messages. As a sociological phenomenon, they do not merely correspond to their cyberspace, nor to their members' interactions, nor to their population of users. Rather, Jones (1997) differentiated between a community and its cyberspace, which constitutes the *virtual settlement*, that is between a social aggregation and its medium or platform, in which participants interact (Lechner & Schmid, 2000). A virtual settlement is defined as a virtual place symbolically delineated by a particular subject, and within which a considerable part of CMC technologies takes place, allowing people to interact (Jones, 1997).

Jones (1997) specified that for a cyberspace, with associated CMC technologies, to be considered a virtual settlement, it needs to satisfy the following four requirements: (a) a minimum level of interactivity, (b) a variety of communicators, (c) a virtual-common-public space where a significant portion of a community's interactive CMCs occurs, and (d) a minimum level of sustained membership. Drawing on Fletcher's (1995) theory, Jones (1997) maintained that the existence of a virtual settlement (and the occurrence of its requisites) implies the presence of a connected community. Thus, the virtual settlement is a precondition for the emergence and existence of a virtual community, and the existence of a virtual settlement is evidence of the existence of an associated virtual community (Jones, 1997).

A Minimum Level of Interactivity

Multidisciplinary literature considers three essential conceptual views of interactivity (Tremayne, 2005): structural, perceptual, and process. The first approach defines interactivity a "characteristic of a medium" (Lombard & Snyder-Duch, 2001; Roehm & Haugtvedt, 1999) or an intrinsic part of new media (Heeter, 1989; Rust & Varki, 1996). It is considered as a multidimensional construct that needs to be investigated through an analysis and a categorization of its features or dimensions (Sohn & Lee, 2005). That is, the characteristics of the communication environment that make it interactive.

The second approach assumes that interactivity is a "perceptual variable that involves communication mediated by technology" (Bucy, 2004, p. 377), that is whether or not users perceive the communication environment to be interactive. Numerous authors used experimental design to examine perceived interactivity (Chung & Zhao, 2004) or developed appropriate attitudinal or emotional scales for its measurement (Jee & Lee, 2002; McMillan, Hwang, & Lee, 2002).

The third approach considers interactivity as a process (the actual activity of interacting) of message exchange. Rafaeli (1986), who is the most cited proponent of this approach, defines interactivity as "an expression of the extent that in a given series of communication exchanges, any third (or later) transmission (or message) is related to the degree to which previous exchanges referred to even earlier transmissions" (p. 111). He describes it as a variable attribute of a communication setting that indicated how reciprocal a specific exchange is. Interactivity, thus, is a process that relies on participants. Therefore, it cannot be characterized as a feature of the medium, but rather as a quality of the communication process (Rafaeli & Sudweeks, 1997).

Jones (1997) pointed out the importance of interaction as a necessary condition for a series of CMC messages to demonstrate the existence of virtual communities. So, interactivity is the prerequisite of communication in which simultaneous and continuous exchanges take place. In a real community the relationship occurs through face-to-face communication, whereas in the virtual one, new technologies offer auxiliary instruments to interact in the group, ensuring the same possibilities of reaction (Rafaeli & Sudweek, 1997). Thus, interactivity can be considered a fundamental measure of group social dynamics as it can facilitate the sociality of a group, highlighting the links within it.

A Variety of Communicators

The presence of a variety of communicators is a requisite strictly related to the condition of interactivity, as a single person in contact with another one through CMC technologies does not produce an interactive relationship. Therefore, any possible interaction between a user and a database is excluded from virtual communities (Jones, 1997). This requisite is also discussed by many authors (e.g., Ko & Kim, 2003) when considering the necessary dimensions for a sense of virtual community. For example, people who feel a sense of belonging, people who influence other participants, and people who experience the state of "flow" during virtual communication. Generally the number of communicators in CMC

technologies is higher than in real communities, thanks to their ease of access.

Furthermore, Porter (2004) considers virtual communities according to their population interaction structure: virtual communities as computer-supported social networks (CSSNs), virtual communities as small groups or networks, and virtual communities as virtual publics. According to the type of CSSNs and publics, members can have strong, weak, or stressful social ties in virtual communities. Strong ties are a consequence of regular and supportive communication among socially connected participants; weak ties are a consequence of expressly supportive and reciprocal contacts, even though members are socially and/or physically distant; stressful ties are anti-social communication (e.g., flaming, spamming) (Wellman, Salaff, Dimitrova, Garton, Gulia, & Haythornthwaite, 1996). Small-group-based virtual communities are characterized by strong ties and socially close relationships among participants; weak and likely stressful ties typify networked-based virtual communities. Members are geographically and socially dispersed and directed at the utilitarian advantages of a community, and relationships are frequently of brief extent and propelled by functional needs.

A Virtual-Common-Public Space

In Jones' (1997) theory, a virtual-common-public space denotes a symbolically delineated place, that is a virtual space shared by participants to interact and to form relationships. Considering a virtual-common-public space as an essential requirement for the virtual settlement emphasizes the definition of a community as allocated in the cyberspace (Fernback & Thompson, 1995; Smith, 1992). It "distinguishes a virtual settlement from private communication where postings are directly exchanged from an individual to another with no common virtual place" (Jones, 1997, p. 8). They do not simply correspond to a community subset, but represent a different approach to classifying cyberspace into private virtual places and public ones. Public virtual places can be defined as "places created for CMC conversations using different technologies ... their value depends on the quantity of its (sic) population and on the quantity and quality of their users' contributions" (Jones, 2001, p. 1). Private virtual places, unknown to the mass of the public, allow access only through the insertion of a password. Recent studies (e.g., Blanchard, 2004) refer to CMCs as social or conceptual spaces that members feel is a place, and consider the factors which play a role in the development of a sense of place in virtual communities: the social exchanges that happen in virtual communities, and the "individual cognition of the computers' functioning, because individuals create mental models to help them understand what is going on inside the computer" (Weick, 1990, p. 14).

A Minimum Level of Sustained Membership

A group using CMC technologies is classified as a virtual community when it has a certain degree of sustained membership (Jones, 1997), which is related to the density of messages, defined as the message posting in a group per-unity of time. This condition is emphasized by describing virtual communities as "relatively stable groups of people who interact primarily over CMC and who have developed a sense of community" (Blanchard, 2004, p. 3). Sense of community can be defined as "the members' feeling of shared emotional attachment belonging, influence, and the integration of fulfillment of needs that makes the community different from simply a group of individuals" (Mcmillian & Chavis, 1986, p. 4). Membership is mainly voluntary. Usually participants search for virtual communities sharing the same interests (Wellman & Gulia, 1999), and join them on the basis of their individual interest in a sustaining membership (Blanchard, 2004).

The minimum level of sustained membership required for reaching the stability of the association between members changes according to the CMC medium. Some of them, such as IRCs and forums, produce a higher level of interactivity and of exchange density than other ones due to their structural characteristics.

Table 1. Conditions for the development of virtual communities in asynchronous and synchronous communications

BELOW-THE-WEB TECHNOLOGIES	A Minimum Level of Interactivity	A Variety of Communicators	A Virtual-Common-Public Space	A Minimum Level of Sustained Membership
Asynchronous communications (e-mails, discussion lists, BBSs, newsgroups, forums, peer-to-peer)	*High correlation between all written messages*	*Presence of more than two participants*	*Presence of a virtual delineated space*	*Active participation of each member*
Synchronous communications (IRCs, MUDs, MOOs)	*Message targeting, relatedness of message content*	*Increasing number of members, stability of nicknames*	*Presence of a virtual delineated space, free accessibility to each member*	*Active participation of each member, co-appearance of a substantial number of participants*

COMMUNITIES THROUGH ASYNCHRONOUS AND SYNCHRONOUS BELOW-THE-WEB TECHNOLOGIES

We investigated when the various below-the-Web technologies satisfy the four requirements for the existence of a virtual settlement and its related community, as stated by Jones (1997). Being a community more than a series of CMC messages, we verified when these technologies meet the attributes of a community: analogous to an archaeological perspective, cultural artifacts—such as the characteristics of the place they occupy, and physical traces of life left and created around them by their inhabitants—were analyzed. This is summarized in Table 1. In order to show when below-the-Web technologies hold the four necessary characteristics to create a virtual settlement and its related community, we discriminated between asynchronous below-the-Web technologies, in which messages are read at a later point in time (e-mail, discussion lists, newsgroups, forums, BBSs, and P2P), and synchronous below-the-Web technologies, in which messages appear on users' screens as they are typed (IRCs, MUDs, and MOOs).

A Minimum Level of Interactivity

In asynchronous communications, conversations on particular subjects can last a considerable period of time (some weeks or months); consequently, the verification of a minimum level of interactivity consists in the analysis of the correlation degree between all written messages. In synchronous communications the degree of interactivity in a session is studied analyzing verbal messages, which allow users to "talk" to the other components of the group, and action stimulating messages, which allow users to "act out" imagined actions, and are lines of text sent by participant and describing "what he or she is doing, or rather what his or her virtual being is doing or wishes to do, had it been given a physical body" (Liu, 1999, p. 9). Interactivity is investigated considering message directing/targeting and relatedness of message content. Each communication can be sent to the entire group or sub-group (untargeted), or directed to a specific individual (targeted), specifying the nickname (although every user can see the message). Targeted messages can be unidirectional, that is, directed but not responded to by the targeted recipient, or dyadic, that is responded to by the targeted recipient. Recognition of higher-order patterns of message directing (dyads, triads, and quadruples among others) allows specifying the intensity of within-group interaction (Liu, 1999). As regards message content, it needs to distinguish between the referring message and the referred-to message, and, consequently, messages referring to postings in the same sequence/session (within-session reference) and ones referring to postings in earlier sessions (cross-session reference). Cross-section reference may indicate how long persons have been socializing and the intensity of their interest in each other people (Liu).

A Variety of Communicators

This condition is also definitely found both in synchronous and in asynchronous channels, as Jones (1997) referred to the presence of more than two participants, that is, the minimum number of users needed for any occurring interaction. Generally, asynchronous communication is characterized by a large number of users that allow a continuing mass interaction; in the Internet there are also users (*lurkers*) that read messages without participating actively in the conversation. They cannot be considered members of the virtual community as defined by Jones (1997), as they do not establish interactive relationships (Whittaker, Terveen, Hill, & Cherny, 1998). With regard to synchronous communications, Liu

(1999) includes further conditions: an increasing number of members, in order to reject groups of very insignificant dimension; a stable virtual place (i.e., a channel) existing for a considerably long period of time, with a non-sporadic presence, and the stability of the *nickname*, because an interpersonal relationship cannot mature without recognizing individual identities, so nicknames allow participants to distinguish themselves in the mass and create a personal identity, developing a reputation within the community, and allowing them to be identified for a long period of time (Bechar-Israeli, 1995).

A Virtual-Common-Public Space

A virtual common place can be considered "public" when it is accessible to each user participating in the group (Jones, 1997). As regards asynchronous communications, private exchange of e-mails does not represent a suitable environment for the development of a community, but when members of an existing community exchange messages through a small number of e-mails, these dialogues can be considered community communications (Whittaker et al., 1998). Newsgroups cannot be considered a single virtual community, as they do not correspond to a delineated space, but to thousands of single environments. On the other hand, a single newsgroup, strictly connected to the others on the basis of the subjects discussed, can represent a virtual settlement (Jones, 1997). Discussion lists represent a below-the-Web technology suitable to the development of a virtual community when several users produce a long-lasting interactive communication (Jones, 1997). As regards synchronous communications, the attribute "public" indicates that the shared space is accessible to everyone in the community and that every participant can interact with the others, although the place is not open to everyone. A virtual community can decisively exclude some individuals from becoming a member. Furthermore, the term "public" does not indicate that

interaction between members always needs to be visible to everyone. In fact, private exchanges between participants are crucial for a relationship to develop in a community (Liu, 1999).

A Minimum Level of Sustained Membership

According to Jones (1997), sustained membership stability is related both to a qualitative dimension, which requires the presence of approximately the same group of members for a significant long period of time, and to a quantitative dimension, which requires the presence of a considerable number of active members. As regards asynchronous communication, discussion lists can hold some inequalities the frequency of messages each member sends. Although all the registered members can post, generally only a minority actively takes part in conversations (Whittaker et al., 1998). A large part of users is not linked to a specific group, but continuously searches for groups discussing matters they are interested in, but, in this way, they do not contribute to the establishment of a community relationship (Jones, 1997). With regard to synchronous communication, a minimum level of sustained membership is found when members of a community actively take part in the conversation in a particular channel, demonstrating a minimum level of participation, and not only visualizing various conversations (Liu, 1999). A group of lurkers cannot be considered a community, but a lurker can be a part of a community only if it already exists. Liu (1999) asserts that monitoring sustained membership stability does not require permanent participation of users in a channel, but the co-appearance of a substantial number of participants over a period of time, introducing the concept of sustained level of co-appearance, which has three aspects: a considerable number of clusters of participants whose co-appearance presents durable patterns, a significant size of such clusters (number of members in a co-appearance group), and long lasting patterns of co-appearance.

Since below-the-Web technologies have the potential to support the development of a trust-based below-the-Web virtual community, they increase antecedents of online trust—that is, perceived competence and perceived goodwill of participants. Goodwill and benevolence are strictly related to the conditions of a minimum level of interactivity and sustained membership. By the creation of conversations, participants give support, enhancing perception of cooperative intentions. Members who post messages in a community frequently wait for a reply. Greater interactivity is a sign of motivation to give information and support other community members; it also intensifies the reciprocal nature of relationships (Ridings et al., 2002). If outcomes are consistent with expectations, namely if the trustee has fulfilled promises made in the past, one of the prerequisites for the development of trust is reached. The existence of a below-the-Web virtual community is centered on conversations and activities between members, thus encouraging interactivity and membership reveals benevolence. In an online environment, perceived goodwill is deeply associated with integrity (Gefen, 1997) concerning the reciprocity in creating and sustaining the communities' dialogues, responding to other members or obtaining replies, and showing adherence to social norms and accepted collective rules.

AN EXPERIMENT ON THE PROPAGATION OF A MARKETING MESSAGE THROUGH BELOW-THE-WEB COMMUNITIES

To verify whether below-the-Web technologies are suitable tools for communicating a trust-based message to numerous users displaced in the Internet environment, we conducted an experiment pertaining to the propagation of a marketing message through selected below-the-Web technologies developing a virtual settlement. In particular, this study had the following three aims:

- To test whether communication of a marketing message by means of below-the-Web technologies can correspond to a trust-based, convenient, and quantifiable way to reach a considerable number of recipients
- To evaluate the level of interest aroused from the marketing message sent and, in particular, the way by which possible attention and curiosity are expressed
- To check if the proposal contained in the message was understood and appreciated by members reading it

Creation of a Trust-Based Marketing Message

The message used for the experimental manipulation was designed, structured, tested, and adapted based on characteristics of perceived competence and perceived goodwill as they relate to encouraging the development of trust. The fundamental design elements considered were: first, for sustaining competence and ability in virtual communities, transparency, high-quality content, motivation and background of senders, and access rights; second, the basic design element for supporting perceived goodwill was the comprehensible specification of the message objective (Leimeister, Ebner, & Krcmar, 2005).

In many different languages the introduction of the term "Euro" in European Union (EU) countries created and continues to cause numerous linguistic dilemmas. In our research the suggestion to assign a familiar name to this currency was advanced, with the aim of finding a familiar term that could be used in all EU countries without perplexities of its phonetics and orthography. After the selection of the word Euro for the new European currency, dated December 15, 1995, the European Council stated that the orthography of Euro had to be identical in all the official languages of the European countries and that the plural form of Euro had to be identical to the singular one (Directive n. 1103/97, June 17th

1997), thus violating the principle of subsidiarity. According to subsidiarity, a fundamental principle of European Union law established in the Treaty of Maastricht (1992), the EU does not take action (i.e., make laws) unless it is more effective than taking action at national, regional or local level. This rule caused uncertainty and ambiguity, for the reason that, normally, in the orthography of European languages the conventional value of Latin, Greek, and Cyrillic letters is adapted to the phonetics of each different language, and furthermore the sequence of letters is adapted to the rules of pronunciation (Everson, 2001). A debate thus developed on the way by which the term Euro could be adapted to pronounced, grammatical rules, and phonetics of the European and worldwide official languages. This language uncertainty also generated ample discussions on the Internet. Numerous Usenet groups discussed this question, involving not only lawgivers and linguists, but also the public. Effectively, searching through Google Groups (a discussion group service that offers an ample archive of Usenet postings, including more than a billion messages) the terms Euro and "Euros" led to about 7,990,000 results shortly after its introduction. The proposal to substitute the term Euros, in common terms, with the new term "Ducks" was suggested, in view of its similarity to the term "Bucks," with which Americans informally call dollars (Bucks also can refer to male deer) A message with the object "Refer to EUROs as Ducks!" was therefore created for our experimental manipulation (see Appendix).

The message contained the essential elements encouraging competence and ability. In particular, it displayed: transparency of senders (name, address, and e-mail address, and function of senders are clearly specified in the message); transparency of purpose of the message (the goal of linguistic experiment aimed to measure the spread of a new term in the Internet community is clearly indicated in the message); transparency of feedback procedures (the chance of adhering to the proposal,

substituting the term Euro with the term Ducks is plainly specified). The quality of the content was testified by the numerous and recent debates on the introduction of the term Euro in the European Community. Motivation and background of senders were clearly indicated in the message. The precise explanation of information concerning identification, purposes, and activities of senders were specified. Access rights regard interaction with senders and other members; the asynchronous exchange of information was encouraged within Forums, Newsgroups, and discussion lists (i.e., Yahoo Groups), and through the indication of the sender e-mail address. The message also includes the basic element for supporting perceived goodwill in virtual communities. The intentions and objectives of the research were clearly declared in the message and the absence of commercial aims was declared.

Selection of Below-the-Web Technologies and Message Recipients

The alternative between synchronous and asynchronous below-the-Web technologies, to be chosen for sending the message, led to the choice of the latter. The reasons were that asynchronous technologies allow each community member to read the messages the member wants, regardless of whether the member is connected or not. Furthermore, all responses to messages can be visualized and studied. So, among below-the-Web asynchronous technologies, e-mails, Forum, Newsgroups, and discussion lists (i.e., Yahoo Groups) were used to carry out the experiment.

The selection of specific message recipients was done separately for each different below-the-Web technology, taking into account their perceived competence on the theme of the message and their perceived goodwill in taking part in the community, as described below. With regard to e-mails, a sampling plan was carried out to gather 12,300 e-mail addresses for the experiment. Subjects were selected on the basis of the message content, considering their technical competencies. Specifically, recipients were economic and financial academicians, and financial operators having an e-mail address published on the Internet. Identification of population units was done using two different methods for the e-mail address search, one for academicians and the other for financial operators. Initially, identification of the academic population having economic and financial competencies was accomplished using the directory of the web *links* of the most important universities in the world, obtainable from the Italian Ministry of University and Scientific Technological Research website, and from the University of Bologna website. The sample contained: several e-mail addresses of Italian university professors of Economics, Management, and related subjects, which were published on their university Web sites, for instance LUISS (Libera Università degli Studi Sociali, Rome) and Bocconi University; numerous e-mail addresses of European and worldwide Universities and Schools of Management professors of Economics, Management, and related subjects, for example Cambridge, Oxford, Harvard, and Yale. Secondly, the identification of financial operators' e-mail addresses was undertaken through a double search on Google (www.google.com) considering both *finance* and *bank* as key-words and analyzing the Web sites included in the first one hundred pages. This technique allowed us to find financial and economic private e-mail addresses of operators, institutions, and companies. Forums were selected on the basis of an inquiry carried out on the Google search engine. The most renowned Forums on financial Web sites and their links to other notable financial sites were identified, especially in the U.S., Great Britain, France, Germany, and Italy. Newsgroups were chosen on the basis of a survey accomplished within the Google Group search engine: two searches were carried out using two different key-words, the first one focused on the Usenet groups considering financial subjects, and

the second one on those concerning European and linguistic general themes. Discussion lists were selected among Yahoo Groups containing conversations on financial topics: after selecting the "Finance" category, ten groups were selected on the account of subjects, number of components, and estimated likelihood to receive a feedback from users visualizing the message.

Procedure

Transmission of the marketing message was preceded by a pilot study. An online pretest was carried out in order to test the content and functionality of the trust-based marketing message, and to record eventual negative reactions to such a non-requested communication. The message with the object "Refer to EUROs as Ducks!" (see Appendix) was sent to 50 e-mail addresses to test the message wording. The message was then transmitted, by means of asynchronous below-the-Web technologies—specifically, e-mail, forums, newsgroups, and discussion lists. According to the antispamming law, the message was sent to directory components containing about 12,300 e-mail addresses, by means of a software—Gammadyne Mailers©—allowing forwarded messages in real time to a considerable number of users. The message was also sent to 50 selected forums. Most of them required a user registration; consequently, only a restricted category of financial operators really interested in the community and inclined to give their personal data could read the message. Two hundred messages concerning the currency name were sent to 35 European newsgroups, selected among those dedicated to linguistic discussions, financial problems, and European themes. In some of the 35 groups the message was not published because moderators, having to read and select a massive quantity of messages, slackened procedures for publication, and some of them rejected the message defining it off-topic. Furthermore, with regard to discussion lists, a Yahoo group labeled "Euro_as_ducks" was created: its description contained the same proposal enclosed in the message. Table 2 highlights the summary of opinions and thoughts of those taking part in various debates.

Analysis and Results

Collected data, which included both replies to e-mails and discussions generated within forums, newsgroups, and discussion lists, were first examined with the purpose of studying reactions and effects and afterward classified on the basis of observations and remarks to the proposal. Although the message merely required using the term "Ducks" in Web communications, further observations and remarks gushed from a specific interest to this theme, expressed by recipients replying to the message. The debate rose within several forums, discussion lists, and, in particular, in Newsgroups where linguistic discussions on the term "Euro" dated back to 2000. Numerous replies to e-mails demonstrated that the proposal, even though everybody did not accept it, produced curiosity and interest towards the experiment. Table 2 highlights the summary of opinions and thoughts of those taking part in various debates.

Over 21% of subjects replying to the message declared they would substitute the term Euro with Ducks, and 8.1% of them would like to receive further information on this theme. Nineteen percent of subjects replied, both by the e-mail and in the groups, using vulgar and explicit messages. They were openly unwilling to adopt the term Ducks,

Table 2. Summary of replies to messages

Comments on the Use of the Term "*Ducks*"	Total %
They will use the term "*Ducks*" (8.1% of them asked for further information on the research)	21.5%
They will not use the term "*Ducks*" and expressed their disagreement in an explicit or vulgar manner	19.0%
They will not use the term "*Ducks*", without motivating it	13.5%
They stated the term "*Ducks*" can be easily confused with the term *Loony*, the Canadian dollar nickname	8.1%
They will not use the term "*Ducks*" because the French translation for *duck* is *canard* (the French term for newspapers)	5.4%
They will not use the term "*Ducks*" because it has a negative meaning	5.4%
They will not use the term "*Ducks*" because they stated the importance of establishing a regional European culture, refusing to use a similar name used for the American dollar	5.4%
They will not use the term "*Ducks*" because a European language does not exist and consequently a European nickname cannot be used; a single Country could have its own nickname, actually in Germany *Euro* is already called *Teuer*, that is expensive	2.7%
They will not use the term "*Ducks*" because it is not a worthy nickname for a currency: in cricket language the word "*Ducks*" has a negative meaning because it is similar to zero	2.7%
They will not use the term "*Ducks*" for the reason that it is not linguistically correct	2.7%
They will not use the term "*Ducks*" because it is not an appropriate nickname for a currency: it also means *failure*	2.7%
They will not use the term "*Ducks*" because Euro is a single currency, whereas various kind of *Ducks* exist (black ducks, brown ducks); they indicate the negative meaning related to *failure*	2.7%
They will not use the term "*Ducks*" for the reason that it could be confused with the term "*Bucks*", especially from buyers and sellers of different currencies	2.7%
They stated that in France *Euro* is already called *Balles*, nickname of the old Franc	2.7%
They stated that in one EU state *Euro* is already called *Neuro*	2.7%

considering it a very singular proposal and an ironic and coarse way of communication. This is a remarkable and interesting aspect of the research that can be considered not merely anecdotal. This phenomenon could be due to anonymity, which characterizes impersonal communication, and it would be presumably excluded in other forms of interpersonal contact. Thirteen and a half percent of subjects answering the message absolutely refused the proposal, and 46% of them did not agree, giving specific reasons. In particular, the term "Ducks" was not considered an appropriate substitution for the word Euro, for many reasons: the assonance with the term "Bucks" is not considered a reasonable and legitimate motivation to change a word already used in every language. Furthermore in some comments the similarity with the term "Bucks" was the principal reason for rejecting the word "Ducks", underlining the independence of the European language from the American influence; the term "Ducks" does not derive from a specific historical or cultural context, the same as "Bucks" which is associated with Native American Indians, and Loonies, the nickname of the Canadian dollar, which derives from the Canadian loon impressed on the currency. Some comments were focused on the fact that European countries do not have a common language; therefore it could be very problematic to use a common nickname. It was also significant to know that in some regions different nicknames are used for the term Euro. In Germany it is called Teuer (expensive), in France Balles (balls) (nickname given to the old Franc), and in another European country Neuro. Other remarks also stated that different nicknames, linked to historical or cultural motivation, could emerge in the future in each single region. Further observations were focused on the fact that the term "Ducks" has a negative financial meaning, since it is used to identify a failure or a financial catastrophe. Thus from this point of view, attributing this name to a currency is not of good omen.

GENERAL DISCUSSION AND IMPLICATIONS

With regard to the first objective of the experiment—specifically, investigating whether below-the-Web technologies correspond to a trusting, convenient, and quantifiable method of contacting a significant number of recipients—the directory created by means of the e-mail addresses found on the Internet was appropriate to the communication sending, because the comments received brought new information about the proposal. Below-the-Web technologies thus allowed for the propagation of a message with a small amount of resources—more affordable than those needed for the spread of a message through traditional media—and, at the same time, contributed to enlarging the number of involved users. This exploited the low price or free of charge Internet access, making available communication and interaction of individuals, companies, and institutions situated in geographically dispersed areas.

As to the second objective—evaluating the level of interest aroused from the proposal and, specifically, the way by which attention and interest are expressed—the curiosity produced from the message sent demonstrated the desire to take part in the discussion through advice, suggestions, and opinions on the proposed subject. Even though the message did not require a reply, numerous answers were sent to the e-mail box, and within various forums, Newsgroups, and discussion lists a debate occurred, even though comments were not required. Most of the subjects replying to the message requested additional information, or was available to exchange their knowledge and personal thoughts, opinions, or launch new ideas on the issue, also demonstrating competence and goodwill. According to Ridings et al. (2002), there is a robust association between trust and desire to provide and acquire information. The transmission of the message allows for the activation of a trust-based communication flow, making possible well-timed responses to specific issues, their evaluations and judgments, in absence of a stable organization.

Marketing implications for both customers and companies are relevant as below-the-Web communities allow:

- Customers to find what they need on products and companies, to choose freely which information they desire, in which period of time, and in how much detail.
- Companies to enhance customers' relations and trust, in various ways. They can generate data by means of: (1) systematic *online* marketing research tools—online customer panels and online customer surveys and questionnaires; and (2) unsystematic *online* marketing research—the evaluation of e-mail correspondence, feedback forms (mainly in the occurrence of complaints), newsgroups, and the evaluation of online consulting sessions (with customers permission). Companies can also make available customized information upon customers' explicit requests (information on demand), providing information on various products/services and stimulating communication—online customer advice, customer tuition in the form of web-based training, tuition and learning forums, Internet discussions, video-conferencing.

With reference to the specific content of the experiment, checking if recipients appreciated the trust-based proposal contained in the message, controversial results emerged. A significant part of subjects replying to the message (21.4%) gave a positive response, declaring to replace the word "Euro" with the term "Ducks." This result demonstrated that recipients trusted the proposal contained in the message; they had confidence both in the quality and in the content of the proposed message (perceived competence) and revealed a willingness to adhere to it (perceived goodwill). A considerable part of recipients (19.4%) replying to the proposal declared explicitly or in a vulgar manner to be unwilling to adopt the term ducks. This result showed that, in contrast to face-to-face or verbal communication, below-the-Web technologies diminish users' psychological inhibitions in complaining, rendering them more open in voicing objections and criticisms.

These findings imply consumers can use the Internet and below-the-Web communities not solely to seek advices, suggestions, and opinions on product/service and brands, but also to share personal information: communities exist even if companies do not encourage them, do not know them, or even do not have a Web site (Antognazza & Moeder, 2002). Companies have to consider opportunities and threats deriving from this situation, in particular:

- Using information which allows them to develop a more intimate relationship with customers
- Offering products and services tailored to their individual expectancies and desires (Reichheld & Schefter, 2000)
- Preventing the diffusion of negative information and negative online word-of-mouth in the Internet (Urban et al., 1999)

This research confirmed the importance of below-the-Web technologies in intensifying and enlarging trusting relations between companies propagating messages through below-the-Web communities and their potential customers. The substitution of a word used in the common language was neither straightforward nor immediate; nevertheless, the analysis completed proved the existence of concrete basis for future positive evidence.

FUTURE TRENDS AND CONCLUSION

The present experiment on the transmission of a trust-based marketing message by means of below-the-Web communities represents a relevant but initial step toward the investigation of building trust developing computer-mediated relationships. There are numerous unexplored areas of study and prolific opportunities regarding the creation and increase of online trust. New information technologies will intensify and facilitate, in an exponential way, the use of below-the-Web technologies, allowing a considerable number of individuals and companies worldwide to use devices and services not yet designed. Thanks to multiple technologies—such as broadband, wireless fidelity, and mobile applications—billions of people will have high-speed wireless Internet access in the future and will obtain new contents and services. In particular:

- Broadband service will provide high-speed data transmissions, allowing access to numerous high quality Internet services, resources, and products, thus stimulating interactivity and membership between companies, individual consumers, businesses, and institutions. Broadband can surmount geographical and financial barriers providing access to a broad variety of educational, cultural, and recreational opportunities and resources (i.e., video, music); it can encourage companies' growth by the means of e-commerce, creating new jobs and providing access both to local and global markets. Furthermore, it will make available new telecommunications technologies, such as the voice over Internet protocol (VoIP), which allows voice and video communication using the Internet.
- Wireless fidelity (Wi-fi), a high-speed wireless technology, will connect homes and businesses—for example, cafes, hotels, airports—using a radio link through the Internet between the customer's location and the service provider's facility.
- Mobile applications, such as Next Generation 3G cellular services, provide a long-range wireless coverage for data access across wide geographic areas, assuring the maximum mobility for voice communications and Internet connectivity.

High-speed wireless technologies will work together allowing individuals and companies for mobile computing and communications worldwide, offering original and stimulating opportunities for end users, application developers, content providers, and network operators. These technologies will support the development of trust-based virtual communities. Billions of people all over the world will be encouraged to stay connected virtually anytime and anywhere (variety of communicators) and to connect wirelessly using devices and services not yet designed (minimum level of interactivity and membership), combining and matching wireless technologies and mobile platforms (virtual-common public space).

From a theoretical point of view, trust is a multidimensional construct. Future steps for a comprehensive and detailed examination should consider the process of building trust, and, in particular, the identification of elements influencing its antecedents—perceived goodwill and perceived competence. Potential factors may be the trustee's reputation, which can be defined as an expectation of individual's actions on the basis of its past behavior (Abdul-Rahaman & Hailes, 2000), and the trustor's propensity to trust. Furthermore, antecedents and consequences of trust may be different in various types of communities. Trust building elements and their effects may differ in communities of transaction, in which individuals buy, sell, or find information about products and services, in communities of fantasy, in which member explore new identities, in communities of interest, in which individuals share common interests, and in communities of relationships, in which members develop social relations (Hagel & Armstrong, 1997). Within the same kind of community, trust can be different depending on the type of information that members get or desire to give. The area of cross-cultural and international differences in trust perceptions can also be examined, principally race, ethnicity, and culture. Antecedents of online trust may change in distinct cultural environments or may have diverse influences in high than in low context culture.

In conclusion, this experiment demonstrated that a marketing message can earn consumer trust; consequently, the Internet, and particularly below-the-Web communities, could become a new channel for trusting relationships. With reference to the first objective of the experiment, below-the-Web technologies can be considered a trusted and suitable tool for communicating and interacting with individuals, institutions, and companies all over the world. Below-the-Web communities have a high capacity of directing and targeting: trust-based messages—characterized by perceived competence and perceived goodwill—can reach particular market segments defined on the basis of the sociocultural variable and life style, of particular interests and specific competences. Considering the second objective of the research, below-the-Web communities are more effective than simple Web sites for coalescing interests and people on the Internet. Trust-based communications motivate members to participate in communities, sharing a large mass of information and allowing for the integration between content (including information from companies and consumers) and communication. Interaction makes a comparison on shared interests from a common perspective possible. Participants search and provide content, thus generating a collective competence. With regard to the third objective of the experiment, results show that trust-based below-the-Web communities could be appreciated by participants, and they have an influence—positive or negative—on community members. Consequently, companies and institutions have to consider their potential impact, even if they do not directly insert specific information on the corporation itself, other Web sites and below-the-Web communities could contain information on their products and activities.

REFERENCES

Abdul-Rahman, A., & Hailes, S. (2000). Supporting trust in virtual communities. In *Proceedings of the 33rd Annual Hawaii International Conference on System Sciences (HICSS 33), 6,* 6007. IEEE CS Press. Retrieved March 11, 2007, from http://citeseer.nj.nec.com/235466.html

Adams, L., Toomey, L., & Churchill, E. (1999). Distributed research team: Meeting asynchronously in virtual space. *Journal of Computer-Mediated Communication, 4*(4). Retrieved March 11, 2007, from http://www.ascusc.org/jcmc/vol4/issue4/adams.html

Anderson, H., & Kanuka, T. (1997). On-line forums: New platforms for professional development and group collaboration. *Journal of Computer-Mediated Communication, 3*(3). Retrieved March 11, 2007, from http://www.ascus.org/jcmc/vol3/issue3/anderson.html

Antognazza, E., & Moeder, P. (1999). *Web marketing per le piccole e medie imprese.* Milano: Hops Libri.

Baym, N. (1995). The emergence of community in computer-mediated communication. In S. Jones (Ed.), *CyberSociety* (pp. 138-163). Newbury Park, CA: Sage.

Bechar-Israeli, H. (1995). From <Bonehead> to <cLoNehEAd>: Nicknames, play, and identity on Internet relay chat. *Journal of Computer-Mediated Communication, 1*(2). Retrieved March 11, 2007, from http://www.ascusc.org/jcmc/vol1/issue2/bechar.html

Bhattacherjee, A. (2002). Individual trust in online firms: Scale development and initial test. *Journal of Management Information Systems, 19*(1), 211-241.

Blanchard, A. (2004). Virtual behavior setting: An application of behavior setting theories to virtual communities. *Journal of Computer-Mediated Communication, 9*(2). Retrieved March 11, 2007, from http://jcmc.indiana.edu/vol9/issue2/blanchard.html

Bucy, E.P. (2004) Interactivity in society: Locating an elusive concept. *The Information Society, 20*(5), 373-383.

Butler, J.K. (1991). Towards understanding and measuring conditions of trust: Evolution of a condition of trust inventory. *Journal of Management, 17*(3), 643-663.

Castaldo, S. (2002). *Fiducia e relazioni di mercato.* Bologna: Il Mulino.

Chung, H., & Zhao, X. (2004). Effects of perceived interactivity on Web site preference and memory: Role of personal motivation. *Journal of Computer-Mediated Communication, 10*(1), 7. Retrieved March 11, 2007, from http://jcmc.indiana.edu/vol10/issue1/chung.html

Coon, D.A. (1996). *An investigation of friend Internet relay chat as a community.* Unpublished master's dissertation, Kansas State University. Retrieved March 11, 2007, from http://www.davidcoon.com/thesis.txt

Directive n. 1103/97, June 17th (1997), Retrieved June 29, 2007, from http://eurlex.eruopa.eu/LexUriServ/LexUriServ.do?uri-OJ:C:2007:056:0009:0010:EN:PDF

Everson, M. (2001). The Name of the Euro in European languages. *An Aimsir, 2.*

Fernback, J., & Thompson, B. (2000, May). Virtual communities: Abort, retry or failure? Online version of *Computer Mediated Communication and the American Collectivity: The Dimensions of a Community Within Cyberplace.* Paper presented at the Annual Conference of the International Communication Association. Retrieved March 11, 2007, from http://www.well.com/user/hlr/texts/Vccivil.html

Fishbein, M., & Ajzen, I. (1975). *Belief, attitude, intention, and behavior: An introduction to theory and research.* Reading, MA: Addison-Wesley.

Fletcher, R.J. (1995). *The limits of settlement growth: A theoretical outline.* Cambridge: Cambridge University Press.

Fukuyama, F. (1995). *Trust: The social virtues & the creation of prosperity.* New York: The Free Press.

Gabarro, J.J. (1978). The development of trust, influence, and expectations. In A.G. Athos & J.J. Gabarro (Eds.), *Interpersonal behavior: Communication and understanding in relationships* (pp. 290-303). Englewood Cliffs, NJ: Prentice Hall.

Gefen, D. (1997). *Building users' trust in freeware providers and the effects of this trust on users' perception of usefulness, easy of use and intended use of freeware.* Unpublished doctoral dissertation, Georgia State University.

Hagel, J., & Armstrong, A. (1997). *Net gain: Expanding markets through virtual communities.* Boston: Harvard Business School Press.

Handy, C. (1995). Trust and the virtual organization. *Harvard Business Review, 73*(3), 40-50.

Heeter, C. (1989). Implications of new interactive technologies for conceptualizing communication. In J.L. Salvaggio & J. Bryant (Eds.), *Media use in the information age: Emerging patterns of adoption and consumer use* (pp. 221-225). Hillsdale, NJ: Lawrence Erlbaum Associates.

Hoffman, D.L., & Novak, T.P. (1996). Marketing in hypermedia computer-mediated environments: Conceptual foundation. *Journal of Marketing, 60*(3), 50-68.

Jarvenpaa, S. L., Tractinsky, N., Saarinen, L., & Vitale, M. (1999). Consumer trust in an Internet store: A cross-cultural validation. *Journal of Computer Mediated Communication, 5*(2). Retrieved November 5, 2005 from http://www.ascuusc. org/jcmc/vol5/issue2/jarvenpaa.html

Jee, J., & Lee, W-N. (2002). Antecedents and consequences of perceived interactivity: An exploratory study. *Journal of Interactive Advertising, 3*(1). Retrieved March 11, 2007, from http://jiad.org/vol3/no1/jee/index.htm

Jones, Q. (1997). Virtual-communities, virtual settlements & cyber-archaeology: A theoretical outline. *Journal of Computer-Mediated Communication, 3*(12). Retrieved March 11, 2007, from http://www.ascusc.org/jcmc/vol3/issue3/jones. html

Jones, Q. (2000). Time to split virtually: Expanding virtual publics into vibrant virtual metropolis. In *Proceedings of the 33rd Hawaii International Conference on System Sciences*. Retrieved March 11, 2007, from http://www.computer.org/proceedings/hicss/0493/04936/ 04936003.pdf

Jones, Q. (2001). *The boundaries of virtual communities: From virtual settlements to the discourse of dynamics of virtual publics*. Unpublished PhD thesis, Graduate School of Business, University of Haifa, Israel.

Jones, Q., Ravid, G., & Rafaeli, S. (2001). *Empirical evidence for information overload in mass interaction*. Paper presented at the Conference on Human Factors and in Computing Systems, CHI 2001. Retrieved March 11, 2007, from http://gsb. haifa.ac.il/~sheizaf/publications/chi2001.pdf

Jones, S. (1998). *Cybersociety 2.0: Revisiting computer mediated communication and community*. London: Thousand Oaks.

Ko, J., & Kim, Y. G. (2003). Sense of virtual community: Determinants and moderating role of the virtual community origin. *International Journal of Electronic Commerce, 8*(2), 75-88.

Kollok, P., & Smith, M. (1999). *Communities in cyberplace*. London: Routledge.

Lechner, U., & Schmid, B.F. (2000). Communities and media: Towards a reconstruction of communities on media. In *Proceedings of the 33rd Hawai International Conference on System Sciences*. Retrieved March 11, 2007, from http://ieeexplore. ieee.org/xpl/freeabs-alljsp?arnumber=926817

Leimcister, J.M., Ebner, W., & Krcmar, H. (2005). Design, implementation, and evaluation of trust-supporting components in virtual communities for patients. *Journal of Management Information System, 21*(4), 101-131.

Liu, Z.G. (1999). Virtual community presence in the Internet relay chatting. *Journal of Computer-Mediated Communication, 5*(1). Retrieved March 11, 2007, from http://www.ascusc.org/jcmc/ vol5/issue1/liu.html

Lombard, M., & Snyder-Duch, J. (2001). Interactive advertising and presence: A framework. *Journal of Interactive Advertising, 1*(2).

Mayer, R.J., Davis, J.H., & Schoorman, F.D. (1995). An integrative model of organizational trust. *Academy of Management Review, 20*(3), 709-734.

McMillan, D.W., & Chavis, D.M. (1986). Sense of community. A definition and theory. *Journal of Community Psychology, 14*(1), 6-23.

McMillan, S.J., Hwang, J., & Lee, G. (2003). Effects of structural and perceptual factors toward the website. *Journal of Advertising Research, 43*(4), 400-409.

Menichella, E. (2000). Chat r u ready. *Internet News.* Retrieved March 11, 2007, from http://inews.tecnet.it/Articoli/2000/2000-12/dossier0012.html

Morgan, R.M., & Hunt, S.D. (1994). The commitment-trust theory of relationship marketing. *Journal of Marketing, 58*, 20-38.

Pastore, A., & Vernuccio M. (2004). *Marketing, innovazione e tecnologie digitali.* Padova: Cedam.

Porter, C.E. (2004). A typology of virtual communities: A multidisciplinary foundation for future research. *Journal of Computer-Mediated Communication, 10*(1). Retrieved March 11, 2007, from http://jcmc.indiana.edu/vol10/issue1/porter.html

Rafaeli, S. (1986). The electronic bulletin board: A computer driven mass medium. *Computers and the Social Sciences, 2*(3), 123-136.

Rafaeli, S., & Sudweeks, F. (1997). Networked interactivity. *Journal of Computer-Mediated Communications, 2*(4). Retrieved March 11, 2007, from http://www.ascusc.org/jmc/vol2/issue4/rafaeli.sudweeks.html

Reichheld, F., & Shefter, P. (*2000).* E-loyalty: Your secret weapon on the Web. *Harvard Business Review, 78*(4), 105-114.

Rheingold, H. (1993). The virtual community: Homesteading on the electronic frontier. Reading, MA: Addison-Wesley. Retrieved March 11, 2007, from http://www.rheingold.com/vc/book/

Ridings, C.M., Gefen, D., & Arinze, D. (2002). Some antecedents and effects of trust in virtual communities. *Journal of Strategic Information Systems, 11*(3-4), 271-295.

Roehm, H.A., & Haugtvedt, C.P. (1999). Understanding interactivity of cyberspace advertising. In D.W. Schumann & E. Thorson (Eds.), *Advertising and the World Wide Web* (pp. 27-39). New Jersey: Lawrence Erlbaum Associates.

Rust, R.T., & Varki, S. (1996). Rising from the ashes of advertising. *Journal of Business Research, 37*(3), 173-181.

Schollmeier, R. (2002). A definition of peer-to-peer networking for the classification of peer-to-peer architecture and applications. In *Proceedings of the First International Conference on Peer-to-Peer Computing.* IEEE Press.

Shankar, V., Sultan, F., & Urban, G.L. (2002). Online trust: A stakeholder perspective, concepts, implications, and future directions. *Journal of Strategic Information Systems, 11*(4), 325-344.

Sheppard, B., & Sherman, D. (1998). The grammar of trust: A model and general implications. *The Academy of Management Review, 23*(3), 422-437.

Smith, M.A.. (1992), *Voices from the WELL: The logic of the virtual commons.* Unpublished master's thesis, University of California, Los Angeles.

Sohn, D., & Lee, B. (2005). Dimensions of interactivity: Differential effects of social and psychological factors. *Journal of Computer-Mediated Communication, 10*(3). Retrieved March 11, 2007, from http://jcmc.indiana.edu/vol10/issue3/sohn.html

Sultan, F., Urban, G. L., Shankar, V., & Bart, I. Y. (2002). *Determinants and role of trust in e-business: A large scale empirical study.* MIT Sloan Working Paper, 4282.

Treaty of Maastricht (1992), Retrieved June 29, 2007, from http://www.eurotreaties/com/masstrichtext.html

Tremayne, M. (2005). Lessons learned from experiments with interactivity on the web. *Journal of Interactive Advertising, 5*(2). Retrieved November 5, 2005 from http://www.jiad.org/vol5/no2/tremayne/

Urban, G.L. (2003). *The trust imperative* (MIT CeB Working Paper, 175).

Urban, G.L., Sultan, F., & Qualls, W.J. (2000). Placing trust at the center of your Internet strategy. *Sloan Management Review, 42*(1), 38-48.

Wasko, M., & Faraj, S. (2000). It is what one does; Why people participate and help others in electronic communities of practices. *Journal of Strategic Information Systems, 9*(2/3), 155-173.

Weick, K.E. (1990). P. Goodman, L. Sproull and Associates (Eds.) Technology as equivoque: Sense making in new technologies. In *Technology and organizations* (pp. 1-44). San Francisco: Jossey-Bass.

Wellman, B., & Gulia, M. (1999). Net surfers don't ride alone: Virtual communities as communities. In P. Kollock & M. Smith (Eds.), *Communities in cyberspace*. Berkeley: University of California Press.

Wellman, B., Salaff, J., Dimitrova, D., Garton, L., Gulia, M., & Haythornthwaite, C. (1996). Computer networks as social networks: Collaborative work, telework, and virtual community. *Annual Review of Sociology, 22*(1), 213-238.

Whittaker, S., Terveen L., Hill, W., & Cherny, L. (1998). *The dynamics of mass interaction*. In Proceedings of CSCW, Seattle, Washington. Retrieved March 11, 2007, from http://citeseer.nj.nec.com/cache/papers/cs/14002/http:zSzzSzwww.research.att.comzSz~stevewzSzcscw98-published.pdf/the-dynamics-of-mass.pdf

Zuboff, S. (1988). *In the age of the smart machine. The future of work and power.* New York: Basic Books.

APPENDIX

Marketing Message

In your electronic mail and communications, we invite you to refer to Euros (the new European currency) as "ducks". Please, call them familiarly "ducks", as you would refer to U.S. dollars as "bucks." "Ducks and bucks" could be an easy and memorable pair to be used within financial, academic, and social communities. This is a linguistic experiment carried out by a research group at the Chair of Marketing at the University of Lecce, Italy. We are measuring the spread of a new term in the Internet community. This message has no commercial aim and will be sent to you only once. We thank you in advance for promoting in your communications the use of this new word. We shall periodically check the use of this term in search engines.

Cordially,

The Marketing Research Group at the University of Lecce, Italy
Faculty of Economics, Palazzo Ecotekne, Via per Monteroni, 73100, Lecce, ITALY
E-mail address: mktg-group.lecce@libero.it

Chapter XIV
Overcoming Hurdles to Trust:
Infomediaries and Public Records

Robert M. Easter
Qwest Group, LLC

Linda L. Brennan
Mercer University, USA

ABSTRACT

This chapter discusses paradoxes that have arisen in response to society's dependence on the Internet and the new business models that attempt to resolve the contradictions. The authors present a Trust Model illustrating a series of hurdles that must be overcome in order for an individual or an organization to exhibit trust, that is, risk-taking behavior. The chapter addresses the formation of new information providers, "infomediaries," which can inform risk-taking behavior in computer-mediated relationships.

INTRODUCTION

In times past, a "man's word was as good as his bond." Deals were made based on reputations and handshakes. Kings' decrees were validated by the seal of a signet ring. Marriages were arranged by families—or at least introductions were made by mutual acquaintances.

With the proliferation of the Internet, an interesting paradox has emerged: we have become more interconnected, and, at the same time, more isolated. The technology that enables us to transcend the limitations of time and space also limits our ability to judge a person's character, evaluate his or her trustworthiness, or establish a long-term relationship. Yet we are becoming increasingly more dependent on the technology for our social interactions.

Naisbitt (1982) identified this paradox as a "high-tech backlash" (p. 43). He suggested that:

... when institutions introduce a new technology to customers or employees, they should build in a high-touch component; if they don't, people will

try to create their own or reject the new technology .... When high tech and high touch are out of balance ... dissonance results. (pp. 43-44)

As a result, new business models are emerging. Kelly (1998) predicted the need for intermediaries that could help negotiate the web of information that is available. He described relationship technology that is applied towards forging—and binding—the connections among individuals, and between individuals and organizations. He warned that:

The advent of relationships technologies on the net creates a larger role .... None of this enlargement of relationships can happen unless there are vast amounts of trust all around .... Trust is a peculiar quality .... It can't be instant – a startling fact in an instant culture .... But it can disappear in a blink. (pp. 132-133)

Kelly goes on to say that trust cannot be purchased and has to accumulate slowly, which takes us to another paradox: how can we accumulate trust slowly, yet have it instantly?

This chapter examines these paradoxes and the emerging business models that attempt to resolve these contradictions. It starts with background information on trust elements, public records, and how trust can be built using public records. The chapter continues with descriptions of emerging business models and illustrations of how they work. It concludes with a discussion of future and emerging trends.

BACKGROUND

To frame this analysis, is helpful to have a common understanding of how trust is defined and established. The literature on trust is extensive, drawing from economic theory and the social sciences. Rather than present an exhaustive analysis

of this work, the following model is offered as a succinct—and salient—summary of key factors to consider.

Trust Model

Hardin (2006) distinguishes between the quality of trustworthiness and the expectation of trust:

To say that I trust you in some context is to say that I think you are or will be trustworthy toward me in that context. You might not be trustworthy toward others and you might not be trustworthy in other contexts. (p. 1)

Mayer, Davis, and Schoorman (1995) perceive trustworthiness as a function of three broad factors: ability, benevolence, and integrity:

Ability is that group of skills, competencies, and characteristics that enable a party to have influence within some specific domain .... Benevolence is the extent to which a trustee is believed to want to do good to the trustor, aside from an egocentric profit motive .... Integrity involves the trustor's perception that the trustee adheres to a set of principles that the trustor finds acceptable ... the consistency of ... past actions, credible communications ... a strong sense of justice, and the extent to which ... actions are congruent with ... words all affect the degree to which the [trustee] is perceived to have integrity. (pp. 717-719)

In their view, trust is a function of these factors, moderated by the trustor's propensity to trust. Elsewhere, this characteristic is referred to as "generalized" trust, relatively positive expectations of the trustworthiness, cooperativeness, or helpfulness of others (Hardin, 2006, p. 125). This factor becomes especially important in temporary relationships with time pressures, as might occur in Internet-based interactions. Meyerson, Weick, and Kramer (1996) suggest that without the time to build expectations from experience with a relationship, trustors rely on their predispositions and category-driven assumptions (i.e., generalized trust).

Whether the trustor acts on this trust depends on one more factor, the perception of risk. Without risk, there is no need for trust. One does not need to risk anything to trust; however, one must take a risk in order to engage in trusting action (Mayer et al., 1995, p. 724). Risk is a function of both the probability for failure and the cost of that failure. To reduce the perception of risk, the trustor might limit the exposure of failed expectations. This perception of risk can be mitigated by contractual specifications (Gulati, 1995; Lui & Ngo, 2004), explicit structuring processes (Walther & Bunz, 2005), and frequent and open communications (Sydow, 1998).

In summary, the trust model implicit in this chapter is based on:

- Perceptions of trustworthiness (a function of ability (a), benevolence (b), and integrity (i))
- Predisposition to trust (i.e., generalized trust, T)
- Perceptions of risk (R)

as determinants of risk-taking behavior (RTB). In a generalized form, we might represent this model as:

$$RTB = \{\ F\ (a, b, i)\ \}*T*R$$

where each value is non-negative, and predisposition to trust and perceptions of risk are moderating variables that range from [0,1]. This is depicted graphically in Figure 1 as a series of hurdles.

Understanding trust as a series of hurdles will help to underscore how the creative use of public records can engender trusting actions in a

Figure 1. The trust model

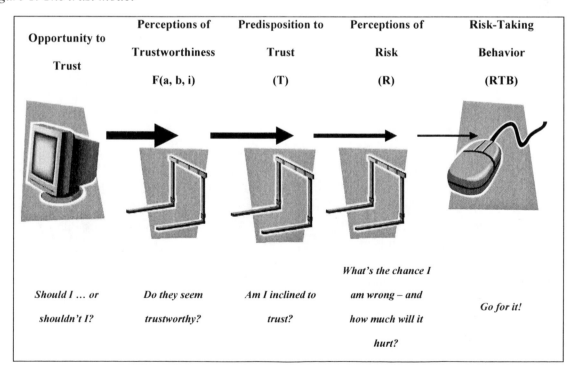

Opportunity to Trust	Perceptions of Trustworthiness F(a, b, i)	Predisposition to Trust (T)	Perceptions of Risk (R)	Risk-Taking Behavior (RTB)
Should I ... or shouldn't I?	*Do they seem trustworthy?*	*Am I inclined to trust?*	*What's the chance I am wrong – and how much will it hurt?*	*Go for it!*

computer-mediated environment. The following section provides background information about this special kind of data.

Public Records

Public records document and register changes of status and the transfer of assets. They are pieces of information that have been filed or recorded by public agencies and can be created by the government (vital records such as birth, marriage, and death certificates; property records such as deeds; professional licenses; driving history; and criminal and civil court records, as examples); by companies (corporate filings, contact lists, real estate transactions, and mortgages); or by individuals (magazine subscriptions, voter registrations, and phone numbers). These records can be accessed directly from the government or from companies that aggregate the data to deliver risk management services or the raw data itself.

The use of public records is an essential activity for companies and individuals to make informed decisions and to determine the risk/reward relationship for many decisions that are frequently made. Does a retailer want to acquire goods from this particular company? Do I as a homeowner want to allow this repairman into my house? Should this person be permitted to teach children?

Use of public records to make informed decisions are a center of controversy. One perspective maintains that the "wheels of commerce" could quickly grind to a halt without the ability to access publicly available information, simply because continual risk management decisions would not be able to be made. Offsetting that viewpoint is the one that says that others should not be able to access information about another, even if it already exists in the public realm. Current discussions in our neighborhoods and in our state and federal legislative houses generally acknowledge the need to use public records, but differ in opinions about how much to control access and/or use.

Access and use can be controlled by various authentication systems and by organizations or individuals. These systems use biometrics like voiceprints or fingerprints to determine specific identification (ID), verify that ID, and confirm entitlement for that ID. Examples: Is there a John Smith? Is this John Smith with whom I am dealing? Is this the John Smith who should be entitled access to this Web site?

Framework

In the context of the trust hurdles, public records might be used to evaluate the perceived trustworthiness of the trustee, at least in terms of integrity, variable (i). Has Mary Doe been convicted of a crime? Is Mary Doe licensed to do business? Have there been complaints or liens filed against Mary Doe, Inc.?

Public records might also undergird a trustor's generalized trust, variable (T). For example, in the case of online dating services, if the trustor is able to verify that the representations made by the trustee are true (e.g., single female with a college degree), then there is a predisposition to trust that the trustee is worth getting to know.

With respect to the third factor of the trust model, perception of risk (R), the trustor might rely on public records to be able to seek redress in the case of failed expectations. An interesting example comes from an automobile manufacturer offering dealer incentives. To receive payment on the incentives, the dealers submit the vehicle identification numbers (VINs) of the automobiles sold under the program. The manufacturer can consult public records to ensure that, in fact, the title on these VINs was transferred. Being able to verify this information with public records limits the exposure the manufacturer has in this incentive program.

BUSINESS MODELS

Business models describe the value proposition that a firm uses to make money. This framework stems from Porter's (1985) concept of a value chain. Simply put, organizations produce outputs by adding value to inputs in a transformation system. Meredith and Shafer (2007) describe four major ways to add value: alteration (physical, sensual, or psychological), transportation, storage, and inspection.

To address the need for trust in a computer-mediated relationship, several business models have emerged. They are generally referred to as "infomediaries," or information intermediaries. A key to the success of these infomediaries is their use of public records. They add value to their customers by some transformation public records as a key input. Specific applications are described in the following models.

Data Aggregators

Public data are fragmented, available at the municipal, state, and federal levels. One business model, the data aggregator, profits by combining these disparate sources of information to consolidate individuals' public records and sell them. They enable the data users, that is, the trustors, to transcend time and space by accessing the public records where and when they choose.

Such companies not only aggregate the data but also sell services to mine that data. These services can result in powerful applications of public records. Individuals can learn specific information that is available about them, companies can analyze backgrounds of job applicants, governments can verify identities and qualifications for benefits, and companies can determine appropriate insurance rates to charge businesses and households. All are able to make better personal and business decisions, as a result.

In this model, the aggregators are adding value by transporting and storing the public records. By mining the data, the aggregators are also altering it and presenting it in a new, presumably more valuable, way. They are providing access to this information to prospective trustors, informing their perceptions of trustworthiness, applications of generalized trust, and assessments of risk.

ChoicePoint's (www.choicepoint.com) Insurance Services division uses the data it accumulates and combines to help insurance companies make appropriate underwriting decisions. They also host and manage services to determine rates and issue quotes for their customers. For example, while reviewing a new homeowner's application from a former Allstate customer, it is helpful for the State Farm underwriter to know the types and numbers of claims filed on this particular home over the years. The ability to either fuel customer processes, or provide the actual services, are common to the data aggregators.

Thomson West (www.west.thomson.com) and LexisNexis (www.lexisnexis.com) each support the legal, corporate, government, and academic markets. Both offer Web-based and dedicated access to legal, regulatory, and business information that can be integrated into customer processes, as well as a myriad of services to help law practices operate their businesses. As an illustration, consider how a law firm frequently accesses legal records to research past court rulings. The firm also needs to view public records for backgrounds of prospective clients or witnesses to ensure no conflicts of interest.

Acxiom (www.acxiom.com) is another aggregator that promotes its strength as helping its clients maximize the value of their customers to build and reinforce relationships. The most visible of these services would be delivering specific messages to targeted customers to maximize exposure and response. In this way, this data aggregator could be viewed as enhancing its clients' trustworthiness by enhancing the perception of their benevolence.

Trust Brokers

Many businesses have emerged by building reputations and brands to be trusted in a gradual way, so that if they endorse a transaction or an individual, that trust will be transferred instantaneously. These "trust brokers" are much like the "Good Housekeeping Seal of Approval" endorsement in a networked society. In the trust model, these brokers add value by inspection, validating, and assuring the quality of the data being presented to the trustor. In this manner the trust brokers are reducing the trustor's perception of risk, (R).

Trufina (www.trufina.com) is one of these companies helping individuals to determine whether others with whom they are dealing across the Internet are being truthful. A person simply submits personal information about himself; Trufina then verifies that this information is accurate and issues a Trufina Identity Card that indicates that this verification has been performed. It is important to note that each "Trufina'd" individual completely controls how much information and with whom this verification is shared. Trufina can be used for electronic interactions such as dating, other social interactions, auctions, blogs, employment sites, and job contracting.

Verisign (www.verisign.com) is another trust broker, and focuses on securing specific Web sites, intranets, and extranets to ensure confidential e-commerce transactions and communications. In addition, Verisign provides infrastructure services to ensure security across any Internet Protocol (IP)-based network, citing examples that range from supply chain transactions that employ wireless technologies to voice-over-Internet communications.

Identity Proofers

One side effect of this reliance on public records is the consumers' growing awareness of the data that are available—and realization that these data are error-prone. Data entry mistakes are common.

Mismatches between identities also occur. These are benign, yet troubling sources of inaccuracies. Added to them is the issue of identity theft, and we find a need for yet another infomediary, "identity proofers," troubleshooters who correct errors in public records or resolve abuses of them. They add value to the public records by altering them. This level of assurance can enhance the perception of trustworthiness by the trustor.

Historically, credit bureaus have existed as infomediaries in the financial sector. Companies are using automated credit-checking applications such as Dun & Bradstreet's Global DecisionMaker to establish credit limits for online transactions (Violino, 2002).

Other businesses have emerged to address the broader sphere of public records. For example, MyPublicInfo (MPI; www.mypublicinfo.com) was founded to provide consumers with their own background check information and identity theft-prevention tools. MPI's Personal Information Profile scours billions of public records to display information available about persons who register. It is important to note that MPI uses a multi-step authentication process to ensure that individuals can only request information about themselves. This Profile allows one to determine if there are any discrepancies that need to be corrected and suggests procedures to make those corrections. These discrepancies can be unintentional data-entry errors, or the signs of identity fraud being committed.

On an ongoing basis, MPI's IdentitySweep (www.identitysweep.com) and Intersections' IdentityGuard (www.intersections.com) continually track billions of records, looking for manipulations to individuals' Social Security numbers. Currently, identity thieves are creating new identities by gradually attaching new names and addresses to existing Social Security numbers, saddling their victims with compromised identity exposure and financial liabilities.

For this service to have value, timeliness is a key element in the business model. The identity proofer will issue an alert to its subscribers *as*

Table 1. Transformation of public records for perceptions of trustworthiness

Public Records Transformation	Transport	Store	Alter	Inspect
Infomediary Business Model	DATA AGGREGATORS	DATA AGGREGATORS	IDENTITY PROOFERS	TRUST BROKERS

soon as a manipulation is detected. In contrast, the credit bureaus summarize *past* creditworthiness of individuals. In fact, using credit reports alone can give one a false sense of confidence—often many SSN mismatches are viewed as acceptable errors and are within the tolerance of the systems that accumulate records.

Infomediaries as Trust Brokers

To summarize, these infomediaries add value by the way they transform public records for trustors, transcending time and space limitations. This value is manifest in how the information can be used to overcome hurdles to trust, especially the perceptions of trustworthiness. The synthesis of this framework is presented in Table 1.

Other Business Models

Certainly, public records are not the only means to enhancing trust in the online world. Technology vendors are offering tools to safeguard the computer-mediated communication. Digital certificates and public-key encryption are often used to verify identities, for example.

Third parties, referred to as Internet business rating services (IBRS) aggregate consumer feedback and ratings of online retailers. Examples are Bizrate (www.bizrate.com) and Gomez (www. gomez.com). Similarly, eBay (www.ebay.com) is renowned for the assurance provided by customer ratings of the online auction's suppliers.

Another type of third party provides seals of approval by ensuring that recipients adhere to standards for privacy and security. For example,

"TRUSTe (www.truste.com) grants its seal to sites that adopt its standards for privacy and comply with its audits. Similarly, VeriSign (www.verisign. com) grants its seal of approval to sites that use its encryption and authentication services" (Urban, Sultan, & Qualls, 2000, p. 41).

A variation of that model in B2B commerce is the online industry trading exchanges. Violino (2002) likens this model to a gated residential community. By vetting members before they join and providing a secured marketplace behind a firewall, companies such as Trade-Ranger (www. trade-ranger.com) and Elemica (www.elimica. com) offer trustworthy exchanges.

FUTURE TRENDS

Future trends begin in the past. Poole (1999, p. 469) identified organizational challenges specific to businesses working in the virtual realm: their tight coupling of complex business systems, fluidity, limits on interactivity, stresses on individuals, and technology as a Trojan horse. Keen (1999, p. 19) agreed, noting that "business and its customers are being pushed to *having* to trust by complexity, interdependence, and telecommunications networks, which are the basis for new and ever-tighter business relationships." These challenges continue today and represent a significant managerial need for the future. New infomediary business models that simplify and enable tight *and* fluid relationships will emerge.

Existing infomediary models will continue to be in demand. The trend towards an increasing reliance on computer-mediated communications shows no sign of abating in either business or personal relationships. As we examine the existing forms of infomediaries in the context of trust hurdles, we see that they do little to enhance perceptions of risk and almost nothing to impact the trustor's propensity to trust. Considering online dating services as a bellwether also indicates that there is still plenty of room for enhancement within the trust model. Baig (2005) reported that, for online dating services:

... success is getting harder .... JupiterResearch says online-dating revenue hit $473 million in 2004, up from $396 million in 2003. But Jupiter expects revenue growth to slow by the end of the decade .... A survey released exclusively to USA Today by Keynote Systems found that customer satisfaction with dating sites is lower than for online venues as a whole. A key reason: 61% of those surveyed said they feared that online daters were misrepresenting themselves (for example, the "bachelor" who is not a bachelor).

This would suggest that there would be increasing expectations of services from trust brokers. Shapiro and Varian (1998) describe a layered approach to extend information-based services that they call "versioning," offering different versions for different market segments. They point to several dimensions for versioning, such as varying the information product by timing, convenience, quality, flexibility, capability, support, and user interface. It is not hard to imagine the extension of the informediaries' services along these dimensions.

Anthropomorphisizing the trust agent is one possible extension. An emerging model is described and evaluated by Urban et al. (2000) as a virtual advisor:

Such advisors may range from full personas (for example, Jill at CompUSA.com by Kana and Ida at etown.com by Ask Jeeves) to simple utility calculations that identify the most preferred

brand (for example, software from Frictionless Commerce or from PersonalLogic ...). If trust is a criterion and resources are limited, we feel that virtual advisors represent the best approach to helping consumers make the correct decision on the basis of full information, learning, and shopping help. (p. 44)

It is not hard to imagine the scenario in which computer-mediated communicators have a virtual alter ego, an online persona that represents them in a nonverbal, perhaps even animated way. This could be preferable to video teleconferencing and open new questions as to validation—does Michael Jones *really* look like that?

Hardware developments are making computing ever more pervasive. We use numerical sequences as key codes, special cards with magnetic strips, and biometric devices for authenticating physical access. Standardization of their use can extend their application. In another example, pet owners can have a radio frequency identification chip implanted in their pets; if Fido or Fluffy gets lost, the owner can use global positioning technology to locate the missing pet. The technology is available to do this for humans, but its application is limited by personal and public policy concerns.

CONCLUSION

Brown and Duguid (2000, p. 21) noted an "over reliance on information," and questioned whether it "can really be useful ... to redefine complex issues such as trust as 'simply information?'" We believe that the answer is "yes, but ..."

Organizational Effects

We have presented a model in terms of a series of hurdles that must be overcome in order for an individual or an agent of an organization to exhibit risk-taking behavior, that is, to trust. Information can be used to address the hurdles, particularly perceptions of trustworthiness (which we present as a function of ability, benevolence, and integrity. Information may also help communicators evaluate the risk inherent in their relationship.

The need for such information has led to the formation of new information providers, which we call "infomediaries." The essence of the value these infomediaries (i.e., data aggregators, trust brokers, and identity proofers) provide is the transformation of public records. By transporting, storing, altering, and inspecting the data that they gather, they are able to provide information that can inform risk-taking behavior in computer-mediated relationships.

From the standpoint of the infomediary organization, there are clear opportunities, or "space," available for their business development. How might public records be used to validate ability (a), for example? How can an organization aggressively use technology to transcend limitations of access and availability to provide accurate and timely information for perceptions of risk? What other creative uses of public data can add value to decision-makers?

Managerial Effects

From a decision-maker's vantage point, it is important to be intentional about the decision to trust. This is always the case, but is even more important in computer-mediated relationships where many of the cues managers typically rely on are either missing—or virtual. Contrast hiring decisions made 20 years ago, where personal introductions, interviews, and references were key with the current-day decision to contract with a team of programmers the manager has never met. You may be giving the team access to systems that are critical to the organization—can the team members be trusted?

Before engaging in risk-taking behavior, managers should be deliberate in their evaluation of the perceptions of trustworthiness and risk. In

this way, infomediaries and public records can be used to overcome these hurdles.

REFERENCES

Baig, E.C. (2005, February 13). Love growing strong on the Web. *USA Today.* Retrieved March 12, 2007, from http://www.usatoday.com/money/industries/technology/2005-02-13-online-dating-usat_x.htm

Brown, J.S., & Duguid, P. (2000). *The social life of information.* Cambridge, MA: Harvard Business School Press.

Gulati, R. (1995). Does familiarity breed trust? The implications of repeated ties for contractual choice in alliances. *Academy of Management Journal, 38*(1), 85-112.

Hardin, R. (2006). *Trust.* United Kingdom: Polity.

Keen, P.G.W. (1999) Transforming intellectual property into intellectual capital: Competing in the trust economy. In N. Imparato (Ed.), *Capital for our time: The economic, legal, and management challenges of intellectual capital.* Stanford, CA: Hoover Institution Press.

Kelly, K. (1998). *New rules for the new economy: 10 radical strategies for a connected world.* New York: Viking/Penguin Putnam.

Lui, S.S., & Ngo, H. (2004). The role of trust and contractual safeguards on cooperation in non-equity alliances. *Journal of Management, 30*(4), 471-485.

Mayer, R.C., Davis, J.H., & Schoorman, F.D. (1995). An integrative model of organizational trust. *Academy of Management Review, 20*(3), 709-734.

Meredith, J.R., & Shafer, S.M. (2007). *Operations management for MBAs* (3rd ed.). Hoboken, NJ: John Wiley & Sons.

Meyerson, D., Weick, K., & Kramer, R. (1996). Swift trust and temporary groups. In R. Kramer & R. Tyler (Eds.), *Trust in organizations: Frontiers of research and theory* (pp. 166-195). Thousand Oaks, CA: Sage.

Naisbitt, J. (1982). *Megatrends: Ten new directions transforming our lives.* New York: Time Warner.

Poole, M.S. (1999). Organizational challenges for the new forms. In G. DeSanctis & J. Fulk (Eds.), *Shaping organizational form: Communication, connection, and community.* Thousand Oaks, CA: Sage Publications.

Porter, M. (1985). *Competitive advantage.* New York: Free Press.

Shapiro, C., & Varian, H. (1998). *Information rules: A strategic guide to the network economy.* Cambridge, MA: Harvard Business School Press.

Sydow, J. (1998). Understanding the constitution of interorganizational trust. In C. Lane & R. Bachman (Eds.), *Trust within and between organizations.* United Kingdom: Oxford University Press.

Urban, G.L., Sultan, F., & Qualls, W.J. (2000). Placing trust at the center of your Internet strategy. *Sloan Management Review, 42*(1), 39-48.

Violino, B. (2002a, June 17). Building B2B trust. *Computerworld.* Retrieved March 12, 2007, from http://www.computerworld.com/management-topics/ebusiness/story/0,10801,71988,00.html

Violino, B. (2002b, June 17). Gated communities yield B2B trust. *Computerworld.* Retrieved March 12, 2007, from http://www.computer-world.com/managementtopics/ebusiness/story/0,10801,71929,00.html

Walther, J.B., & Bunz, U. (2005, December). The rules of virtual groups: Trust, liking, and performance in computer-mediated communication. *Journal of Communication, 55*(4), 828-846.

Chapter XV
Knowledge Management and Trust in E–Networks

G. Scott Erickson
Ithaca College, USA

Helen N. Rothberg
Marist College, USA

ABSTRACT

This chapter focuses on trust issues relating to knowledge management. Knowledge management is increasingly reliant on information systems to identify, collect, and disperse information and knowledge. Moreover, such systems are stretching across the borders of the firm to include collaborators and their knowledge assets in e-networks. This scenario has important implications for trust between the organization and individuals who contribute to and/or use knowledge management systems. Organization-to-organization trust issues are also apparent as valuable, proprietary information and knowledge are shared across the borders of firms. The authors hope that with an increased awareness of the trust issues implicit in the burgeoning field of knowledge management, executives and managers will be better prepared to employ some of our suggestions for dealing with this complex problem.

INTRODUCTION

Over the past decade, a school of thought has developed that the only enduring competitive advantage for an organization is found in its people, especially what its people know. Further, some of us believe that firms better at acquiring and then managing knowledge resources will be the winners as we increasingly move into a knowledge-based economy. To manage intellectual capital effectively, however, organizations need to realize the trust issues implicit in knowledge sharing and knowledge protection. These issues involve fairly clear organization-individual trust dynamics, which we'll discuss, and some less obvious organization-organization matters also deserving of further analysis.

BACKGROUND: DEFINITIONS

Knowledge management (KM) has become such a buzzword that the general business world sometimes doesn't understand the true nature of the concept or its power. Once a phrase winds up in Dilbert, it can be hard to take it seriously. But knowledge is increasingly seen by strategists as not only a unique, sustainable source of competitive advantage but perhaps as the only source that lasts. From Drucker's (1991) knowledge workers to the resource-based theory of the firm (Dierickx & Cool, 1989), a group of influential strategists have increasingly moved toward the idea that the only competitive advantage that cannot be eventually copied is the one that is continuously reinvented through the skills and knowledge of an organization's people (Zack, 1999a; Grant, 1996). Indeed, most identifiable core competencies in firms that lead to competitive advantage can be attributed to individuals in the firm having unique knowledge or insight into their jobs, be they operations, management, marketing, finance, R&D, or elsewhere.

Knowledge in this context is usually differentiated from basic data or information. Knowledge, some of which is referred to as know-how, implies that some reflection and/or learning has taken place based on data, information, or experience. The basic idea is that individuals develop knowledge about their duties and the context of those duties. Thus, they perform better, yielding better results for the organization. Li & Fung Chair Victor Fung, for example, has noted that value is found not in his firm's list of sourcing and manufacturing partners or its customer lists (information), but rather in the personal knowledge he holds concerning strengths, weaknesses, and preferences of all the members of Li and Fung's widespread virtual network (Magretta, 1998). The information could be stolen tomorrow without any advantage to the thief. The personal knowledge of relationships behind the information is the key to Li and Fung's competitive advantage.

Several other definitions are probably appropriate. Intellectual capital (IC) is a companion field to knowledge management. IC generally grew out of attempts to measure knowledge-related, intangible assets. KM refers more to attempts to better manage these assets. Since both deal with identifying and managing knowledge assets, the terms are interchangeable for our purposes. Another related term is intellectual property (IP). The idea of intellectual capital grew out of an interest to manage knowledge assets not structured enough or innovative enough to rise to the level required to protect them with intellectual property mechanisms. Well-defined knowledge assets can be identified, described, and protected with IP mechanisms such as patents, copyrights, and trademarks. The challenge of IC and KM is to take less well-defined knowledge assets and identify, describe, and protect them as well. Within a pharmaceutical firm, for example, data and information can come from experiments and clinical trials, from production and quality control, from marketing and sales, and elsewhere. Intellectual property would include the patents filed on new drug discoveries. And knowledge, or intellectual capital, would be found in how to optimize production processes, how to conduct more effective research, sales expertise regarding a particular product or particular customer, expertise in dealing with regulators such as the FDA, insight into competitor strategies and tactics, and similar circumstances.

To summarize, KM refers to attempts to better manage intangible knowledge assets. These knowledge assets include both intellectual property and less well-defined knowledge (or intellectual capital) such as know-how. Most of the discussion centers on non-IP assets as IP has previously received a lot of attention and its management is better understood. The less well-defined knowledge (non-IP) assets are the newer topic, have less existing scholarship concerning their management, and are not as well understood.

Table 1. Definitions of terms

Data	"Observations or facts out of context" (Zack, 1999b, p. 46).
Information	"Data within some meaningful context" (Zack, 1999b, p. 46).
Knowledge	"That which we come to believe and value on the basis of the meaningfully organized accumulation of information (messages) through experience, communication, or inference" (Zack, 1999b, p. 46). Also sometimes termed know-how, knowledge is learning that takes place leading to individual expertise (Zander & Kogut, 1995).
Knowledge assets	Valuable, intangible assets of the firm. Personal knowledge, corporate culture, intellectual property or any other valuable organizational knowledge.
Intellectual property	Formalized knowledge assets, qualifying for a patent, copyright, trademark or other institutionalized protection mechanism.
Intellectual capital (IC)	Knowledge assets of the firm. The field of intellectual capital focuses on the identification, measurement, and management of these intangible assets. IC includes both IP and less formalized knowledge (Edvinsson & Malone, 1997).
Knowledge management	The practice of managing knowledge assets, focused on identification, capture, organization, sharing, and analysis. Closely related to IC, the differences are more in emphasis on measurement (IC) and management (KM).
Tacit knowledge	Knowledge assets that are personalized and hard (perhaps impossible) to communicate (Nonaka & Takeuchi, 1995; Polanyi, 1967).
Explicit knowledge	Knowledge assets that are captured and codified by the organization, more easily communicated, perhaps stored in a formalized manner in an IT system or elsewhere (Choi & Lee, 2003).

Thus, although KM and IC include both categories of knowledge assets, IP receives considerably less attention in the literature. The basic themes can be further extended to include the management of information and data, pre-knowledge that can become valuable knowledge assets with appropriate use and/or analysis.

BACKGROUND: OTHER CONSIDERATIONS

In order to better manage knowledge assets, firms seek to identify knowledge, capture it as an organizational (rather than individual) asset, leverage it through sharing, and expand it through learning, analysis, and acquisition. Consequently, the field has antecedents in areas as disparate as organizational learning, accounting and performance measurement, strategy, and innovation theory.

One of the critical early insights concerning knowledge was that it is of two types: tacit and explicit (Nonaka & Takeuchi, 1995; Polanyi, 1967). Explicit knowledge is structured and explainable, easy to communicate to others and/or store in information systems (Choi & Lee, 2003). Tacit knowledge is more nebulous, less easy to define, less easy to explain or communicate. Indeed, tacit knowledge may be something that individuals don't even realize they know or that they may find impossible to explain or pass on to someone else. Although there are different prescriptions for knowledge management depending on the type of knowledge (a tacit to tacit exchange, for example, can be carried out through an apprenticeship program, storytelling, or some other personal relationship), managing knowledge is a lot easier if the knowledge assets are explicit or can be made explicit. Once codified as explicit knowledge, knowledge assets can be more easily stored in information technology (IT) systems, analyzed for further insights, and/or shared throughout the organization. As a result, many of the most visible KM installations during its consulting pinnacle in the late 1990s involved IT systems. Often disappointing because they didn't include the human (and tacit) element as well, such systems do remain an important part of most KM systems.

Indeed, once organizations have a handle on the knowledge at their disposal, managing it is a much easier task. With IT systems, individuals with a problem can easily search for in-house expertise. This may be cataloged within the system. Alternatively, the system may be able to recommend someone else within the organization that has the necessary knowledge, encouraging the inquiring party to contact him or her. Stephen Denning likes to tell a story from his World Bank experience, for example, where the organization faced a problem with a road project in Pakistan. The details of the situation were posted on the internal knowledge system, and the project leaders soon had specific suggestions on appropriate mixes and conditions for application from other parts of the world with similar temperatures, humidity, and so forth (Brown, Denning, Groh, & Prusak, 2004). Once captured, knowledge assets can be analyzed for further insights (business intelligence and data mining are related aspects of this process). Individuals or the organization as a whole can then also identify knowledge deficiencies, heading outside the organization to acquire new knowledge to fill in the gaps, as necessary. So not all knowledge can be codified or managed through a computer system but those looking to best manage their existing or potential knowledge benefit from employing IT as a KM tool.

One other important aspect of KM's conceptualization is that valuable knowledge exists throughout the organization, in all workers. Intellectual capital theory suggests that knowledge assets can be grouped into four categories (Rothberg & Erickson, 2002; Bontis, 1999; Edvinsson &

Malone, 1997). Human capital refers to know-how related to doing a particular job. As managers or employers gain education, training, and/or experience, they learn how to better perform their work, building their human capital. Structural capital has to do with knowledge embedded in the organization, whether in its structure, in its culture, or some other aspect that maintains the knowledge. Southwest Airline's unique culture, for example, illustrates an example of structural capital routinely passed on to individual employees. Relational or collaborative capital is knowledge having to do with those outside the firm. Knowledge about a specific customer's likes and dislikes, about dealing with a regulator, about a supplier, or about a research partner all qualify as relational capital. Finally, competitive capital is knowledge about a specific competitor. Whether a salesperson clued in to competitor offerings, a scientist aware of competitive R&D projects, or a finance staffer who has studied a competitor's SEC filings, all have potentially useful knowledge about a competitor's strategies and activities. The key point is that any employee in the organization, from the CEO to the overnight cleanup crew may have valuable knowledge in any of these areas that can help a company compete.

The difficulty in managing knowledge assets is in the process stages. An effective system identifies existing knowledge (the Balanced Scorecard (Kaplan & Norton, 1992), for example, is related to efforts to identify and manage such assets), seeks to capture it for the organization (informally, through interpersonal connections, or, increasingly, through formal mechanisms including installed KM software, Wiki-type systems, and so forth (Schulz & Jobe, 2001)), subjects it to analysis for further insights and identification of gaps, leverages it through sharing (Boisot, 1995), and further builds it through dedicated search and acquisition (Kogut & Zander, 1992). Each step can be problematic as it includes not only management challenges but personal elements. Individuals must reveal that they hold knowledge, they must agree to contribute it to the system, they must be willing to help with analysis (as appropriate), they should access the system when knowledge may be helpful, and they should help by adding to the system to fill in gaps. This is where the issue of trust enters in, as the personal dynamics of contributing to, utilizing, and protecting the knowledge assets are critical to the system's success.

Finally, note that knowledge is increasingly not just an organizational issue but a network issue. Knowledge assets are available not just to a single firm but to all of the collaborators in its network. Further, with the interconnections provided by the Internet, not only is data flowing between members of the extended enterprise in increasing quantities but knowledge is, as well, since knowledge management systems are often open to network members. And remember that today's information can become tomorrow's knowledge, so valuable information and knowledge are increasingly passing beyond the control of the core firm into the hands of its network partners.

Overall, firms are increasingly aware that better management of knowledge assets can yield unique, sustainable competitive advantage in today's economy. As a result, with the help of IT systems, organizations are installing tools to help all types of employees to build and leverage their intellectual capital, both tacit and explicit. Further, they are extending these systems to encompass their entire e-networks, providing collaborator access not only to day-to-day data but more also more developed knowledge assets.

ORGANIZATION-INDIVIDUAL TRUST

For KM systems to work, organizations have to identify and, ideally, take ownership of the knowledge of their individual employees. As noted earlier, this can be done with tacit knowledge through face-to-face interactions or tools such as storytelling or communities of practice. But many organizations also want to identify

and exploit their explicit knowledge and this practice is most efficiently accomplished by encouraging employees to contribute to and use an IT system. Most major consulting firms have standard packages for managing knowledge, but the basics include storage, search, and perhaps ratings/rankings of the most popular or most useful pieces of knowledge (Matson, Patiath, & Shavers, 2003). Employees feed knowledge into the system (troubleshooting tips, case histories, customer likes/dislikes, supplier idiosyncrasies, competitor activities, etc.) which stores it, yielding it (or the contact information for even more in-depth knowledge) when requested.

The surrender of knowledge is the first issue related to trust in KM systems. Individual employees, whether line, staff, or management, are giving up something of professional or personal value. Once the organization has identified and codified the knowledge, it essentially becomes part of the property of the firm, so the employee is unequivocally surrendering something. But the surrender goes beyond property rights; it often involves power as well.

Employees who have the know-how to do a job particularly well possess personal power. The salesperson who knows how to analyze a potential client's business better, the engineer who knows how to build cost-effective prototypes, or the purchasing agent who knows how to get a quick response from a supplier all have ways to perform their job better than similar employees in the firm. By giving up such knowledge, individual employees surrender their unique insights that lead to superior performance. If they keep them secret, they possess power because the firm cannot move them or replace them without losing that knowledge and that superior performance. Indeed, some authors have suggested that the importance of knowledge signals a major shift in power from the organization to the individual employee (Belasco & Sayer, 1994).

At a Michigan plant producing industrial pumps, for example, establishing procedures to explain best practices on a variety of machines proved to be extremely difficult because of varying perceptions of power (Aeppel, 2002). One machine operator "known for the accuracy of his cuts" and able to do machine setups "hours faster than anyone else" refuses to contribute his best knowledge. He sees keeping his secrets to himself as the best way to preserve his job, his preferred machine assignment, and his preferred pace of work, among other things. Having sole access to the knowledge, keeping it tacit, gives him a sense of power.

Individuals also surrender time and effort when contributing to a KM system. As with many initiatives, the startup phase can make demands on participants that are not paid back until sometime far in the future. In the beginning, a KM system will have very little available knowledge for those looking for insights. All the initial use of the system is found in inputting, with very little immediate payback other than pride of contribution.

So individuals give up power, time, and effort in building a KM system. Management has to consider what the return might be. One could argue that the knowledge available to employees and managers on the system will benefit them, but the nature of KM is such that whether a specific individual ever has need of the system is hit or miss. Simply at its face value, potential contributors may not be convinced a KM contribution is worthwhile.

This is where the issue of trust starts to build. Individuals contributing to a KM system have to trust that the organization will reward them for participating or, at the very least, not punish them. Proper incentives have to be installed for becoming an active member of the knowledge community. The authors know one manager at a major pharmaceutical firm, for example, who believes in the KM system and wants to contribute to it. But when time gets tight, proper FDA documentation is always a more important concern, so when time allows only certain work, the required regulatory work gets done (and is rewarded) while the optional

KM work is ignored (with no one really noticing). Not that the FDA should be ignored, but if some activities have proper incentives and some don't, we all know which will be completed. Individuals looking to participate in KM systems must trust the organization to have the proper incentives in place, be they pay, recognition, status, or some other reward.

And it's especially important that the incentives not be counterproductive. In the industrial pump example discussed earlier, employees are worried about being shifted to another job once they have revealed the best way to perform their current duties. They are worried about being replaced by scabs if they strike. They are worried about the organization shipping their jobs and knowledge overseas to a lower-cost location. Essentially, once you have created a detailed blueprint of how to do your job, you have to trust your management team not to use that knowledge base to make you redundant, whether you are a line worker, a salesperson, a scientist, or an office manager. Similarly, individuals using the system have to believe that performance expectations won't be ratcheted up unreasonably if they choose to avail themselves of the KM system's knowledge stores.

Much of this is true of any KM system, not just IT-driven ones. But a computerized structure can worsen the problem. It's one thing to share tacit knowledge with another individual, face-to-face. Trust can be established, and helping a single co-worker or two to improve performance may not be a big issue. Once one is potentially sharing the knowledge throughout the organization, perhaps with thousands of people, including those in low-wage overseas locations (many KM systems stretch across firm boundaries to include collaborators), the situation obviously changes dramatically. An individual could be sending a blueprint for outstanding performance to a potential replacement in Bangalore.

On a different topic, the organization also has to have some trust in its individuals. We noted earlier that knowledge assets can be the key to unique, sustainable competitive advantage. If so, firms will generally want to keep the knowledge to themselves (exceptions exist when they have goals such as setting an industry standard). Just as intellectual property is protected through patents, trademarks, copyrights and such, non-IP knowledge assets are usually protected in some way by firms. Trade secret procedures can sometimes be employed, either formally or not, along with legal instruments such as non-compete and nondisclosure agreements. But these can be hard to rely on (non-compete agreements, for example, don't hold up in California courts).

Consequently, an organization with critical proprietary knowledge will establish procedures to protect that knowledge, both technological (passwords and firewalls, encryption, secured connections) and behavioral (what knowledge leaves the company premises, what is allowed to be on personal laptops, what can be talked about with the general public, etc.). Organizations trust individuals to know protection guidelines and to adhere to them in both letter and spirit. Generally, when a major breach in knowledge security occurs, it's often because of carelessness or someone not following procedures. The HP pretexting scandal, for example, illustrated a common competitive intelligence technique, but it's often not necessary to misrepresent oneself to gain knowledge. Simply calling up and asking will often do the trick. Similarly, sitting in on public presentations or keeping one's ears open on particular flights or in particular watering holes are well-known mechanisms for legally and ethically collecting competitive knowledge. Essentially, individuals are often sloppy in keeping proprietary knowledge under wraps. Better awareness, better training, and, again, an appropriate motivation and reward system are key to keeping everyone on board with knowledge protection.

Regarding the organizational-individual relationship, KM raises interesting issues related to

trust. Individuals surrender their personal tacit knowledge to the organization, trusting they will be rewarded in some manner. Effective organizations will fulfill that trust in the manner in which they structure their knowledge system and their reward scheme. Alternatively, organizations leverage their knowledge assets through redistribution, allowing access to individuals. As a result, the organization trusts individuals to be careful with that knowledge and not reveal it in places where competitors may notice. As noted throughout the discussion, computerization has elevated the trust issues in both circumstances and complicated the jobs of organizations and individuals in fulfilling their responsibilities.

ORGANIZATION-ORGANIZATION TRUST

With the advent of the Internet and its impact on communication links between organizations, new and perhaps even more interesting aspects of trust have arisen in organization to organization relationships. As might be expected based on the preceding discussion, these relate to safekeeping of knowledge, especially that belonging to your partners and not to you.

For many large businesses, competition is no longer just company vs. company but network vs. network. A firm and its entire value chain of partners compete with other firms and their entire value chains of partners. Sharing knowledge throughout the e-network can help everyone do their jobs better. Further, KM systems work best when inputs come from more directions, with more diverse points of view. Consequently, a number of organizations have chosen to extend their KM systems to partners outside the core firm. It's uncommon to include all collaborators, but those closest to the core firm are frequently full participants in the most advanced systems.

This step is actually pretty easy to accomplish since any number of large networks are already linked together by information systems running the massive enterprise resource planning, supply chain management, and customer relationship management systems that coordinate contemporary business. To these data flow mechanisms, the addition of knowledge flows is a minor matter. As a result, the plugged-in firm today constantly shares information and knowledge with a range of network partners. Procter & Gamble (P&G) and Wal-Mart, for example, directly share all sorts of information and knowledge, including sales figures, promotional plans, production plans, and so forth. Indeed, when the former let the trademark rights to White Cloud toilet paper lapse in the 1990s, the latter was right there to pick them up and launch the product as a store brand, armed with scads of P&G consumer information and knowledge to back up its efforts (Ellison, Simmerman, & Forelle, 2005).

As with individuals, these relationships create substantive trust issues. We have already talked at length about the potential importance of proprietary knowledge (and, in this instance, we can include proprietary information as well) and so won't belabor the point. But again realize that this can be information and knowledge that are critical to the success of the core firm and that, in many cases, it would prefer to keep the knowledge away from competitors. As we also discussed in the previous section, the core firm can try to do this by employing security systems to protect the knowledge. But we are now talking about the knowledge and information moving beyond the boundaries of the organization and into the hands of network partners and their employees. This has always been the case, of course, but the Web-based IT systems that are now employed in these activities have driven up the number of bytes exchanged between e-network partners exponentially. Tremendous amounts of knowledge and information are now passed routinely to collaborators.

All of this takes place in a contemporary environment that also contains substantial competitive

intelligence (CI) activity. CI has been growing as an organized practice for the past couple of decades. The legitimate practice of CI does not include the industrial espionage practices often associated with it but practitioners have been known to go right up to, if not past, that line. Our discussion includes only legal and ethical activities. The more questionable sort, of course, only add to the threat of incursion by competitors.

Competitive intelligence operations seek to gather knowledge about competitors, looking to gain insights into competitive strategies, tactics, and activities (Rothberg & Erickson, 2005; Fleisher & Bensoussan, 2002). In terms of our previous discussion, CI practice builds competitive capital, a form of IC. Although the tools and techniques are varied and continuously growing, they can be grouped into a few categories. Initially, a good amount of CI is simply reviewing publicly available information, essentially a library function. Published articles, regulatory filings, patent filings, publicity concerning executive speeches, Web logs, and other such sources can provide a wealth of insight to those knowing how to use them. Internal sources, the employees of the firm, can also be valuable to a CI operation. Managers and employees may have worked for a competitor; salespeople, scientists, purchasing personnel, and others may have information on what competitors are doing; and other staff may simply have picked up something somewhere (overheard conversations, information from a friend of a friend, etc.) that could be useful to the core firm. Indeed, there is a truism in the field that everyone in the organization should be part of the competitive intelligence operation. Similarly, external sources can provide the same sorts of knowledge as network partners or other external organizations contribute what they know. Finally, active gathering techniques such as observation of competitive facilities, requests for information, trolling at trade shows, and such can also be effective. Put them all together, with skilled analysts, and a CI operation can predict competitive initiatives and long-term strategies. CI can uncover the valuable proprietary knowledge we've been discussing this entire chapter.

Further, CI looks for the weak spots in a knowledge system. While a core firm may have effective protection structure in place, the multiple hands of collaborators into which information and knowledge passes may not be quite so secure. When Oracle was looking for information on Microsoft, it targeted a public relations nonprofit that worked with the Redmond company (Simpson & Bridis, 2000). When a generic toner cartridge maker sought information on a new Hewlett Packard (HP) technique to slow copying of its cartridges, it talked with some HP suppliers (Tam, 2002). E-network participants are quite likely to be targeted by CI efforts, particularly when the information or knowledge is important and the core firm's protection scheme is strong.

And the computerization trend also contributes to this situation. When critical knowledge (or raw data) is digitized, it is, of course, easier to share with collaborators along the e-network. But digitized knowledge is also more vulnerable. It does not need to be physically removed to be useful. Digitized knowledge can be removed with no trace of the acquisition and it can be easily moved en mass remotely via the Web or on site with portable storage devices. Further, KM systems provide access to virtually the full knowledge assets of an organization to anyone who taps into them, a potentially huge number of members of the extended e-network. Hence, there are many more attack points and the volume of knowledge to be gained from incursion is not limited just to the targeted individual. It is the full knowledge base of the e-network.

Given this admittedly lengthy buildup, the point is quite simple. When knowledge is shared among organizations in this manner, in this CI-filled environment, trust becomes an issue of paramount importance. How can organizations share their knowledge and information with partners, as must often happen to be competitive, if they don't

trust them to protect the assets in an appropriate manner? To establish trust, e-network partners need to show they have established effective protection procedures. These include proper IT security measures, proper manager and employee training procedures, and proper incentives to encourage individuals to follow the procedures (those we noted in the previous section).

One approach is certification. Just as organizations certify partners for quality or for environmental standards, so mechanisms can be developed to better secure knowledge protection. If you give a partner your critical knowledge, you need to have trust in them to keep it safe. They need a plan, a structure, and execution to guarantee knowledge security. Both computer security and employee procedures need to be addressed. Another alternative is levels of access. Organizations have been known to separate knowledge assets into those stored in IT systems and freely shared, those only passed along in hard copy with appropriate "confidential" designations, and those shared only orally, with not visible record at all. The more sensitive the knowledge, the lower the level of documentation, access, and sharing. Depending on the type of knowledge and the nature of the collaborator, sharing would only take place in a given manner.

FUTURE TRENDS

All the factors raising trust issues vis a vis knowledge management appear likely to accelerate in coming years. As we noted at the beginning of this discussion, there is more and more agreement that sustainable competitive advantage tends to come from some unique competency of the firm—and that old standbys such as quality, service, technology, and efficiency are things that can be copied unless unique, specific knowledge accompanies them. And organizations can't stand still in this regard. Those that learn most quickly and most effectively will be those best able to stay one step ahead of competitors regardless of the nature of their current advantage.

In terms of the organization-individual relationship, the trends in employee longevity (shorter) will undoubtedly create ever greater complications in terms of who owns what knowledge. Even as organizations try to harvest employee knowledge, individuals will undoubtedly question what parts of that knowledge are the firm's and what part are the employee's. And as employees move on to other firms, what knowledge they are entitled to take with them will become a bigger issue. With computers holding ever bigger pieces of the organization's knowledge assets, employees will have access not only to their own knowledge but that of others. If they choose to review and remove knowledge assets upon departure, they can do so easily if proper safeguards are not in place.

Employee mobility and computerization, combined with other trends such as outsourcing, are likely to raise the levels of distrust between organization and employees unless the relationship is well-managed. Firms need to give careful thought to how they will structure their KM systems in terms of contributions, access, motivation and reward systems, and security processes. The stakes will only get higher.

Similarly, the Web-based economy will continue increasing ties between organizations, including across borders. Firms will find tightening relationships with an ever increasing network of collaborators, sharing information and knowledge across computer systems with other entities that may be two, three, four, or more links removed (i.e., sharing with a supplier of a supplier of a supplier). It is quite possible for an organization to routinely share its knowledge crown jewels with an entity with which it is not familiar. The trends are such that, again, attention to appropriate KM structures, particularly in relation to security, are critical. One could imagine ISO standards for knowledge

security or proprietary certification systems (such as those used by Toyota for collaborator quality standards) coming into vogue. Once again, the issues look to get only more complicated and more important in coming decades.

CONCLUSION

This chapter has dealt with the basics of knowledge management and the implications for trust in both organization-individual and organization-organization relationships. Because effective knowledge management depends on sharing, trust is a preeminent issue in the field. In spite of this, little has been written in this area.

On the organization-individual level, KM systems ask individual employees to surrender their personal knowledge to an organizational structure looking to codify it, analyze it, and share it with others, usually through a computerized network. Employees must give up power and time to do so. Thus, they trust the organization to properly motivate and reward them, and then to use the knowledge appropriately. Alternatively, the organization that shares its knowledge back out among individual employees trusts them to follow proper procedures to safeguard these important assets.

On the organization-organization level, e-networks are also increasingly sharing information and knowledge with one another. In these situations, each entity trusts its collaborators to use the knowledge effectively and to protect it as if it were their own proprietary asset. In an age when competitive advantage increasingly comes from what you know that your competitor doesn't, the ability of the network to develop, share, and protect knowledge is critical. And it is highly likely that it will become more so in the future.

REFERENCES

Aeppel, T. (2003, July 1). On factory floors, top workers hide secrets to success. *The Wall Street Journal*, pp. A1, A10.

Belasco, J.A., & Sayer, R.C. (1994, March-April). Why empowerment doesn't empower: The bankruptcy of current paradigms. *Business Horizons, 37*(2), 29-41.

Boisot, M. (1995). Is your firm a creative destroyer? Competitive learning and knowledge flows in the technological strategies of firms. *Research Policy, 24*, 489-506.

Bontis, N. (1999). Managing organizational knowledge by diagnosing intellectual capital: Framing and advancing the state of the field. *International Journal of Technology Management, 18*(5-8), 433-462.

Brown, J.S., Denning S., Groh, K., & Prusak, L. (2004). *Storytelling in organizations*. Woburn, MA: Butterworth-Heinemann.

Choi, B., & Lee, H. (2003). An empirical investigation of KM styles and their effect on corporate performance. *Information & Management, 40*, 403-417.

Dierickx, I., & Cool, K. (1989). Asset stock accumulation and the sustainability of competitive advantage. *Management Science, 35*, 1504-1513.

Drucker, P.F. (1991, November-December). The new productivity challenge. *Harvard Business Review, 69*, 69-76.

Edvinsson, L., & Malone, M.S. (1997). *Intellectual capital: Realizing your company's true value by finding its hidden brainpower*. New York: Harper Business.

Ellison, S., Simmerman, A., & Forelle, C. (2005, January 31). P&G's Gillette edge: The playbook it honed at Wal-Mart. *The Wall Street Journal*, pp. A1, A18.

Fleisher, C., & Bensoussan, B. (2002). *Strategic and competitive analysis: Methods and techniques for analyzing business competition*. Upper Saddle River, NJ: Prentice Hall.

Grant, R.M. (1996, Winter). Toward a knowledge-based theory of the firm. *Strategic Management Journal, 17*, 109-122.

Kaplan R.S., & Norton, D.P. (1992, January-February). The balanced scorecard: Measures that drive performance. *Harvard Business Review, 70*(1), 71-79.

Kogut, B., & Zander, U. (1992, August). Knowledge of the firm, combinative capabilities, and the replication of technology. *Organization Science, 3*(3), 383-397.

Magretta, J. (1998, September-October). Fast, global, and entrepreneurial: Supply chain management Hong Kong style. *Harvard Business Review, 76*(5), 102-114.

Matson, E., Patiath, P., & Shavers, T. (2003). Strengthening your organization's internal knowledge market. *Organizational Dynamics, 32*(3), 275-285.

Nonaka, I., & Takeuchi, H. (1995). *The knowledge-creating company*. New York: Oxford University Press.

Polanyi, M. (1967). *The tacit dimension*. New York: Anchor Day Books.

Rothberg, H.N., & Erickson, G.S. (2002). Competitive capital: A fourth pillar of intellectual capital? In N. Bontis (Ed.), *World congress on intellectual capital readings* (pp. 94-103). Woburn, MA: Butterworth-Heinemann.

Rothberg, H.N., & Erickson, G.S. (2005). *From knowledge to intelligence: Creating competitive advantage in the next economy.* Woburn, MA: Elsevier Butterworth-Heinemann.

Schulz, M., & Jobe, L.A. (2001). Codification and tacitness as knowledge management strategies: An empirical exploration. *Journal of High Technology Management Research, 12*, 139-165.

Simpson, G.R., & Bridis, T. (2000, June 19). Oracle hired firm to probe Microsoft allies. *The Wall Street Journal*, p. A48.

Tam, P-W. (2002, September 25). High-technology giant duels with nimble knock-off artists. *The Wall Street Journal*, pp. A1, A10.

Zack, M.A. (1999a, Spring). Developing a knowledge strategy. *California Management Review, 41*(3), 125-145.

Zack, M.A. (1999b, Summer). Managing codified knowledge. *Sloan Management Review, 40*(4), 45-58.

Zander, U., & Kogut, B. (1995, January-February). Knowledge and the speed of transfer and imitation of organizational capabilities: An empirical test. *Organization Science, 6*(1), 76-92.

Chapter XVI
When Trust Does Not Matter:
The Study of Communication Practices Between High–Tech Companies and Their Clients in the Environment of Distrust

Dominika Latusek
Leon Kozminski Academy of Entrepreneurship and Management, Poland

ABSTRACT

The chapter focuses on the dynamics of trust and distrust by presenting a qualitative field study of interorganizational collaboration between customers and providers in the Polish IT industry that illustrates practices of communication between parties engaged in collaboration within IT projects. The chapter is intended to merge two perspectives: the academic viewpoint on the theorizing of trust and distrust and the practitioners' reflections on the reality of relationships in business. The author hopes that the study may further our understanding of the process of cooperation in project work, provide an interesting insight into the role of trust in cooperation, and offer a reflective account of actual practice of cooperation in a distrustful environment.

INTRODUCTION

Trust, seen as "indispensable in social relationships" (Lewis & Weigert, 1985, 1968) and "vital for the maintenance of cooperation in society and necessary as grounds for even the most routine, everyday interactions" (Zucker, 1968, p. 56), is also claimed to be of importance for the cooperation between organizations (e.g., Sako, 1992, 1998). This is especially in the case of buyer-supplier relations that "trust plays an important part because the threat of opportunism and vulnerability is acute even when there is a mutual interest in continuing the relationship" (Möllering, 2006, p. 156).

In the following chapter, the author presents excerpts of a qualitative field study of interorganizational collaboration between customers and providers in the Polish IT industry. Although it is based on an empirical study that was conducted in Poland and therefore is confined by time and space, it offers valuable insights into the interplay of communication and trust that may broaden our understanding of both phenomena. Poland is still a reality of transition, although the very moment of transformation already belongs to the distant past; the feeling of change still pervades almost every aspect of social life – be it politics, economy, or culture (Koźmiński & Sztompka, 2004; Staniszkis, 2001; Sztompka, 1996). As Sztompka (1996, 1999), one of the most prominent sociologists engaged in research on trust in the transition societies, points out:

The vicissitudes and fluctuations of trust and distrust during the last fifty years of Polish history, as well as the condition of trust in the present turbulent period of post communist transformations, have proven to be an excellent 'strategic research site' (Merton, 173, p. 373), a kind of useful laboratory for applying and testing viability of theoretical concepts and models. (Sztompka, 1999, p. xi)

The project was originally intended to be the study of the process of creation and maintenance of trust in the relations between organizations in a knowledge-intensive sector. It was conducted in three software providers based in Poland, as well as their clients from both the private and the public sector. Here, the analysis puts emphasis on the communication between organizations engaged in projects. The main idea of this chapter originated from a reflection that occurred to me after an initial reading of the field data. Yet, the

language of all the parties did not appear as a language of cooperation built on the foundation of trust; indeed, what I encountered was not as much about trust itself as it was about the converse: it was the discourse of distrust. Considering this, the more general question emerged: how cooperation, in such a deep climate of distrust, can take place at all? The existing literature on distrust, however limited, may support the conjecture that cooperation without trust is also possible, and, even more, it may be valuable (Hardin, 2004; Cook, Hardin & Levi, 2005).

In this light, the goal of the chapter is threefold. First, the interest of the chapter lies in the actual communication practices between the parties as they are employed throughout the projects. What do the everyday encounters of customers and providers look like and what are the specific tools intended to facilitate this cooperation? As we are in a high-tech business, how are the e-tools involved in this communication? How do the communication practices build up the cooperation process; are they relatednd if so, then howo trust?

Second, it is intended to investigate the process of interorganizational cooperation in IT projects from the perspective of trust. It focuses on the perspective of two major parties involved: the providers and the customers. Through a qualitative, strictly local study, I hope to capture the dynamics of collaboration, and identify the building blocks of trust, or, alternatively, other elements that constitute the "glue" of cooperation.

Finally, taking into account both the perspective of the communication and trust, this chapter aims at exploring the issue of external environment in which particular cases of cooperation take place. In the specific Polish context the following question comes up: how does the environment with a low level of trust, or rather pervaded with distrust, influence the communication practices and, consequently, cooperation? The project takes on the perspective of practitioners, representatives of both parties engaged in cooperation. This standpoint is particularly important in the qualitative research projects, as Möllering (2006) writes:

In dyadic relationships, both sides of the dyad should be interviewed (Huemer, 1998) not so much as a means of confirmation or triangulation (Altheide & Johnson, 1994) but rather as a means of taking in multiple perspectives that allow reflection on the idiosyncrasy of trust experiences. (p. 153)

The research presented here is qualitative and ethnographical. It is also performative, not ostensive (Latour, 1986); therefore, what is to be presented are the viewpoints of the actors, without attempting to fit their stories into the preconceived theoretical framework. Quite certainly, these opinions are under the influence of group biases (Gill, 2003). They should not be perceived as a flaw, though. The way organizational actors describe their social worlds gives the picture of how they enact their roles and re-establish the reality they believe in (Berger & Luckman, 1966). Thus, through the presentation and analysis of their stories, it is possible to obtain a better understanding of the phenomena under study. The conclusions provided in the chapter are grounded in field data, but, due to the obvious spatial limitations, I was not able to provide enough stories from the field to make the picture complete. In this regard I totally share the observation made by Möllering (2006) that it is a challenge to fit qualitative research into the journal or article format, as it requires much more space to provide an exhaustive account of the field study. The interpretation I provide in the conclusive part of the chapter is just one of the possible readings of the data. But, as I am also talking openly about the limitations of the study (Kostera, 2003; Silvermann, 2001), I hope the reader will see the chapter as an inspiring piece that broadens the understanding of the multifaceted nature of trust and distrust.

The chapter begins with a clarification of the notions of trust and distrust. I adopt the rather narrow understanding of trust, with the concept of "suspension" (leap of faith) at its heart (Möllering, 2006). A brief discussion of cooperation in the conditions of trust and distrust is also included here. In the main part, I will provide an account of the empirical research project; subsequently follows the practitioner's perspective on the cooperation in buyer-supplier relationships; then, the process of communication, especially the role of computer-mediated communication, is presented. The chapter concludes with confronting the academic viewpoint on theorizing about trust and communication with the image emerging from the field work.

NOTIONS OF TRUST AND DISTRUST

According to Lane and Bachmann (1998), there are two main levels in analyzing trust: micro and macro. Within the macro perspective, we focus on interpersonal and interorganizational relationships; while on the macro level we consider the "impersonal" dimension of trust, that is, its social functioning, roles, antecedents and consequences, and so forth. This chapter concentrates on the micro level, that is, on buyer-seller organizational relations.

As far as the definition of trust is concerned, the statement made 11 years ago by Hosmer (1995) that "there appears to be widespread agreement on the importance of trust in human conduct, but unfortunately there also appears to be equally widespread lack of agreement on a suitable definition of the concept" (p. 380), still holds true today, although significant effort has been undertaken by academics to integrate various conceptualizations. Generally, these endeavors fall into two broad categories (Huemer, 1998): the calculative view of trust that is the feature of the research within the organizational economics and the social and affective view of trust characteristic of the organizational sociology. Despite being set in two distinct traditions, these conceptions share a common core consisting of three convictions (Lane & Bachmann, 1998). First of all, there is an interdependence between the trustor and the trustee: the trustor may be harmed by the trustee but he cannot estimate whether this might happen or not, therefore he remains vulnerable and uncertain towards the trustee. Hence, trust is seen as a way of dealing with this inherent uncertainty—as Möllering (2006) notes, there is here a fine, but important difference between the notions of risk and uncertainty; trust is "indeed 'risky' (Luhmann, 1979, p. 24) in a general sense of the word, but it is irreducible to calculation and therefore more than simply a probabilistic investment decision under risk" (p. 8). Furthermore, there is a trustor's positive expectation (and upon it actual actions are undertaken) that the other actor will not exploit the vulnerability, and so will not fail the trust vested in him – to put it simply, that no harm would be done (Baier, 1986).

However, Möllering (2006, pp. 105-121) also indicates the element that many theoretical considerations tend to miss, but which constitutes the crux of the experience of trust. Referring to the works of Georg Simmel he writes that trust is "a state of mind which has nothing to do with knowledge, which is both less and more than knowledge" (Möllering, 2006), and, furthermore, he writes that "Complete knowledge or ignorance would eliminate the need for or possibility of trust" (p. 109). The essence of the concept of trust is, hence, that element of "suspension" (or the leap of faith) that ultimately makes the concept of trust so unique and powerful. This is the very element that ultimately allows for reaching the state of positive expectation.

The notions of the trustor and the trustee have long been subject of an academic debate.[1] Avoiding this discussion here, I resort to the distinction set out by Sztompka (1999) between the primary and secondary objects of trust. Secondary objects may be, for instance, organizational roles (positional trust), brands (commercial trust), infrastructure (technological trust), institutions (institutional trust), and whole social systems (systemic trust). However, it is the human being that is ultimately the primary object of trust:

Behind all of them there looms the primordial form of trust n people, and their actions. Appearances nothwithstanding, all of the above objects of trust are reducible to human actions. We ultimately trust human actions, and derivatively their effects, or products. (Sztompka, 1999, p. 46)

Distrust, unlike trust, has rarely been posed as an autonomous research problem (Cook et al., 2005). Initially defined as a converse of trust, it recently began to be seen as a separate, independent construct (Lewicki, McAlliser, & Bies, 1998). Researchers have indicated several powerful differences that actually call into question the parsimonious view of these notions as opposite sides of a single continuum (Lewicki et al., 1998; Ullman-Margalit, 2004).

First, trust usually builds up in a troublesome and long-term process, while distrust may be a matter of single action: hurting conversation, misconduct, unfortunate actions, and so forth. As Six (2005) wrote (quoting Dutch statesman J. Thorbecke), "trust comes on foot, but leaves on horseback" (p. 5). Second, the experience of distrust, in contrast to trust, is usually more pronounced and more readily experienced (Ross & LaCroix, 1996). Third, misplaced trust tends to bring about much more harmful results than undue distrust then in doubt, the distrustful attitude may then seen as a wisely taken precaution (Cook et al., 2005). Fourth, trust may be falsified in action, while it is hardly possible in the case of distrust (Gambetta, 1998; Luhmann, 1979; Nooteboom, 2002). Finally, the elements that reduce distrust do not necessarily build trust.

The concept of continuum, however, directs our attention to the fact that actually, apart from trusting or distrusting attitudes, we may also be simply indifferent, that is, a state in which there is actually neither trust nor distrust (Ullman-Margalit, 2004). It allows us to see a huge similarity between the notions: yet, they both involve agency (Cook et al., 2005). Inasmuch as trust involves actions taken on the ground of positive expectations, distrust calls for an active seeking of safeguards against the opportunism of the other party. Therefore, it might be easier to build trust when we simply begin at the point of the mere lack of it, rather than when we have to overcome the condition of (still active) distrust.

As several authors claim, collaboration is possible both in the conditions of trust and distrust. Distrust may also be functional precisely because the human civilization has developed alternative ways of securing the reliability of our partners, for example, law (Cook et al., 2005). Moreover, as Lewicki et al. (1998) point out, in reality human relationships have the quality of "thickness," that is, they may simultaneously involve trust and distrust. "Just as it is possible to experience attraction and disattraction, to like and dislike, and

to love and hate, it may be possible to both trust and distrust others" (Lewicki et al., p. 449).

Referring to market relations, Beckert (2005) makes a powerful statement that "not all market relations depend on trust" (p. 21). Specifically, when we consider "suspicion" being at the heart of the concept, we cannot talk of trust when one party can "either calculate the actions of the exchange partner, or integrate them through power" (Beckert, p. 21). Calculation or power seem to be alternative arrangements that substitute for trust and meet "universal cravings for certainty, predictability, order, and the like. These are the functional substitutes for trust" (Sztompka, 1999, p. 115). The concept of functional substitutes for trust was developed as a result of studies of trust on the macro-level. However, it remains an open question as to whether this concept could be applied also on the micro level, to the relations on market. Table 1 gives a brief description of such functional substitutes for trust.

Table 1. The functional substitutes for trust (Source: Compilation based on Sztompka, 1999.

Providentialism	"[T]he regression from the discourse of agency toward the discourse of fate" (Sztompka, 1999, p. 116).
Corruption	"[I]t provides a misleading sense of orderliness and predictability, some feeling of control over a chaotic environment, some way to manipulate others into doing what we want them to do" (Sztompka, 1999, p. 116).
Overgrowth of vigilance	"[T]aking into private hands the direct supervision and control of others, whose competence or integrity is put into doubt, or whose accountability is seen as weak, due to inefficiency or lax standards of the enforcing agencies" (Sztompka, 1999, p. 117).
Excessive litigiousness	"[The attempt to] safeguard all relationships formally: draw up meticulous contracts, insist on collaterals and bank guarantees, employ witnesses and notaries public, and resort to litigation in any, even the most minuscule, even of breaching trust" (Sztompka, 1999, p. 117).
Ghettoization	"[C]losing in, building impenetrable boundaries around a group in an alien and threatening environment" (Sztompka, 1999, p. 117).
Paternalization	"[Dreaming] about a father figure, a strong autocratic leader, a charismatic personality…who would purge with an iron hand all untrustworthy…persons, organizations or institutions" (Sztompka, 1999, p. 118).
Externalization of trust	"[Tendency] to turn to foreign societies, and deposit their trust in their leaders, organizations, or goods…By contrast with locally targeted distrust, such foreign targets of trust are often blindly idealized" (Sztompka, 1999, p. 118).

THE FIELD STUDY OF BUYER-SUPPLIER RELATIONS

Buyer-supplier relationships are one of the most common kinds of relations between organizations. It has been widely recognized that they are actually "thick" relationships and thus trust plays an important role in them, as they require rather close cooperation (e.g., Lane & Bachmann, 1996; Sako, 1992). That particularly holds true here, as the research is focused on relations with clients for whom the companies were providing tailor-made solutions, produced exactly to match the order; they are build upon collaboration of the provider and the buyer. Furthermore, the potential role of trust becomes salient when we consider the mutual reliance on information, the vulnerability of both parties, and the threat of opportunism, further reinforced by the rather volatile environment. Beckert (2005) substantiates that trust is an issue only in exchange situations that share the three following features:

1. Uncertainty regarding the characteristics of the goods/services. The trust-giver cannot perfectly judge their quality and value.
2. There is competition in the market; that is, the trustor can choose between several trustees.
3. The relations are short-term, so that tradition and identity (mechanisms that reduce uncertainty, and, hence, the need for trust).

From the analytical point of view, all of these qualities are fulfilled in the case of IT project work that is studied here, so it seemed particularly relevant for the study of the concept of trust.

From the methodological point of view, the presented research project was inspired by the need of local, interpretative studies of trust, pointedly expressed in the recent literature (see, for example, Möllering, 2006, pp. 151-154). As Wicks, Berman, and Jones (1999) put it, "empirical work [should] begin with localized studies (e.g., keeping country and industry the same). Researchers could then, based on multiple studies, determine if (and under what conditions) any broader generalizations about trust are true" (Wicks et al., 1999, pp. 113).

Thanks to keeping the empirical focus and keeping the study highly localized I was able to explore the buyer-supplier relations with the superior goal being explorative (understanding the process within one fragment of the social reality), much rather than normative (looking for general rules applicable under any circumstances).

The first part of the study involved in-depth, nonstructured interviews conducted in three IT companies in Poland, and, subsequently, in four organizations representing their clients (together 28 people were interviewed and the interviews usually took 1.5-2.0 hours). The interviews were tape-recorded (with two exceptions) and transcribed. This in-company phase of the research project was then complemented by direct observation of everyday interaction between the client and the supplier as well as the analysis of selected professional literature and professional press. The choice of specific sources was guided by interviewees themselves, as they mentioned their usage in their professional life.

THE FIELD: IT INDUSTRY

The value of the Polish IT market in 2005 is estimated to be $49.5 billion USD, continuing the growth evidenced in Table 2.

Table 2. Polish IT market in 2001-2004 (Source: Based on Młynarczyk, 2004 and IDC, 2005)

[Mln USD]	2001	2002	2003	2004
Services	727	819	918	1150
Software	512	587	680	750
Hardware	1648	1703	1860	2250
Total	2887	3109	3458	4150

Out of the three segments, the software part has been certainly the one growing most rapidly. Experts reckon this rising tendency to be stable, as the software market created by business in Poland might be considered saturated only in regard to the least sophisticated products. As the Polish enterprises are entering the world of global competition, they tend to look for more sophisticated solutions. This, together with the effect of the European Union (EU) accession (transfer of significant funds supporting IT modernization for both business and public institutions), indicates that the sector of highly specialized custom-made software solutions will be on the rise in the years to come. Additionally, EU imposes high operation standards on the public administration and it particularly encourages the small and medium-sized companies to implement electronic tools in order to change the business culture into a more modern one. On the other hand, the EU also provides funds assigned specifically for the implementation of electronic support systems.

There are two traits of the IT industry that intuitively induced me to take up research of this particular business. First, the truly global nature of this industry (Obłój, 2002) may provide a good springboard for comparative research, and the local, ethnographic studies would indeed constitute a good beginning for such a project. Second, the whole business of IT heavily draws upon technological knowledge. As it seems to me (or at least that was the case at the outset of the project, by and large an IT-layman's perception) there exists a substantial asymmetry between the supplier and the client. In business relations, this asymmetry is accompanied by a relatively high disclosure of key information about the organization on the part of the client. This, along with the oftentimes particular character of the service (like, for example, software design) that eludes precise evaluation, forms an environment conducive to the emergence of trust.

THE ENVIRONMENT OF DISTRUST

Actors engaged in any relationship within the society do not exist in a vacuum; indeed, they are embedded in a social context that exerts impact on how they define themselves, the other party, and how they enact their own agency (Berger & Luckmann, 1966). In relation to trust, Möllering (2006) formulated this in the following way: "Trust is practically never a purely dyadic phenomenon between two isolated actors; there is usually always a context and a history, and there are also other actors that mater" (p. 9). There were two such contexts most often mentioned by my informants in the field, by both the customers and the providers.

First, all the organizations operate in Poland and they originated as Polish companies.[2] For the interviewees it is often not taken for granted, but highlighted as an important feature of their work environment:

You know, I was working in the U.S., and then in Germany for several years, so I have the perspective. I'm telling you, it's a whole world of difference... how the companies see me, as a person who manages IT in the bank. It's something different in Poland, although I'm Polish ... but I see that difference because I was working somewhere else before.[3] (MK, buyer, personal communication)
The words of people like MK, who have previous record of work abroad, are particularly drawing attention to this issue of national context. One woman, COO of a medium-size company reflected upon her previous experience:

I came back to work in Poland after 18 years in the UK. I was first a programmer, and then a manager in a company .... If you asked me to compare between the UK and Poland, I just .... I'm discussing it often with my husband. It is not that we regret the comeback, we enjoy it a lot, but it is like being in some other reality. Here it is crazy. The rules are flawed, the speed of project is incredible. It is just unfeasible, or at least it seemed so to me when I came here. (GK, provider, personal communication)

Second, the frame of reference distinguished by the interviewees themselves is the IT industry itself. They refer to the characteristic features such as: the project-oriented organization of work, the lengthiness of the projects, and the intangible character of the software product and accompanying services. Relating to the trust research, Bart Nooteboom (2002) introduced the notion of "industry recipes" (Spender, 1989, cited in Nooteboom, 2002, p. 60), claiming that we may also attempt to assess the trustworthiness of the entire business sectors. Admittedly, although this distinction does not appear to refer directly to distrust, it may contribute to the understanding of the condition of the IT industry that is in question here. For instance, the industries with fierce, intense competition between players can simply not afford "trust building" initiatives, as they require the deployment of unavailable resources. That is why such industries will be usually characterized by the overall lower level of trust than businesses that are stable, even maybe quasimonopolistic. There, firms can afford more benevolence; their moves are more predictable, adding to the general feeling of security, and, therefore, fostering trust.

According to this nomenclature, on the basis of the previously described characteristics of the industry, with the reference to the buyer-supplier relations, the IT sector in Poland appears as a rather low-trust industry. It is interesting that although the interviewees were not asked directly about trust (I worded the question in terms of cooperation, collaboration, relationships with the buyer/supplier), they, unprompted, characterized the climate of the relations within the industry as distrustful. MK was the most memorable to me, as while he was telling a longer story and mentioned trust I snatched at it:

And could I believe them, how could I trust them at all? What do you think?

DL: So, you basically are saying that you didn't trust this guy?

Trust? Please, don't make fun of me.

DL: I don't. I'm serious.

Oh, my God, I never trust them. If I ever trusted them I'd be dead [laugh]. (MK, buyer, personal communication)

Actually, it was quite often that the interviewees were just laughing at the word of trust. This made me realize that the conversations were actually not about trust itself, but rather about the profound lack of it, or even about explicitly expressed distrust, as in the case of JK, who stated, "I never trust the customers. I do not need it, and I do not want it. I should make them believe in what I am saying, but … in my position, I am rather distrustful towards them" (JK, provider, personal communication).

COMMUNICATION

Against the background of the explicitly stated distrust, it may be interesting to look at how the parties indeed do collaborate on the projects. Yet, despite the suspicious attitude towards each other, they still make deals. When we take a look at the tools that the actors deploy in their every-day communication we see a rather surprisingly (considering that we are in a high-tech business) traditional catalogue. Among the e-tools used in every-day communication between providers and their clients are e-mails and instant messengers; in a very few cases the voice over Internet protocol (VoIP) technologies were mentioned. But, when the interviewees talk about CMC communication with the provider, or respectively, the customer, these usually appear in a negative context: as misspent time, as unwanted effort, or even as impediments.

E-mails, well ... it happens that you send something and they never reply ... or they keep telling you they have never actually got it. We don't have time for that in my team .... If I want to know something I send my people there, or sometimes ... you know, sometimes it may be necessary to ask the boss to go to talk to the boss there ... and I talk to somebody here over the phone also ... writing a message would not work. (DT, provider, personal communication)

You must understand that business, at least here, but I think that also in other sectors, is made on face-to-face basis. Naturally that we have all the cute things: e-mails ... even that ... Skype⁴ ... but I wouldn't rather do that.

DL: So, what's your way?

MW: I just call or ... I also would rather meet them ... somehow. If the person I want to talk to doesn't have time I'd go for a lunch, or wait. (MW, provider, personal communication)

Also, computer-mediated communication to some managers is a kind of warning, a sign of trouble. The statement made by PK seems pretty exemplary. PK is now a head of an implementation unit at the large IT company. He previously worked on a position of a business consultant for the same firm, and then for several years worked as a project manager. Finally he was promoted to a senior managerial position and now he leads the team of project managersn the passage below he talks about how troubling any communication other than face-to-face is to him as a person in charge of the project.

Well ... I become suspicious when I see my people writing too much. Doesn't matter what, e-mails, *or letters .... I almost see the end then .... That's the tragedy of the project. When they don't meet in person, don't talk ... then as a manager I have to take over, and I have to be decisive .... Usually I have to change the teamleader, or even all the teammembers.* (PK, provider, personal communication)

The act of "writing" (in traditional or electronic form), as PK puts it, draws attention to the broader, and still a bit paradoxical, subject raised by the interviewees. On one hand, there is the issue of recording, meaning that once the words are put in writing they gain long-lastingness, as if a life of their own. Written on a piece of paper, in the e-mail, or even typed on the electronic chat (where it is possible to record the conversation) they may be brought up afterwards, or may be quoted to in a further conversation. What was once recorded cannot be easily erased and its existence cannot be openly denied as it might be in the case of words that were "only" spoken.

I prefer not to write.

DL: Why?

You see ... mhm ... when you are angry ... and ... well ... with our providers I am almost always angry, I feel ... irritated because something happened, or didn't happened, and then I want to talk to them ... And then it is better to talk, because when you write a message then you cannot avoid writing something emotional.

DL: Such as?

I do not know ... something like: "finally", or "at last", or ... I do not now. But you can say actually something really worse, and then they feel you are angry ... but it does not last ... and the e-mail, or the letter is always here ... you know. (DC, buyer, personal communication)

But there is also the other side of the coin, and here is where the paradox, and perhaps the distrustful environment, comes clearly into sight. Both the customers and the providers claim thats far as the partner's opinions, decisions or commitments are concernednly the written documents have a binding meaning to them. They directly express misgivings about the partner's words in a direct conversation. Taking into account the previous statements about the importance of personal contact such opinions seem rather surprising. On one hand they seek personal communication and claim that do not like written, or computer-mediated communication while one the other hand they openly declare to rely only on words that were written or signed by the other party. The providers, as GK is saying, want to protect themselves against the denial on the part of the customer.

You know, over time, we learned... that you can't trust what the customer is saying us.... And now we always want them to confirm what we've agreed upon. To sign a note, or a report. They don't want it. And they try to get away. When we want them to write, or to confirm... simply to sign... they try to avoid it usually. I know there may be several reasons.... But at the end we are left with no decisions.... And we have to find a way out. (GK, provider, personal communication)

Customers actually talk along the same lines, they want to get written statements from the seller, as these also serve as a clear point of reference.

Well, we know very well when the salesperson is going to take us in... simply sell us the moon. It's the old rule of salespeople all over the world in every single firm: "we can promise you everything". But it works like that all the time. Later, when we work on the solution, the provider comes here and promises us literally everything.... But you know... that happens only face-to-face. When we tell them to e-mail us, or to write and sign it

they suddenly change. And it turns out that things are not so rosy. It's always good to ask them to record... somehow... what they have just told. (KK, buyer, personal communication)

The last commonly mentioned feature of electronic communication brought up by the actors in the field is its impersonality. You cannot, essentially, see what emotions it evokes, you cannot argue with it, you can push or delay issues with it. Conversely, in the face-to-face communication you talk until the issues are solved. In this light, all interviewees were emphasizing the importance of routines established to maintain personal communication, such as kick-off sessions or regular team meetings. It seems to be popular and recognized especially among the managers, while computer-mediated communication is treated rather suspiciously.

I want my people and their people meet face-to-face. I do not want them to waste time chatting on their computers with each other... writing these e-mails ... thinking how to phrase the issue .... I want them to meet together, here or anywhere else .... They should sit there, and talk, and fight ... of course ... I am always there with them to watch over .... But they should sit together and talk until the issue is solved .... I am now the project manager for one project and I made it very clear form the beginning. The people from the software company must be here every week. Just be here, not sending me reports, or calling me, or anything. That's my condition ... necessary condition. To talk with my people, to see how things are going, to see where the problems are. (MK, buyer, personal communication)

Many people said that while they actually feel free in using VoIP technologies, or regular e-mails in communicating with the colleagues, they would be much more wary of doing that with the customer. It seems that the possibility of being misunderstood, of missing the point is too high.

And there too much at stake not to see the person they are talking to, not to be able to react on the spot to their body language, facial expressions, immediate words, the way the person speaks.

There were, however, several strikingly similar stories where interviewees talk about the projects where, surprisingly, the trustworthiness of either the buyer or the supplier seemed not to matter at all. That were the few cases of projects carried out under the auspices of the European Union; the EU also provided funds for the implementation of the solution. The EU, in the accounts of both clients and providers, seemed to serve as a powerful patron, a point of reference, to whom all actions must ultimately be carefully accounted for, who demands reports not only restricted to financial settlements, but also enforces its standards and procedures. Then, also the communication practices are often carefully specified and the room for manipulation is highly limited. As WS, CEO of one of the provider companies put it:

When there's the EU project, it makes all the difference. Then there are clear restrictions and they can't push us, as they normally do. You know ... when something isn't OK with the EU, Brussels will just withdraw funding, they don't have any problems doing that ... when something is not conforming with the rules. We all lose then and the customers also know that, so they would rather behave. (WS, provider, personal communication)

People on both sides also declare that they value very high personal competencies and trustworthiness of single individuals in the partner organization. The organizational context matters a lot, because it dictates how the organizational roles are performed and there is an interplay between the trustworthiness of the person and of the organization (Nooteboom, 2002). However, as the interviewees keep saying, the dense network of intimate, personal relationships and just knowledge about the people in the business are

the ultimate safeguards that can be relied upon.

I never believe the provider when they tell me "we're going to send you our best people for the project." I just don't believe that. I demand them to give me all the names and résumés of these people. And I look if I know them, or I know somebody who knows them .... Then I can check up. Or ... at least I can verify the competencies .... But I'd rather wait and take the people I know personally. (MK, buyer, personal communication)

On the part of the provider, DT confirms the above attitude which underlies the importance of reputation and the dense network of personal connections in the field. The name seems to matter much more than the brand of the organization. According to DT, "People go around, they change jobs, they quit and take up new positions in new organizations, but once you're a reliable person that reputation sticks to you whenever you go" (DT, provider, personal communication).

CONCLUSION

The field data supports the proposition that cooperation without trust, or even in sheer distrust, is possible; in other words, distrust may be functional and is not devoid of the cooperation potential (Cook et al., 2005; Möllering, 2006). In the words of the people in the field, the nature of the relations that have been studied constitute everything but trust, especially understood as having "suspension" at its heart (Möllering, 2006). There are not any positive expectations towards the other, but there is, indeed, an explicitly negative expectation. From the very outset of cooperation both parties pay great attention to designing safeguards and take precautions to protect themselves against potential, and evidently expected, opportunism.

The notion of the substitutes of trust is not completely supported by the field study; however, it may still be attractive as an interesting way of framing the ways of how people deal with cooperation in the lack of trust. It might be that Sztompka's concept of the functional substitutes for trust was developed in the analysis of trust on the macro level and its application to interpersonal or interorganizational collaboration may be limited. But in regard to this, note that in the empirical cases people do not declare that they strive for overcoming distrust – they just take it for granted and learn how to function within its boundaries. Part of the explanation of this phenomenon may be the concept of the "culture of trust," also offered by Sztompka (1999, 2005). Trust, or, respectively, distrust, may be inculcated into human experience to the point of becoming a cultural rule. Such rules cannot be changed overnight, as, for example, is the case of political systems. Consequently, even when the newly established political order may actually support the attitude of trust, this is in fact unattainable for people who were brought up in the culture of distrust. Then, the previously functional distrust becomes dysfunctional, but it remains as a cultural rule. Regardless of the new rules, the parties re-create the patterns of the reality of distrust, as it is, paradoxically, the world most familiar to them, the reality with established rules they were accustomed to (Kostera, 1996). In this light it may perhaps be more accurate to talk about alternatives rather than substitutes for trust. Yet, the interviewees have not expressed the wish to make things different, to have "more" trust, that is, trust does not seem to be perceived as inherently better. Instead, they rather focus on establishing alternative mechanisms that would secure cooperation.

Another proposition (to be confirmed in further research) would be that one of the mechanisms governing successful cooperation is the balance between the engaged parties. In buyer-supplier relations, customer and providers know they are mutually interdependent in bringing a project to the successful conclusion. When there is no third, strong supervising subject, both parties have a power of breaching the contract and they are ready to resort to it any time. For sure, they all have high stakes in the projects, and perhaps it is also that everybody has the goods on somebody—and that is the link that forces the partners to cooperate, regardless of the distrustful and disregarding attitude towards each other.

The complex practice of formalization of communication between parties (prefer personal contact, but actually want the other party to sign virtually everything) may be interpreted as one of the manifestations of potential extensive litigiousness as Sztompka (1999) formulated it, or of the desire to exercise control over the other party, or at least provide oneself with an illusion of such control. Control, nonetheless, in the literature on trust is seen as one of the basic ways of reducing uncertainty and, respectively, the need for trust.

Also, striving for face-to-face contact and reliance on personal relationships may constitute another manifestation related to control. The personal reputation and trustworthiness, as well as interpersonal connections, are seen as relatively stable bases for cooperation. It may bring to mind the notion of corruption, one of Sztompka's substitutes of trust; however on the basis of data collected within this study I am not able to assess how much the actual practices in the field are in fact corruption. It is significant, though, that in a few cases where the topic of corruption was mentioned the interviewees either openly refused to talk on this subject or asked to turn the tape-recorder off. Two interviewees acknowledged that corruption in various guises is indeed a common phenomenon of their work but they refused to talk about it.

Finally, there is a rather clear indication of the mechanism of the externalization of trust. The case of project realized in partnerships with the EU indicates that it serves as a powerful, external point of reference that provides a relatively reliable framework of cooperation.

As far as communication is concerned, there is clearly a need for face-to-face, personal interaction. The emphasis and recurrence of this issue may be surprising when we take into account the environment of the studye are, after all, in a high-tech business. Quite intuitively, high-tech should be in the vanguard of implementing the virtual tools of communication and maintaining computer-mediated relationships. Yet, in my opinion, what may be seen in the field is that it does not matter much whether we are in the context of distrust, as it practically overshadows all the other aspects circumstances of the relation. I would venture a hypothesis that computer-mediated communication may serve well when there is a basic consensus between people collaborating, or, in other words, when at least distrust is suspended, that is, we have the condition of the lack of trust (or, obviously, trust). It is the case of, for example, the intraorganizational communication, where the interviewees declared to use e-communication more willingly. Electronic tools of communication, still, in the environment of distrust, do not appear to have a potential to secure cooperation between organizational partners. In the condition of distrust, people retreat to basics, the traditional ways of communication and they want to feel and see the other.

REFERENCES

Baier, A. C. (1986). Trust and antitrust. *Ethics, 96*, 231-260.

Beckert, J. (2005). *Trust and the performative construction of markets* (MPIfG Discussion Paper 05/8). Cologne: Max Planck Institute for Studies of Societies.

Berger, P. L., & Luckmann, T. (1966). *The social construction of reality.* Garden City: Double-day.

Cook, K., Hardin, R., & Levi M. (Eds.). (2005). *Cooperation without trust.* New York: Russell Sage Foundation.

Gambetta, D. (Ed.). (1988). *Trust: Making and breaking cooperative relations.* Oxford: Basic Blackwell.

Gill, M. J. (2003). Biased against "them" more than "him": Stereotype use in group-directed and individual-directed judgments. *Social Cognition, 21*(5), 321-348.

Hardin, R. (Ed.). (2004). *Distrust.* New York: Russell Sage Foundation.

Hosmer, L. T. (1995). Trust: The connecting link between organizational theory and philosophical ethics. *Academy of Management Review, 20*(2), 379-403.

Huemer, L. (1998). *Trust in business relations: Economic logic or social interaction?* Umea: Borea bokvorlag.

IDC (2005). Poland IT services 2005-2009: Forecast and 2004 vendor shares. Retrieved March 14, 2007, from http://idc.com/getdoc.jsp?containerId=ES05M

Kostera, M. (1996). *Postmodernizm w zarządzaniu* [Postmodernism in management]. Warszawa: PWE.

Kostera, M. (2003). *Antropologia organizacji* [Organizational anthropology]. Warszawa: PWN.

Koźmiński, A. K., & Sztompka P. (2004). *O wielkiej przemianie* [On the great change]. Warszawa: Wydawnictwo WSPiZ.

Lane, C., & Bachmann, R. (Eds.). (1998). *Trust within and between organizations: Conceptual issues and empirical applications.* Oxford: Oxford University Press.

Latour, B. (1986). The powers of association. In: J. Law (Ed.), *Power, action and belief: A new sociology of knowledge?* (pp. 264-280). London: Routledge & Kegan Paul.

Lewicki, R., McAlliser, D., & Bies, R. (1998). Trust and distrust: New relationships and realities. *Academy of Management Review, 23*(3), 438-458.

Lewis, J. D., & Weigert, A. (1985). Trust as a social reality. *Social Forces, 63,* 967-985.

Luhmann, N. (1979). *Trust and power.* New York: John Wiley.

Mlynarczyk, D. (2004). *Analiza stanu obecnego i perspektywy rozwoju rynku IT* [Analysis of the current condition and the perspectives of the development of the IT market]. Warszawa.

Möllering, G. (2006). *Trust: Reason, routine, reflexivity.* Oxford: Elsevier.

Nooteboom, B. (2002). *Trust: Forms, foundations, functions, failures and figures.* Cheltenham: Edward Elgar.

Obłój, K. (2002). *Tworzywo skutecznych strategii* [Material of effective strategies]. Warszawa: Polskie Wydawnictwo Ekonomiczne.

Poznań: Polska Agencja Rozwoju Przedsiębiorczości.

Ross, W., & LaCroix, J. (1996). Multiple meanings of trust in negotiation theory and research: A literature review and integrative model. *International Journal of Conflict Management, 7(*4), 314-360.

Sako, M. (1992). *Prices, quality, and trust: Inter-firm relations in Britain and Japan.* Cambridge: Cambridge University Press.

Sako, M. (1998). Does trust improve business performance? In C. Lane & R. Bachmann (Eds.), *Trust within and between organizations: Conceptual issues and empirical applications* (pp. 88-117). Oxford: Oxford University Press.

Silverman, D. (2001). *Interpreting qualitative data.* Thousand Oaks, CA: Sage.

Six, F. (2005). *The trouble with trust: The dynamics of interpersonal trust building.* Cheltenham: Edward Elgar.

Staniszkis, J. (2001). *Postkomunizm: Próba Opisu* [Postcommunism: Attempt of description]. Gdańsk: Słowo/Obraz Terytoria.

Sztompka, P. (1996). Trust and emerging democracy: Lessons from Poland. *International Sociology, 11*(1), 37-62.

Sztompka, P. (1999). *Trust: A sociological theory.* Cambridge: Cambridge University Press.

Sztompka, P. (2005). *Socjologia: analiza społeczeństwa* [Sociology: Analysis of society]. Kraków: Znak.

Ullman-Margalit, E. (2004). Trust, distrust, and in between. In R. Hardin (Ed.), *Distrust* (pp. 60-83). New York: Russell Sage Foundation.

Wicks, A. C., Berman, S. L., & Jones, T. M. (1999). The structure of optimal trust: Moral and strategic implications. *Academy of Management Review, 24*(1), 99-116.

Zucker, L. G. (1986). Production of trust: Institutional sources of economic structure. In. S. Barry & L. Cummings (Eds.), *Research in organizational behavior* (pp. 53-111). Greenwich: JAI Press.

ENDNOTES

[1] Even collective or nonhuman entities may be identified as trustors and trustees, as long as it is possible to ascribe actions (responsibilities), or expectations to them (Nooteboom, 2002).

[2] To be precise, some of them attracted foreign investment capital, but it still constitutes a minor share. Two companies also carry out operations abroad, but it is a marginal activity.

[3] All interviews translated from Polish by the author.

[4] Skype - Software that enables communication over the Internet for free.

Chapter XVII
Establishing Trust in Offshore Outsourcing of Information Systems and Technology (IST) Development

Rachna Kumar
Alliant International University, USA

ABSTRACT

This chapter explores the issues and challenges faced in establishing trust among individuals and teams participating in offshore outsourcing of software development projects. While technical and project management aspects have been recognized as important for the success of offshore software outsourcing, the issue of establishing trust among the participants has not received specific recognition. The chapter discusses the special characteristics of offshore software outsourcing relationships which make the establishment of trust a challenge. The discussion emphasizes that a specific and planned approach of utilizing communication and coordination technology in software offshoring relationships will contribute towards trust formation. Use of communication and coordination technology in offshoring environments is recommended to be designed to increase the culture of communication, to establish a culture of transparency in communication, and to systemically maintain a trail and evidence of the communication.

INTRODUCTION

Over the past few years several organizations have embarked on ambitious offshoring projects as a way to respond to pressures for keeping profits up and keeping costs down. "Offshoring" refers to the practice of organizations transitioning part of their business operations to lower cost overseas destinations. The basic idea entails utilizing equivalent skill levels at lower wages in the destination country.

The information systems and technology (IST) services industry segment has been the forerunner in establishing offshore bases. Offshore IST services, involving applications development, maintenance, and R&D services, is approximately 75% of the total global offshoring market today (Hatch, 2005). Several other industry segments have been increasingly utilizing offshore destinations, notably among them hardware and software maintenance, network administration, help desk services, call centers, and telemarketing organizations. The latest entrant in the offshoring race is back-office processing for a myriad of industries ranging from banking and insurance to retail banking, deposits and lending, credit card processing, and mortgage processing to corporate finance and accounting.

Irrespective of the industry or organizational operations, offshored projects entail distance coordination between virtual teams. As a result, one major characteristic shared by all offshored projects is that they are, in large part, IST-mediated. Unlike in offshoring relationships of the manufacturing era, the offshoring relationships in the IST-mediated era (sometimes referred to as computer-mediated communication) require very frequent coordination, sometimes multiple times a day. The ironic part of this close and frequent coordination is the fact that offshored relationships are usually managed by employees in the home-base country, who often harbor ill-will and insecurity against the offshore partners because offshore projects almost always threaten jobs in the home-base country. In addition, differences in culture and work practices between the two participating teams from two different countries also increase the tension between the two teams in the offshoring context.

With this backdrop, establishing trust between participating individuals and teams in offshoring relationships is critical for success in the project but presents several challenges. In this chapter we will explore some of the structural and procedural mechanisms that can be utilized and established as antecedents for trust in offshoring relationships. A short case study of offshoring relationships in a large Fortune 100 company in the U.S. is utilized for illustrating a trust building framework in action. In the case study discussed here, the offshoring destination nation is India, an increasingly popular venue.

BACKGROUND

The explosion of the Internet along with a boom in telecommunications capacity has made it feasible to get IST projects completed remotely. The communications between the U.S. and offshore locations became not only feasible but also efficient and cheap. At the same time, foreign offshore locations produced a worker pool that was well-trained in a wide array of technology skills, who also worked at much lower wages compared to the U.S. So, with available supply of technology skills, favorable economics of IST production, and a means to accomplish the projects, offshoring became a natural business initiative.

Experts assess the global offshore market to be close to a $300 billion opportunity and the size of offshored IST services and business processes is regarded to have almost tripled since 2001 (Chakrabarty, Gandhi, & Kaka, 2006). They estimate that the market has grown by nearly 21% a year in the past five years and over the next five years it will grow by an additional $80 billion. Offshoring is a relatively young market and widely different statistics and trends are quoted. For example, according to Robinson and Kalakota (2004), a McKinsey and National Association of Software and Service Companies (NASSCOM) study estimates that the information technology and enterprise solutions market in India is likely to reach $142 billion in 2009. This estimate contrasts with the current price tag of $532 billion to provide these services in the United States. The difference of $390 billion would be the net savings for the U.S. economy due to offshoring. The information technology and enterprise solutions sector has the potential to generate job opportunities for more than 1.1 million Indians by 2008 (Robinson & Kalakota, 2004). By every account though, the offshoring market segment is large and is growing at an impressive rate.

In the past 10 to 15 years, the vast majority of offshore service jobs have gone to just a handful of cities in India, Eastern Europe, and Russia. Only a handful of cities, such as Hyderabad, Bangalore, Delhi, Mumbai, Budapest, Prague, and Moscow, have been popular as offshore destinations. Currently, new destinations in China, Dubai, South Africa, Morocco, Argentina, and Brazil are growing in popularity (Farrell, 2006). India has been in a leadership position as an offshore destination and has captured two-thirds of the current global market for offshored IST services and almost half of the global market for all offshored business processes.

The major motivation for going offshore is popularly understood as cost savings (Bennatan, 2002; Stiffler, 2006). But a survey of articles published over the past 2 to 3 years finds several other advantages being recognized (Bennatan, 2002; Daga & Kaka, 2006; Hayes, 2003; Jones, 2003; Lacity & Willcocks, 2001; Overby, 2003; Pfannenstein & Tsai, 2004). Other advantages include availability of scarce and cutting-edge software development skills in destination countries, increased responsiveness to business needs and customer service in the home base country, and ability to shorten time to market in the home

base country. At the same time, with the benefit of experience from almost half a decade of offshoring in the IST arena, several risks and concerns are being recognized as well. Chief among these concerns are loss of intellectual property, loss of core business knowledge, vendor delivery failure, scope creep, turnover of key personnel, and political instability (Davison, 2004). As a result, several studies have tried to estimate success factors as well as success rates of offshore projects. A study of 116 outsourcing projects found that 38% of outsourcing arrangements were successful, 35% were failures, and 27% had mixed results relative to cost, quality, flexibility, and other considerations (Kern & Willcocks, 2003). Market watchers, AMR Research Inc., surveyed more than 220 companies and found less than one-third were satisfied with the amount of money they had saved by outsourcing (Travis & Durocher, 2003). More recently, a survey by McKinsey & Co. found that offshore outsourcing can reduce an organization's costs by anywhere from 45% to 55% (Daga & Kaka, 2006). But Gartner reports that most customer-service offshore outsourcing not only fails, but could end up costing companies one-third more than keeping them in-house (Huntley, 2006). Their study concludes that almost 80% of companies that outsource customer service operations to cut costs won't be successful in realizing the targeted cost savings.

Thus, while offshore outsourcing appears to be an unstoppable tidal wave, the success of any offshoring endeavor is far from certain. Following this, the IST industry has come to a point where there is a general realization that the soft skills of people and project management are as important to the success of offshore outsourcing as the hard skills of technical and computer system implementation (Doh, 2005; Foote, 2004; Kishore, Rao, Nam, Rajagopalan, & Chaudhury, 2003; Robinson & Kalakota, 2004). Success factors for offshore outsourcing projects have been emphasizing organizational and human factors in managing projects in addition to factors such as contracts, technology, vendor selection, and infrastructure standards.

Given this context, building trust among the members of the offshore outsourcing team is one of the most important people and project management factors to impact the success of offshoring projects. However, building trust in this environment has several challenges and pitfalls. For one, the team is a virtual team located in two separate locations and time zones and the team works and interacts primarily with IST and computer mediation. In addition, while the team has to adjust to cultural differences and boundaries, they start off by being naturally pitted against each other since the onshore members' work has been intentionally transitioned offshore for cost savings. The next section will further discuss the nature of these challenges with a view to understand their implications for designing systems and procedures for building trust among the members.

ISSUES AND CHALLENGES

IST offshore outsourcing teams have six separate components contributing to the character of their relationships. The "outsourcing" component of the relationship gives rise to (1) virtual teams in separate locations, and, (2) in-house vs. outsource employee rivalry and insecurity. The "offshore" component of the relationship gives rise to (3) teams working across cultural differences and geographical boundaries, and, (4) teams working under separate time zones. The IST component of the relationship gives rise to (5) teams being required to be in close coordination with frequent work hand-offs due to the compatibility requirements of IST projects, and, (6) teams' work and interaction is primarily conducted with computer and other IST mediation. This section will attempt to help the reader understand the interplay of each of these characteristics with trust building among the team members.

1. **Virtual teams:** Virtual teams can cross boundaries related to time, distance, and organization. As members of a virtual team come together on projects, integration of work methods, organizational policies, norms and traditions, workplace cultures, and task technologies becomes difficult. Sometimes establishing an unambiguously shared goal is itself an exercise and increases coordination and collaboration overheads. This leads to the possibility of miscommunication, misunderstanding, and alienation that are the antecedents of mistrust in relationships.

2. **In-house vs. outsource employee rivalry and insecurity:** In outsourcing initiatives, varying degrees of competition between the in-house and offshore teams exist. The in-house team is trying to prove the organization still needs them and the outsource team postures for greater responsibilities. This leads to possibilities of sabotaging success, low employee morale, and employees being unwilling and unsupportive participants of the endeavor. Establishing trust among the members in such an environment is a difficult task.

3. **Teams work across cultural differences and geographical boundaries:** Overcoming cultural differences is not merely differences in language and social interaction. Cultural differences also include differences in work practices, work conduct and behavior, manager roles and responsibilities, professional values and assumptions. For example, cultures that function with large power distances and hierarchical distance between boss and employees find it hard to trust colleagues from individualistic cultures where individuals are given more control over their work and output.

4. **Teams work under separate time zones:** Depending on the geographical distance between the target and destination locations,

especially in offshoring situations, the time difference could be between 8 to 15 hours. In fact organizations prefer this because then their offshore outsourced teams are working "24/7" round the clock and the implications for customer service or for time to market are very beneficial. However, it also means that real-time coordination and communication is almost non-existent and most communication is IST mediated or asynchronous. Building trust among members in such situations requires presence of several structural mechanisms which produce hard facts of trustworthiness.

5. **Teams in close coordination and frequent work hand-offs:** When offshore outsourced teams work on IST projects, the projects typically require much coordination between the in-house and offshore teams. Compatibility issues necessitate that work originated by members in one team often gets re-worked and commented upon by members of the other team. If the roles and responsibilities are not very well defined and the processes are not entirely transparent throughout the team, the possibility of ruffled egos and of misconstrued work requests could cause problems. This excessive interdependence requires ever present evidence of the competence and commitment of team members to establish trust within the team.

6. **Teams work and interact primarily with computer and other IST mediation:** The absence of face-to-face mediation presents a special challenge which has to be made up by the technology used to complete work as well as communicate and coordinate. The richness of media and cues in personal interaction has been widely documented as leading to establishing rapport, understanding, working relationships, and trust. Video or phone conferencing, net meeting, e-mails are all used but fall short in providing social presence and data richness at

the same time. Establishing trust in such settings will require specific promotion of opportunities and incentives that go beyond the naturally occurring work mediation. And yet, since major portions of work is done on the computer and computer-mediation is both the tool for work as well as the media for monitoring and coordinating the relationship, in some senses the openness and complete sharing of work components can actually be harnessed for increasing the trust in offshoring relationships.

CRITICAL SUCCESS FACTORS: IT OFFSHORE OUTSOURCING AND TRUST BUILDING

Several issues and challenges were discussed in the context of offshored outsourcing of IST projects. This section will synthesize their implications and consolidate them into a common framework for trust building in the offshoring context. This framework can then be utilized in analyzing the trust building procedures and mechanisms in place at a successful Fortune 100 organization engaged in extensive offshoring.

Role of Communication and Information Sharing in Building Trust

Trust is a very complex area of interpersonal communication within the context of any type of business transactions, but when trust formation is needed among members of different cultural backgrounds, it can be even more complex. Building trust in business relationships and economic transactions requires business partners or teammates to establish continuity and predictability through cooperative transactions. Additionally, the formation of trust among virtual team members emphasizes open, consistent, transparent information and communication. In the virtual environment of online shopping, Kim, Ferrin, and Rao (2003) found that trust among members is positively related to satisfaction and confirmation of expected trust feeds back and leads to increased trust. Similar mechanisms for trust formation in ecommerce environments have been observed in several other studies (Ba, Whinston, & Zhang, 1999; Gefen, 2000, 2002). Zolin and Hinds (2002) looked at how trust would be formed in geographically dispersed work teams. They found that trust was based on an individual's perceptions about the team member's perceived performance and perceived trustworthiness. A study done by Jarvenpaa and Leidner in 1999 looked at the area of trust and communication in global virtual teams. These teams used e-mail and message boards to communicate and came from different countries and cultural backgrounds. Jarvenpaa and Leidner (1999) found that frequency and richness of communication influenced trust levels. Even when communication levels were high, trust was hard

Figure 1. Synthesizing critical success factors for building trust in offshoring situations

	ISSUES	IMPLICATIONS	CRITICAL TRUST FACTOR	TRUST BUILDING MECHANISM
IT	**Teams Work and Interact Primarily with Computer and other IST mediation**	Easy Access to modes and options for communication providing rich social cues and informational base for relationships.	Easy access to communication and coordination technology.	*COMMUNICATION & COORDINATION TECHNOLOGY*
IT	**Teams in Close Coordination and Frequent Work Handoffs**	Specific, transparent, systemic procedures for work flow needed which don't assume trust.	Specific, transparent communication of procedural mechanisms without assuming trust and for evidence of trustworthy interaction.	*TRANSPARENT COMMUNICATION WITH BODY OF EVIDENCE*
OFFSHORE	**Teams Work under Separate Time Zones**	Structural mechanisms to: (1) allow easy access to team members' work & output; (2) document evidence of trustworthy interaction history.	Specific, transparent communication of procedural mechanisms without assuming trust and for evidence of trustworthy interaction.	*TRANSPARENT COMMUNICATION WITH BODY OF EVIDENCE*
OFFSHORE	**Teams Work across Cultural Differences and Geographical Boundaries**	Training for cultural sensitivity and inclusion.	Cultural sensitivity training.	*CULTURAL SENSITIVITY TRAINING*
OUTSOURCING	**In-house versus Outsource Employee Rivalry and Insecurity**	(1) Transparent communication; (2) Documentation of roles and work output of participants.	Specific, transparent communication of procedural mechanisms without assuming trust and for evidence of trustworthy interaction.	*TRANSPARENT COMMUNICATION WITH BODY OF EVIDENCE*
OUTSOURCING	**Virtual Teams**	(1) Specific, transparent, systemic procedures for work flow needed which don't assume trust; (2) Easy access to structural mechanisms for communication and coordination.	Specific, transparent communication of procedural mechanisms without assuming trust and for evidence of trustworthy interaction.	*TRANSPARENT COMMUNICATION WITH BODY OF EVIDENCE*

to maintain in virtual teams with homogenous members; when culturally diverse teams are formed, the trust factor becomes even harder to obtain. In studies where virtual teams have been observed making financial decisions requiring trust, it has been found that teams build trust depending on explicit and implicit communication about how much information disclosure they are exposed to about their team mates' performance, competence, and operations (Jettmar & Rapp, 1996; Johnson-George & Swap, 1982).

The good news is that results of these studies show that the culturally diverse virtual teams which utilized rich, transparent, specific, systematic, and adequate levels of communication built high levels of trust and were more capable of managing the uncertainty, complexity, and expectations of the virtual environment than teams who did not build the same levels of trust. Their performance, where measured, was also better than the low trust teams. Another good point to note here is that several of the issues which were discussed above in the context of building trust in offshoring teams could be well addressed with communication and information sharing. The offshoring enterprise can define what rich, transparent, specific, systematic and adequate levels of communication would mean for their enterprise or project. They would also design procedures and mechanisms to institutionalize and systematize these communication requirements appropriate to their own content.

Figure 1 synthesizes the issues and their link to the factor of communication and information sharing as a trust building mechanism.

Role of Cultural Sensitivity in Trust Building

Culture refers to a set of learned values, attitudes, meanings, and norms that are shared by members of a group. It is a collective programming that gives scripts of behaviors and understanding and sepa-

rates one group of people from another. Culture becomes second nature for members of the group and affects the group members' assumptions, behaviors, and expectations about work habits, practices, norms, and decisions. Interestingly, culture is often partially or totally hidden and usually has to be learned and absorbed over a period of time naturally. However, with the realization that cultural differences and insensitivity to cultural practices can have serious consequences on the success of two groups working together, several training programs have been designed.

The culture of a virtual team is a blend of national, organizational and functional culture (Jarvenpaa, Knoll, & Leidner, 1998). In offshore outsourcing, differences in the national culture and generally accepted norms between the home country and the destination country become a critical factor to counteract. Offshore outsourcing teams are usually multinational from at least two nations, sometimes more. Although organizational culture is also different and can very importantly impact expectations of norms, quality, schedules, values, and so on, national culture differences are accepted as the major hurdle to overcome in offshore outsourcing teams.

The most widely accepted framework for understanding impacts of national culture in the global arena comes from Geert Hofstede's work in 1967, discussed and expanded in his 1980 book (Hofstede, 1980). From responses of IBM employees in many countries, he discerned patterns of national behavior and derived four dimensions along which national cultures can be classified. The first, *power distance*, refers to degree of hierarchical difference and inequity among people that the population expects and accepts. Nations with low power distance have a culture where team members are more participative and managers seek inputs from members and get challenged on decisions routinely. In a high power distance nation such as India, Mexico or Singapore, offshore team members might expect to be told their task responsibilities rather than be self-starters. On the other hand, offshore team members from low power distance nations such as the U.S. or Germany might openly challenge their superior if given very restrictive steps for task performance. The second culture dimension, *uncertainty avoidance*, is the extent to which people from the culture are comfortable with uncertainty. Team members from nations (such as Mexico and Japan) with high uncertainty avoidance look for detailed plans, need defined rules and procedure, formalized responsibilities and outcomes. Cultures with low uncertainty avoidance, such as India and U.S., are more comfortable with ambiguous situations and open-ended plans. The third dimension for national culture, *individualism-collectivism*, is the degree to which people prefer to act as individuals rather than as members of a group. Teams in cultures with high collectivism expect cohesive work groups, expect team members to put the needs of the team before their individual needs. Examples of high collectivist nations are most of the Asian nations. Low collectivist nations, such as U.S. and Australia, have teams which accept individual needs and identities, act in the job situations with individual preferences. The fourth culture dimension, *masculinity-femininity*, refers to the extent to which the people are oriented towards the traditionally masculine concerns such as earnings, possessions, success achievements as opposed to concerns such as nurturing, caring, sharing. Team members from high masculine cultures such as Mexico and U.S. will tend to be motivated with earnings and achievements much more intensely than team members from low masculine cultures such as Norway and Sweden.

Certainly cultural sensitivity is important for team performance and achievement of task goals. How does cultural sensitivity impact trust building amongst offshore outsourcing teams? Duarte and Snyder (2001) have explained that the cultural dimensions impact team members'

expectations and understanding of each others' integrity, performance and competence, and person quality about concern for the team members. For example, expectations on performance and competence could be affected if a team member from a high power distance culture does not go the extra mile in follow through of the task deliverable. The high power distance culture will expect to be given specific go-aheads and authorizations from the boss before making any decisions, while the low power distance culture would expect that the need for the deliverable would guide the member to take the decision initiative in his/her own hand. The evaluation of performance and competence of members in cultures with high uncertainty avoidance could similarly get affected because it may seem that they cannot work independently and need to be given detailed plans and directions. As another example, members from highly collectivist cultures may evaluate a team member as having doubtful integrity because they actually transfer out of a project due to concern for their own individual career gain while ignoring concern for the team performance. As a final example, over-emphasis on wage earnings and achievement recognition of a team member might appear to other members from low masculine cultures as implying low person quality, having low concern for the welfare of the earnings of other team members. Members from highly masculine cultures see this as merely "looking out for oneself," a moral responsibility in all individualist cultures.

So, cultural dimensions provide us a way to understand differences and address them so that trust building (via expectations of performance and competence, integrity, and person quality) is not diminished. Organizations engaged in offshore outsourcing need to provide cultural sensitivity training in order to allow their team members not only to understand the differences but also to understand how they impact personal trust as well as how to learn ways and means to address the differences such that trust among individuals gets built seamlessly and surely.

Role of Communication and Coordination Technology in Building Trust

Technology and systems' role as it facilitates task completion of IT projects is critical to performance, but not our focus here. This section will discuss its role as it facilitates the dimension of communication and coordination, leading to trust building. Technology mediated communication and coordination impacts the perception and evaluation of integrity, of performance and competence, and of personal quality of other team members.

Technology for communication and coordination can be classified in several different ways (Rice, 1980). Primarily, the technology options can lead to synchronous, real-time communication and coordination or to asynchronous, anytime, anyplace communication and coordination. Synchronous communication and coordination options tend to have higher social presence that refers to promoting a personal connection and individual association leading to better possibilities of trust building. The second dimension to note is whether the media provides rich information in the communication and coordination process involving video, audio and data or lacks one of these types of information. The more the information provided about the team members interaction, the more the cues and data to build trust. Information rich media and technology options are therefore usually preferred in trust building phases of offshore outsourcing endeavors. A third dimension for evaluating technology options for communication and coordination is that of permanence. Permanence is described as the degree to which the technology is capable of creating a historical record of team communications and coordination (Poole & Jackson, 1993). From the point of building trust, technology mediation should be transparent and specific with the ability to provide a body of evidence and history of interactions. When team members utilize history of performance, integrity and personal concern to form expectations of future, it promotes trust. Audio phone conferences are sometimes not high on permanence while e-mails and bulletin boards are high on a interaction history. Figure 2 outlines some technology options in these different categories.

In summary, technology can act as a medium for open, specific, transparent, rich, and adequate communication and coordination between offshore teams, thus contributing to trust building.

Figure 2. Technology mediation options for communication and coordination

Time factor ⇒ Information ⇩ Richness	SYNCHRONOUS	ASYNCHRONOUS
Audio Only	Phone conference	Phone messages
Video and audio	Video conference	Video messages
Data Only	Data conferencing, chats	E-mail, fax, letters, group calendars, bulletin boards,
Video, audio, data	Electronic meeting systems, net meeting	

Technology options can also be utilized to send specific or subtle messages which reinforce expectations about the performance and competence of team members, or of the integrity expectations of the team, or about the personal values of the team. The characteristics of social presence, information richness, and permanence can be utilized to advantage in fostering structural mechanisms such as interaction history (e.g., using e-mails) or fostering social connections (e.g., introducing new recruits in a video conference rather than through e-mails). Organizations engaged in offshore outsourcing should evaluate technologies for their communication needs given the type of project tasks and situational needs they have. In addition, they should carefully lay out mechanisms to be followed when using the technology options in order to specify procedures and mechanisms so that team members can clearly expect specificity and transparency in communication and coordination.

TRUST BUILDING IN OFFSHORE OUTSOURCING ENDEAVORS: A CASE STUDY

A Fortune 50 company that we will call OOE, with revenues of about $80 billion, is currently a big player in the offshore outsourcing of software development. The company is a pioneer in technology and a market leader in its IST product segment. It is headquartered in the U.S., with branches throughout the west coast.

OOE first started exploring the possibility of offshoring software development for one of its $25 million product division in 2000. Cost and performance pressures were cited as the major motivations for considering offshoring. Between 2001 and 2004, the company started small offshoring projects in several destinations, more specifically in India, China, Mexico, Brazil, and Ukraine. The operation in India became their major offshore center and grew in size dramatically.

In 2006, their India center supported software development volume requiring over 1,000 full time equivalent developers. These developers are organized into several teams developing software for the company's product division.

Organization to Organization Level of Trust

The first phase of trust building which the organization engaged in was organization to organization. OOE had to pick a vendor in India to offshore and outsource their software development. The trust at the level of organization-to-organization involved three different dimensions: (1) Will the performance and competence promised actually be delivered? (2) Will our intellectual property (IP) shared in the software development project be sold to competitor companies? And, (3) Will our IP shared in the software development project be exploited for personal gain by the vendor company in other situations? Legal contracting was the means by which the trust in the IP-related issues could be best ensured. In the early phases, organization-to-organization level of trust is often one of faith. Although legal contracting is used to specify several items, it still leaves room for breaks in expectations of performance and failures due to miscommunication and learning. The organization-to-organization level of trust is best built with evidence of performance and history of competence. In later phases, when larger volumes of development are planned, the IP issues are solved by establishing solely owned subsidiaries which employ developers with in-house loyalties.

Team Member to Team Member Level of Trust

In the next phase, the team member to team member trust must be built in order for projects and tasks to be completed and delivered in time, within budget, and of required quality. This phase is where the critical success factors discussed earlier in this chapter come into play. Interestingly, the part of offshore outsourcing team based in the U.S. as well as the ones based in India have not undergone any specific training regarding trust building in computer mediated offshoring situations. However, several of the issues such as cultural sensitivity, communications, and coordination had to be addressed as OOE management gathered experience in this endeavor.

Cultural sensitivity is one area that has been well understood as critically impacting working relationships. All U.S. team members are given brief training on the cultural differences, norms, practices, values, traditions, and expectations of the Indian workplace and the Indian employee. Similarly, the Indian employees are trained in cultural differences in the U.S. workplace. However, a piece of training that was missing was how an understanding of cultural differences can be translated into ways and behaviors to foster trust between team members. The team members learn it over time with varying levels of success. For example, since India is high on the power distance dimension of national culture, one idea would be to train the U.S. team member to honor reporting structures and hierarchy when their U.S. counterpart is supervising a project module. If the U.S. team member continues to make or challenge decisions as is appropriate in the U.S., it could lead to a perception of low trustworthiness by the supervisor as well as by team peers. Cultural sensitivity training from the point of view of trust building is essential to complete the cycle of understanding the issues and then having the tools to address the issues to the advantage of the business goals and the team.

Communication and coordination for trust building is the other success factor and has several procedural and structural mechanisms to systemically implement and institutionalize it. The very nature of software development performed by team members separated by time and distance requires mechanisms for communication and coordination which are often computer-based. This organization has adopted high speed, high bandwidth video conferencing utilities as well as electronic meeting (audio and data) options such as Net Meeting. In addition, some of the other mechanisms such as open and shared team calendars, Web dissemination of minutes and action items, regularly scheduled status meetings, and commonly shared and developed meeting agendas are strongly woven into the organizational culture as well as encouraged. The communication and coordination between team members is emphasized to be frequent, transparent with log and audit trails, specific, and as rich as possible.

Interestingly, the general approach OOE used in the offshore outsourcing management of trust was to employ natural mechanisms to transfer the U.S. organizational culture to the offshore unit. The thinking seemed to be that if the U.S. organizational culture and mechanisms worked for establishing adequate levels of trust among team members, it would work for the offshore unit as well. Although that may be true, it would still be beneficial to systematically announce the specific structural and procedural mechanisms that need to be followed for task performance, thereby maximizing the evidence and opportunity for trust building. For example, if a time limit for responding to any e-mail is publicly set within the offshoring teams, then the expectation for follow-through on commitments and coordination will be ensured. Moreover, if trust building is emphasized as one of the desirable dimensions of performance within teams and specific procedures such as surveys which conduct a trust audit or seminars which specifically talk about trust mechanisms and procedures for fostering trust are conducted, this issue of communication and coordination for building trust would be more intentionally managed. OOE's approach currently is to trust the expertise of individual managers and team members to adopt sound principles of interpersonal business relationships and of communication for team building. No specific customization of expertise is undertaken to the needs of offshoring, to the needs of outsourcing, to the needs to technology mediation, or to the needs of virtual software development.

CONCLUSION AND FUTURE TRENDS

Trust formation in offshore outsourcing teams is one of the most critical needs for project and team success. The teams are virtual and separated by time and distance which makes trust building more challenging. The teams are in two nations separated by cultural differences and that adds complexity to building and maintaining trust relationships. These teams are developing software which has compatibility and modularity issues making coordination and work hand-off requirements very frequent. So, trust building in such teams presents special challenges and pitfalls.

Although critical for success, several organizations deal with trust building in offshoring situations in ways similar to the one home-base organization case. Managerial skills of the managers and interpersonal skills of the team members are often trusted as the mechanisms to ensure trust building in offshoring contexts. Several structural and procedural mechanisms naturally instituted for communication and coordination and for cultural sensitivity training do in fact promote trust formation. But the approach is not specific and several trust fostering opportunities are neglected or ignored. Several offshore outsourcing team members learn by trial and error and are forced to learn on the job as they hit the ground running. This may result in many tensions and pressures being generated and extended growing pain phases when setting up offshore outsourcing operations. As a result, offshore outsourcing teams might be functioning in suboptimal relationships

which impact performance, results, and goal achievement.

Specific training designs that emphasize trust as a desirable characteristic in the workplace and impart knowledge, tips, and ideas about fostering trust in offshore outsourcing teams should be undertaken. In addition, specific structural and procedural mechanisms need to be designed and instituted which make trust building systemic and a natural outcome rather than an art dependent on the team members' individual skills. The use of technology for communication and coordination is already part of the fabric of the workplace today. In offshore IT outsourcing teams these technology options can be harnessed to increase evidence and history of performance and behaviors, thereby improving trust amongst team members. These ideas could possibly contribute to resolving some of the mysteries of mixed performance results and cost savings in offshore IT outsourcing endeavors.

REFERENCES

Ba, S., Whinston, A. W., Zhang, A. (1999, December 12-15). Building trust in the electronic market through an economic incentive mechanism. In *Proceedings of the 20th International Conference on Information Systems* (pp. 208-213). Charlotte, NC.

Bennatan, E. M. (2002, November). Globalization: Boon or bane. *Cutter IT Journal*, 15(11).

Chakrabarty, S. K., Gandhi, P., & Kaka, N. (2006). The untapped market for offshore service. *McKinsey Quarterly*, Issue 2.

Daga, V., & Kaka, N. F. (2006, May). Taking offshoring beyond labor cost savings. *The McKinsey Quarterly.*

Davison, D. (2004). Top 10 risks of offshore outsourcing. *CIO Magazine*. Retrieved March 15, 2007, from http://search.cio.techtarget.cotn/orginalComent/0,289142.sidl9_gci950602,00.html

Doh, J. (2005, May 3). Offshore outsourcing: Implications for international business and strategic management theory and practice. *Journal of Management Studies, 42,* 696-704.

Duarte, L. D., & Snyder, N. T. (2001). Mastering virtual teams: Strategies, tools, and techniques that succeed. Jossey-Bass, a Wiley Company, San Francisco, USA.

Farrell, D. (2006, June), Smarter offshoring. *Harvard Business Review,* 85-92.

Foote, D. (2004, September). Recipe for offshore outsourcing failure: Ignore organization, people issues. *ABA Banking Journal.*

Gefen, D. (2000). E-commerce: The role of familiarity and trust. *The International Journal of Management Science, 28,* 725-737.

Gefen, D. (2002, Summer). Reflections on the dimensions of trust and trustworthiness among online consumers. *ACM SIGMIS Database, 33*(3), 38-53.

Hatch, P. J. (2005). Offshore 2005: Research, findings and conclusions (White Paper). Retrieved March 15, 2007, from http://www.ventoro.com/Offshore2005ResearchFindings.pdf

Hayes, I. S. (2003, August). *Ready or not: Offshoring: Is it a win-win game?* McKinsey Global Institute Research Papers.

Hofstede, G. H. (1980). *Culture's consequences: International differences in work related values.* Beverly Hills, CA: Sage Publications.

Huntley, H. (2006, March). *Offshore deal failure: What providers need to know.* Gartner Research.

Jarvenpaa, S. L., Knoll, K., & Leidner, D. E. (1998). Is anybody out there? Antecedents of trust in global virtual teams. Journal of *Management Information Systems, 14*(4), 29-64.

Jarvenpaa, S. L., & Leidner, D. E. (1999). Communication and trust in global virtual teams. *Organization Science: A Journal of the Institute of Management Sciences, 10*(6), 791-815.

Jettmar, E. M., & Rapp, M. W. (1996). *Computer mediated communication: A relational perspective.* Paper presented at the Annual Convention of the Western States Communication Association, Pasadena, CA.

Johnson-George, C., & Swap, W. C. (1982). Measurement of specific interpersonal trust: Construction and validation of a scale to assess trust in a specific other. *Journal of Personality and Social Psychology, 43*(6), 1306-1317.

Jones, W. (2003). *Offshore outsourcing: Trends, pitfalls, and practices* (Executive Report No. 4(4)). Cutter Business Technology Council.

Kern, T., & Willcocks, L. (2003). Exploring relationships in information technology outsourcing: The interaction approach. *European Journal of Information System, 11*(1), 3-23.

Kim, D. J., Ferrin, D. L., & Rao, H. R. (2003) A study of the effect of consumer trust on consumer expectations and satisfaction: The Korean experience. In *ACM International Conference Proceedings, 50,* 310-315.

Kishore, R., Rao, H. R., Nam, K., Rajagopalan S., & Chaudhury, A. A. (2003, December). Relationship perspective on IT outsourcing. *Communications of the ACM, 46*(12), 87-92.

Lacity, M. C., & Willcocks, L.P. (2001). *Global information technology outsourcing: In search of business advantage.* Chichester, West Sussex, England: John Wiley & Sons.

Overby, S. (2003). The hidden costs of offshore outsourcing. *CIO Magazine.* Retrieved March 15, 2007, from http://www.cio.com/archive/090103/money.html

Pfannenstein, L., & Tsai, R. (2004). Offshore outsourcing: Current and future effects on American IT industry. *Information Systems Management, 2*(4), 72-80.

Poole, M. S., & Jackson, M. H. (1993). Communication theory and group support systems. In L. M. Jessup & J. S. Valacich (Eds.), *Group support systems: New perspectives.* Old Tappan, NJ: Macmillan.

Rice, R. E. (1980). Computer conferencing. In B. Dervin & M. J. Voight (Eds.), *Progress in communication science.* Norwood, NJ: Ablex.

Robinson, M., & Kalakota, R. (2004). *Offshore outsourcing: Business model, ROI, and best practices.* Alpharetta, GA: Mivar Press.

Stiffler, D. (2006). Outsourcing transition case study: Customized communication speeds savings.

Travis, L., & Durocher, C. (2003, November). Strategic use of offshore resources can double labor savings (AMR Research Report).

Zolin, R., & Hinds, P. J. (2002) Trust in context: The development of interpersonal trust in geographically distributed work teams. Center for Integrated Facility Engineering.

Compilation of References

Abdul-Rahman, A., & Hailes, S. (2000). Supporting trust in virtual communities. In *Proceedings of the 33rd Annual Hawaii International Conference on System Sciences (HICSS 33), 6*, 6007. IEEE CS Press. Retrieved March 11, 2007, from http://citeseer.nj.nec.com/235466.html

Adams, L., Toomey, L., & Churchill, E. (1999). Distributed research team: Meeting asynchronously in virtual space. *Journal of Computer-Mediated Communication, 4*(4). Retrieved March 11, 2007, from http://www.ascusc.org/jcmc/vol4/issue4/adams.html

Adler, P. S., & Heckscher, C. (2006). Towards collaborative community. In P. Adler & C. Heckscher (Eds.), *The firm as a collaborative community* (pp. 11-105). New York: Oxford University Press.

Adler, T.R. (2005). The swift trust partnership: A project management exercise investigating the effects of trust and distrust in outsourcing relationships. *Journal of Management Education, 29*, 714-737.

Adler, T.R. (2007, in press). Trust and distrust in long-term contracting. *Journal of Business Strategies.*

Aeppel, T. (2003, July 1). On factory floors, top workers hide secrets to success. *The Wall Street Journal*, pp. A1, A10.

Águila, A. R., Bruque, S., & Padilla, A. (2002). Global information technology management and organizational analysis: Research issues. *Journal of Global Information Technology Management, 5*(4), 18-38.

Ajzen, I., & Fishbein, M. (1980). *Understanding attitude and predicting social behavior.* Englewood Cliffs, NJ: Prentice-Hall.

Amit, R., & Zott, C. (2001). Value creation in e-business. *Strategic Management Journal, 22*(6), 493-520.

Anckar, B., & D'Incau, D. (2002). Value creation in mobile commerce: Findings from a consumer survey. *The Journal of Information Technology Theory and Application (JITTA), 4*(1), 43-64.

Anderson, B. (1991). *Imagined communities: Reflections on the origin and spread of nationalism.* London: Verso.

Anderson, H., & Kanuka, T. (1997). On-line forums: New platforms for professional development and group collaboration. *Journal of Computer-Mediated Communication, 3*(3). Retrieved March 11, 2007, from http://www.ascus.org/jcmc/vol3/issue3/anderson.html

Anderson, J.C., & Naurus, J.A. (1990). A model of distributor firm and manufacturing firm working partnerships. *Journal of Marketing, 54,* 42-58.

Andrews, P. (2006). Virtual teams: Establishing trust is key to successful teamwork. *IBM On Demand Business,* 1-3. Retrieved February 25, 2007, from http://www-935. ibm.com/services/us/imc/ondemand/business/trust_ building.html

Andrews, S., & Shen, A. (2000). *Laws or regulations posing barriers to electronic commerce.* Washington, DC: Electronic Privacy Information Center.

Antognazza, E., & Moeder, P. (1999). *Web marketing per le piccole e medie imprese.* Milano: Hops Libri.

Apfelthaler, G. (1999). *Interkulturelles Management.* Wien: Manz Schulbuch GmbH.

Argyris, C. (1964). *Integrating the individual and the organization.* New York: John Wiley & Sons.

Ariño, A., Abramov, M., Skorobogatykh, I., Rykounina, I., & Vilá, J. (1997). Partner selection and trust building in Western European-Russian joint ventures: To Western perspective. *International Studies of Management and Organization, 27*(1), 19-37.

Arrow, K. J. (1974). *The limits of organization.* New York: W. W. Norton & Co.

Aubert, B.A., & Kelsey, B.L. (2003). Further understanding of trust and performance in virtual teams. *Small Group Research, 34*(5), 575-618.

Avolio, B.J., & Kahaï, S.S. (2003). Adding the "E" to e-leadership: How it may impact your leadership. *Organizational Dynamics, 31*(4), 325-338.

Avolio, B.J., Kahaï, S.S., & Dodge, G.E. (2001). E-leadership: Implications for theory, research, and practice. *Leadership Quarterly, 11*(4), 615-668.

Axelrod, R. (1984). *The evolution of cooperation.* New York: Basic Books.

Axelrod, R. (1990). *The evolution of co-operation.* Harmondsworth, UK: Penguin.

Ba, S., Whinston, A. W., Zhang, A. (1999, December 12-15). Building trust in the electronic market through an economic incentive mechanism. In *Proceedings of the 20th International Conference on Information Systems* (pp. 208-213). Charlotte, NC.

Bachman, T. (2003). Trust and Power as a Means of Co-ordinating the Internal Relations of the Organisation: A Conceptual Framework. In B Nooteboom & F. Six (Eds.) *The Trust Process in Organisations: Empirical Studies of the Determinants and the Process of Trust Development.* Cheltenham, UK: Edward Elgar.

Bachmann, R. (1998). Trust: Conceptual aspects of a complex phenomenon. In C. Lane & R. Bachmann (Eds.), *Trust within and between organisations: Conceptual issues and empirical applications* (pp. 298-322). Oxford: Oxford University Press.

Baier, A. (1986). Trust and antitrust. *Ethics, 96,* 231-260.

Baig, E.C. (2005, February 13). Love growing strong on the Web. *USA Today.* Retrieved March 12, 2007, from http://www.usatoday.com/money/industries/technology/2005-02-13-online-dating-usat_x.htm

Bakos, Y., & Dellarocas, G. N. (2003). Cooperation without enforcement? To comparative analysis of litigation and online reputation ace quality assurance mechanisms (MIT Sloan Working Paper No. 4295-03). Retrieved February 21, 2007, from http://ssrn.com/abstract=393041

Balachander, S. (2001). Warranty signaling and reputation. *Management Science, 47*(9), 1282-1289.

Baldi, S., & Thaung, H. (2002). The Entertaining Way to M-Commerce: Japan's Approach to the Mobile Internet—A Model for Europe? *Electronic Markets, 12*(1).

Barber, B. (1983). *The logic and limits of trust.* New Brunswick, NJ: Rutgers University Press.

Barber, K.S., & Kim, J. (2000). *Belief revision process based on trust: Agents evaluating reputation of information sources.* Retrieved March 9, 2007, from http://www. istc.cnr.it/T3/download/aamas2000/Barber-Kim.pdf

Barney, J. B., & Hansen, M. H. (1994). Trustworthiness as a source of competitive advantage. *Strategic Management Journal, 15*, 175-190.

Barr, T., Knowles, A., & Moore, S. (2003). Trust in transactions: Australian Internet research. Paper presented at the Communications Research Forum 2003. Retrieved March 9, 2007, from http://www.dcita.gov.au/crf/papers03/barr3final.pdf

Baym, N. (1995). The emergence of community in computer-mediated communication. In S. Jones (Ed.), *CyberSociety* (pp. 138-163). Newbury Park, CA: Sage.

Beatty, S. E., Mayer, M., Coleman, J. E., Reynolds, K. E., & Lee, J. (1996). Customer-sales associate retail relationships. *Journal of Retailing, 72*(3), 223-247.

Bechar-Israeli, H. (1995). From <Bonehead> to <cLoNehEAd>: Nicknames, play, and identity on Internet relay chat. *Journal of Computer-Mediated Communication, 1*(2). Retrieved March 11, 2007, from http://www.ascusc.org/jcmc/vol1/issue2/bechar.html

Beck, U., & Beck-Gernsheim, E. (1994). *Individualization.* London: Sage.

Beck, U., Giddens, A., & Lash. S. (1994). *Reflexive modernisation.* Cambridge, UK: Polity.

Beckert, J. (2005). *Trust and the performative construction of markets* (MPIfG Discussion Paper 05/8). Cologne: Max Planck Institute for Studies of Societies.

Bekkering, E., & Shim, J. P. (2006). Trust in videoconferencing. *Communications of the ACM, 49*(7), 103-107.

Belasco, J.A., & Sayer, R.C. (1994, March-April). Why empowerment doesn't empower: The bankruptcy of current paradigms. *Business Horizons, 37*(2), 29-41.

Bell, B., & Kozlowski, S.W. (2002). A typology of virtual teams, implications for effective leadership. *Group & Organization Management, 27*(1), 14-49.

Benassi, P. (1999). TRUSTe: An online privacy seal program. *Communications of the ACM, 42*(2), 56-59.

Bennatan, E. M. (2002, November). Globalization: Boon or bane. *Cutter IT Journal,* 15(11).

Berg, J., Dickhaut, J., & McCabe, K. (1995). *Trust, reciprocity, and social history* (Unpublished Working Paper). University of Minnesota, Minneapolis.

Berger, P. L., & Luckmann, T. (1966). *The social construction of reality.* Garden City: Doubleday.

Bernhardt, H. R., Killworth, P., et al. (1982). Informant accuracy in social network data V. *Social Science Research, 11*, 30-66.

Berson, Y., Dan, O., & Yammarino, F.J. (2006a-d). Attachment style and individual differences in leadership perceptions and emergence. *The Journal of Social Psychology, 146*(2), 165-182.

Bertrand, M., Duflo, E., & Mullainathan, S. (2004). How much should we trust differences-in-differences estimates? *The Quarterly Journal of Economics, 119*(1), 249-275.

Bhattacherjee, A. (2002). Individual trust in online firms: Scale development and initial test. *Journal of Management Information Systems, 19*(1), 211-241.

Bhimani, A. (1996). Securing the commercial Internet. *Communications of the ACM, 39*(6), 29-35.

Blanchard, A. (2004). Virtual behavior setting: An application of behavior setting theories to virtual communities. *Journal of Computer-Mediated Communication, 9*(2). Retrieved March 11, 2007, from http://jcmc.indiana.edu/vol9/issue2/blanchard.html

Blois, K. (1998). A trust interpretation of business to business relationships: A case-based discussion. *Management Decision, 36,* 5.

Blois, K.J. (1999). Trust in business to business relationships: An evaluation of its status. *Journal of Management Studies, 36*(2), 197-215.

Blum, S.D. (2005). Five approaches to explaining "truth" and "deception" in human communication. *Journal of Anthropological Research, 61*(3), 289-315.

Boisot, M. (1995). Is your firm a creative destroyer? Competitive learning and knowledge flows in the technological strategies of firms. *Research Policy, 24*, 489-506.

Boissevain, J. (1974). *Friends of friends. Networks, manipulators and coalitions.* Oxford: Basil Blackwell.

Bolton, G. (1991). A comparative model of bargaining: Theory and evidence. *American Economic Review, 81,* 1096-1135.

Bons, R.W.H., Lee, R.M., & Wagenaar (1998). Obstacles for the development of open electronic commerce. *International Journal of Electronic Commerce, 2*(3), 61-83.

Bontis, N. (1999). Managing organizational knowledge by diagnosing intellectual capital: Framing and advancing the state of the field. *International Journal of Technology Management, 18*(5-8), 433-462.

Boon, S., & Holmes, J. (1991). The dynamics of interpersonal trust: Resolving uncertainty in the face of risk. In R.A. Hinde & J. Grobel (Eds.), *Cooperation and prosocial behavior* (pp. 190-213). Cambridge, UK: Cambridge University Press.

Booz, A. H. (2000). The wireless internet revolution [Electronic Version]. *Insights: Communications, Media & Technology Group, 6*(1). Retrieved from http://www.boozallen.com/media/file/34103.pdf

Borden, G. A. (1971). *An introduction to human communication theory.* Dubuque, IA: W.C. Brown Company Publishers.

Borgatti, S. P., & Foster, P. C. (2003). The network paradigm in organizational research: A review and typology. *Journal of Management, 6*(23), 991-1013.

Borgatti, S. P., Everett, M.G., et al. (2002). *Ucinet 6 for windows.* Harvard: Analytic Technologies.

Bos, N., Olson, J., Gergle, D., Olson, G., & Wright, Z. (2002). Effects of four computer-mediated communications channels on trust development. In *Proceedings of SIGCHI Conference on Human Factors in Computing Systems* (pp. 135-140). New York: ACM Press. Retrieved August 23, 2006, from http://portal.acm.org

Boyd, J. (2003). The rhetorical construction of trust online. *Communication Theory, 13*(4), 392-410.

Bradach, J. L., & Eccles, R. G. (1989). Price, authority and trust: From ideal types to plural forms. *Annual Review of Sociology, 15,* 97-118.

Braudel, F. (1981). *Civilisation and capitalism, 15th-18th century.* London: Collins/Fontana.

Breiger, R. (2004). The analysis of social networks. In M. Hardy & A. Bryman (Eds.), *Handbook of data analysis.* London: Sage.

Bromiley, P., & Cummings, L. L. (1995). Transactions costs in organizations with trust. In R. J. Lewicki, R. J. Bies, & B. H. Sheppard (Eds.), *Research on negotiation in organizations, 5* (pp. 219-247). Greenwich, CT: JAI Press.

Brown, H.G., Poole, M.S., & Rodgers, T.L. (2004). Interpersonal traits, complementarity and trust in virtual collaboration. *Journal of Management Information Systems, 20*(4), 115-137.

Brown, J.S., & Duguid, P. (2000). *The social life of information.* Cambridge, MA: Harvard Business School Press.

Brown, J.S., Denning S., Groh, K., & Prusak, L. (2004). *Storytelling in organizations.* Woburn, MA: Butterworth-Heinemann.

Bruque, S., Moyano, J., & Eisenberg, J. (2006). The effects of social networks on worker's adaptation to a major technological change. *Academy of Management Meeting Proceedings,* Atlanta, GA, electronic version.

Bruque-Cámara, S. (2002). *The paradox of IT productivity. The case of the pharmaceutical distribution industry.* Jaén: University of Jaén.

Brynjolfsson, E., & Smith, M. (2000). Frictionless commerce? A comparison of Internet and conventional retailers. *Management Science, 46*(4), 563-585.

Bucy, E.P. (2004) Interactivity in society: Locating an elusive concept. *The Information Society, 20*(5), 373-383.

Burt, R. (2005). *Brokerage and closure. An introduction to social capital.* New York: Oxford University Press.

Burt, R. S., & Knez, M. (1996). Trust and third-party gossip. In R. M. Kramer & T. R. Tyler (Eds.), *Trust in organizations. Frontiers of theory and research* (pp. 68-89). Thousand Oaks, London, New Delhi: Sage Publications.

Butler, J.K. (1991). Towards understanding and measuring conditions of trust: Evolution of a condition of trust inventory. *Journal of Management, 17*(3), 643-663.

Carlson, J.R., & Zmud, R.W. (1999). Channel expansion theory and the experiential nature of media richness perceptions. *Academy of Management Journal, 42*(2), 153.

Carrier, J., & Miller, D. (1998) *Virtuality: A new political economy.* Oxford, UK: Berg Publishers.

Cascio, W.F., & Shurygailo, S. (2003). E-leadership and virtual teams. *Organizational Dynamics, 31*(4), 362-376.

Casson, M. (1995). *The organisation of international business: Studies in the economics of trust.* Aldershot, UK: Edward Elgar.

Castaldo, S. (2002). *Fiducia e relazioni di mercato.* Bologna: Il Mulino.

Castelfranchi, C., & Tan, Y.H. (2002). The Role of Trust and Deception in Virtual Societies. *International Journal of Electronic Commerce 6*(3), 55-70.

Castelfranchi, C. (2000). Why computers will (necessarily) deceive us and each other. *Ethics and Information Technology, 2*, 113-119.

Castelfranchi, C., & Falcone, R. (2001). *Social trust: A cognitive approach.* In C. Castelfranchi & Y.H. Tan (Eds.), *Deception, fraud and trust in virtual societies* (pp. 55-90). Dodrecht, The Netherlands: Kluwer.

Castelfranchi, C., & Tan, Y. (Eds.). (2001). *Trust in virtual societies.* Amsterdam: Kluwer Academic Publishers.

Castelfranchi, C., & Tan, Y.-H. (2001). *Trust and deception in virtual societies.* Norwell, MA: Kluwer Academic Publishers.

Castells, M. (1998). *The information era: Economy, society and culture.* Madrid: Alianza.

Cavoukian, A., & Gurski, M. (2002). Privacy in a wireless world. *Business Briefing: Wireless Technology.* Retrieved from http://www.ipc.on.ca

Cazier, J. A., Shao, B. B. M., & St. Louis, R. D. (2006). E-business differentiation through value-based trust. *Information and Management, 43*(6), 718-727.

Cellich, C., & Jain, S.C. (2003). *Global business negotiations: A practical guide.* Mason, OH: Thomson/South-Western.

Chakrabarty, S. K., Gandhi, P., & Kaka, N. (2006). The untapped market for offshore service. *McKinsey Quarterly*, Issue 2.

Chen, S. C., & Dhillon, G. S. (2003). Meeting dimensions of to consumer trust in e-commerce. *Information Technology and Management, 4*(2-3), 303-315.

Cheskin Research & Studio Archetype/Sapient (1999). *eCommerce trust study.* Retrieved March 9, 2007, from http://www.sapient.com/cheskin

Child, J. (1998). Trust and international strategic alliances. In C. Lane & R. Bachmann (Eds.), *Trust within and between organizations. Conceptual issues and empirical applications* (pp. 241-272). Oxford: Oxford University Press.

Child, J. (2001). Trust – The fundamental bond in global collaboration. *Organizational Dynamics, 29*(4), 274-288.

Chiles, T. H., & McMackin, J. F. (1996). Integrating variable risk preferences, trust and transaction cost economics. *Academy of Management Review, 21*(1), 73-99.

Choi, B., & Lee, H. (2003). An empirical investigation of KM styles and their effect on corporate performance. *Information & Management, 40*, 403-417.

Chung, H., & Zhao, X. (2004). Effects of perceived interactivity on Web site preference and memory: Role of personal motivation. *Journal of Computer-Mediated Communication, 10*(1), 7. Retrieved March 11, 2007, from http://jcmc.indiana.edu/vol10/issue1/chung.html

272

Clark, B.H., & Montgomery, D.B. (1998). Deterrence, reputations, and competitive cognition. *Management Science, 44,* 62-82.

Clarke, I. (2001). Emerging value propositions for m-commerce. *Journal of Business Strategies, 18*(2), 133-148.

Clayman, S.E. (1993). Booing: The anatomy of a disaffiliative response. *American Sociological Review, 58,* 110-130.

Clothes with a silver lining. (2006, Fall). *Military Officer,* p. 28.

Cohen, D. & Prusak, L. (2001). *In good company: How social capital makes organizations works.* Boston, MA: Harvard Business School Press.

Cohen, S.G., & Gibson, C.B. (2003). In the beginning: Introduction and framework. In C.B. Gibson & S.G. Cohen (Eds.), *Virtual teams that work* (pp. 1-13). San Francisco: Jossey-Bass.

Coleman, J. (1988). Social capital in the creation of human capital. *American Journal of Sociology, 94,* 95-120.

Coleman, J. (1990). *Foundations of social theory.* Boston: Harvard University Press.

Coleman, J. S. (1990). *Foundations of social theory.* Cambridge, MA: Harvard University Press.

Conley Tyler, M., & Raines, S. (2006). The human face of online dispute resolution. *Conflict Resolution Quarterly, 23*(3), 333-342.

Contractor, F. J., & Lorange, P. (Eds.). (1988). *Cooperative strategies in international business.* New York: Lexington Books.

Cook, K., Hardin, R., & Levi M. (Eds.). (2005). *Cooperation without trust.* New York: Russell Sage Foundation.

Coon, D.A. (1996). *An investigation of friend Internet relay chat as a community.* Unpublished master's dissertation, Kansas State University. Retrieved March 11, 2007, from http://www.davidcoon.com/thesis.txt

Coppola, N. W., Hiltz, S. R., & Rotter, N. G. (2004). Building trust in virtual teams. *IEEE Transactions on Professional Communication, 47*(2), 95-104.

Cotton, J.L., Vollrath, D.A., Lengnick-Hall, M.L., & Froggatt, K.L. (1990). Fact: The form of participation does matter – a rebuttal to Leana, Locke, & Schweiger. *Academy of Management Review, 15,* 147-153.

Cox, T. (1993) *Cultural diversity in organizations.* San Francisco: Berrett-Koehler.

Cranor, L.F. (1999). Internet privacy. *Communications of the ACM, 42*(2), 28-31.

Cross, R., & Prusak, L. (2002). The people who make organizations go or stop. *Harvard Business Review, 80*(6), 104-112.

Crow, G., & Allan, G. (1994). *Community life: An introduction to local social relations.* Hemel Hempstead, UK: Harvester Wheatsheaf.

Csikszentmihalyi, M. (1975). *Beyond boredom and anxiety.* San Francisco: Jossey-Bass.

Csikszentmihalyi, M. (1988) *Optimal experience.* New York: Cambridge University Press.

Csikszentmihalyi, M. (1990). *Flow: The psychology of optimal experience.* New York: Harper and Row.

Cummings, J. N. (2004). Work groups, structural diversity, and knowledge sharing in a global organization. *Management Science, 50*(3), 352-364.

Cummings, J. N., & Cross, R. (2003). Structural properties of work groups and their consequences for performance. *Social Networks, 25,* 197-210.

Czepiel, J. A. (1990). Service encounters and service relationships: Implications for research. *Journal of Business Research, 20*(1), 13-21.

Daft, R. L., & Lengel, R. H. (1984). Information richness: A new approach to managerial behavior and organization design. In L. L. Cummings & B. M. Staw (Eds.), *Research in organizational behavior, 6* (pp. 191-223). Greenwich, CT: JAI Press.

Daft, R. L., & Lengel, R. H. (1986). Organizational information requirements, media richness and structural design. *Management Science, 32*(5), 554-571.

Daft, R. L., & Macintosh, N .B. (1981). A tentative exploration into the amount and equivocality of information processing in organizational work units. *Administrative Science Quarterly, 26*(2), 207-224.

Daft, R. L., Lengel, R. H., & Trevino, L. K. (1987). Message equivocality, media selection, and manager performance: Implications for information systems. *MIS Quarterly, 11*(3), 355-366.

Daga, V., & Kaka, N. F. (2006, May). Taking offshoring beyond labor cost savings. *The McKinsey Quarterly.*

Damer, T.E. (2001). *Attacking faulty reasoning: A practical guide to fallacy-free arguments* (4th ed.). Canada: Wadsworth/Thomson Learning, Inc.

Dasgupta, P. (1988). Trust as a commodity. In D. Gambetta (Ed.), *Trust: Making and breaking cooperative relations* (pp. 49-72). New York: Basil Blackwell Ltd.

Davidson, J., & Rees-Mogg, W. (1997). *The sovereign individual.* London: Pan.

Davis, D.D. (2003). The Tao of leadership in virtual teams. *Organizational Dynamics, 33*(1), 47-62.

Davis, J.H., Schoorman, F.D., Mayer, R.C., & Tan, H.H. (2000). The trusted general manager and business unit performance: Empirical evidence of a competitive advantage. *Strategic Management Journal, 21*, 563-576.

Davison, D. (2004). Top 10 risks of offshore outsourcing. *CIO Magazine.* Retrieved March 15, 2007, from http://search.cio.techtarget.cotn/orginalComent/0,289142.sidl9_gci950602,00.html

Dellarocas, C. (2003). The digitization of the word of mouth: Promise and challenges of online feedback mechanisms. *Management Science, 49*(10), 1407-1424.

Dellarocas, C. (2005). Reputation mechanism design in online trading environments with pure moral hazard. *Information Systems Research, 16*(2), 209-230

Dellarocas, C., & Resnick, P. (2003). *Online reputation mechanisms: A roadmap for future research.* Summary Report of the First Interdisciplinary Symposium on Online Reputation Mechanisms. Retrieved March 9, 2007, from http://www2.sims.berkeley.edu/research/conferences/p2pecon/papers/s8-dellarocas

Denison, D.R., Hooijberg, R., & Quinn. R.E. (1995). Paradox and performance: Toward a theory of behavioural complexity in managerial leadership. *Organization Science, 6*(5), 524-540.

Denning, D. E. (1993). A new paradigm for trusted systems. In *Proceedings of the 1993 Association for Computing Management SIGSAC on New Security Paradigms Workshop* (pp. 36-41).

Dennis, A.R., & Kinney, S.T. (1998). Testing media richness theory in the new media: The effects of cues, feedback, and task equivocality. *Information Systems Research, 9*(3), 256.

Dennis, A.R., & Valacich, J.S. (1999). *Rethinking media richness: Towards a theory of media synchronicity.* Paper presented at the 32nd Hawaii International Conference on Systems Sciences.

Dennis, A.R., Valacich, J.S., Carte, T.A., & Garfield, M.J. (1997). Research report: The effectiveness of multiple dialogues in electronic brainstorming. *Information Systems Research, 8*(2), 203.

Deutsch, M. (1952). Trust and suspicion. *Journal of Conflict Resolution, 2*, 265-279.

Deutsch, M. (1958). Trust and suspicion. *The Journal of Conflict Resolution (pre-1986), 2*(4), 265.

Deutsch, M. (1962). Cooperation and trust: Some theoretical notes. In M.R. Jones (Ed.), *Nebraska symposium on motivation* (pp. 275-318). Lincoln: University of Nebraska Press.

Dierickx, I., & Cool, K. (1989). Asset stock accumulation and the sustainability of competitive advantage. *Management Science, 35*, 1504-1513.

Dillon, S. (2004, December 7). What corporate America cannot build: A sentence. *The New York Times*, 23.

Directive n. 1103/97, June 17th (1997), Retrieved June 29, 2007, from http://eurlex.eruopa.eu/LexUriServ/Lex-UriServ.do?uri-OJ:C:2007:056:0009:0010:EN:PDF

Dirks, K. T., & Ferrin, D. L. (2001). The role of trust in organizational settings. *Organization Science, 12*(4), 450-467.

Dirks, K. T., & Ferrin, D. L. (2002). Trust in leadership: Meta-analytic findings and implications for research and practice. *Journal of Applied Psychology, 87*(4), 611-628.

DiStefano, J.J., & Maznevski, M. (2000). Creating value with diverse teams in global management. *Organizational Dynamics, 29*(1), 45-63.

Doh, J. (2005, May 3). Offshore outsourcing: Implications for international business and strategic management theory and practice. *Journal of Management Studies, 42,* 696-704.

DOMINO. (2005). Network information infrastructures management in the construction industry: Emergence and impact on work and management arrangements. Retrieved February 22, 2007, from http://www.ist-domino.net/dynamic/public.php?action=outline4

Doney, P. M., & Cannon, J. P. (1997). An examination of the nature of trust in buyer-seller relationships. *Journal of Marketing, 61*(2), 35-51.

Doney, P. M., Cannon, J. P., & Mullen, M. R. (1998). Understanding the influence of national culture on the development of trust. *Academy of Management Journal, 23*(3), 601-620.

Driskell, J.E., Radtke, P.H., & Salas, E. (2003). Virtual teams: Effects of technological mediation on team performance. *Group Dynamics, 7*(4), 297-323.

Driver, M. (2003). Diversity and learning in groups. *The Learning Organization, 10*(3), 149-166.

Drolet, A.L., & Morris, M.W. (2000). Rapport in conflict resolution: Accounting for how face-to-face contact fosters mutual cooperation in mixed-motive conflicts. *Journal of Experimental Social Psychology, 36*(1), 26-50.

Drucker, P.F. (1991, November-December). The new productivity challenge. *Harvard Business Review, 69,* 69-76.

Duarte, D.L., & Snyder, N.T. (2001). *Mastering virtual teams: Strategies, tools and techniques that succeed* (2nd ed.). New York: Jossey-Bass.

Dubé, L., & Paré, G. (2002). The multi-faceted nature of virtual teams. *Cahier du GreSI, N° 02-11,* 1-33.

Dubrovsky, V.J., Kiesler, S., & Sethna, B.N. (1991). The equalization phenomenon: Status effects in computer-mediated and face-to-face decision-making groups. *Human-Computer Interaction, 6,* 119-146.

Edley, P. P., Hylmö, A., & Newsom, V. A. (2004). Alternative organizing communities: Collectivist organizing, telework, home-based Internet businesses, and online communities. *Communication Yearbook, 28*(1), 87-125.

Edvinsson, L., & Malone, M.S. (1997). *Intellectual capital: Realizing your company's true value by finding its hidden brainpower.* New York: Harper Business.

Eisenhardt, K. M. (1989). Building theories from case study research. *Academy of Management Review, 14*(4), 532-550.

Ekstrom, M. A., Bjornsson, H. C., & Nass, C. I. (2005). A reputation mechanism for business-to-business electronic commerce that accounts for rather credibility. *Journal of Organizational Computing and Electronic Commerce, 15*(1), 1-18.

Elias, N. (1994). *The civilising process.* Oxford: Blackwell.

Ellison, S., Simmerman, A., & Forelle, C. (2005, January 31). P&G's Gillette edge: The playbook it honed at Wal-Mart. *The Wall Street Journal,* pp. A1, A18.

Elron, E., & Vigoda, E. (2003). Influence and political processes in virtual teams. In C.B. Gibson & S.G. Cohen (Eds.), *Virtual teams that work* (pp. 317-334). San Francisco: Jossey-Bass.

El-Shinnawy, M., & Markus, L. (1998). Acceptance of communication media in organizations: Richness or features? *IEEE Transactions on Professional Communication, 41*(4), 242.

Elster, J. (1983). *Sour grapes: Studies in the subversion of rationality.* Cambridge, MA: Cambridge University Press.

Everson, M. (2001). The Name of the Euro in European languages. *An Aimsir, 2.*

Farrell, D. (2006, June), Smarter offshoring. *Harvard Business Review,* 85-92.

Faulkner, D. (1999, November 29). Trust and control in strategic alliances. *Financial Times.* Retrieved February 21, 2007, from http://www.sbs.ox.ac.uk/sbs/newco6jl.html

Felstead, A., Jewson, N., Phizacklea, A., & Walters, S. (2001). Working at home: statistical evidence of seven key hypotheses. *Work, Employment and Society, 15*(2), 215-31.

Feng, J., Lazar, J., & Preece, J. (2004). Empathy and online interpersonal trust: A fragile relationship. *Behavior & Information Technology, 23*(2), 97-106.

Fernandez, R. M., & Gould, R. V. (1994). A dilemma of state power: Brokerage and influence in the national health policy domain. *American Journal of Sociology, 99,* 1455-91.

Fernback, J., & Thompson, B. (2000, May). Virtual communities: Abort, retry or failure? Online version of *Computer Mediated Communication and the American Collectivity: The Dimensions of a Community Within Cyberplace.* Paper presented at the Annual Conference of the International Communication Association. Retrieved March 11, 2007, from http://www.well.com/user/hlr/texts/Vccivil.html

Ferrin, D. L., Dirks, K. T., & Shah, P. P. (2006). Direct and indirect effects of third-party relationships on interpersonal trust. *Journal of Applied Psychology, 91*(4), 870-883.

Fishbein, M., & Ajzen, I. (1975). *Belief, attitude, intention, and behavior: An introduction to theory and research.* Reading, MA: Addison-Wesley.

Fisher, R., Ury, W., & Patton, B. (1991). *Getting to yes* (2nd ed.). New York: Penguin Books.

Fleisher, C., & Bensoussan, B. (2002). *Strategic and competitive analysis: Methods and techniques for analyzing business competition.* Upper Saddle River, NJ: Prentice Hall.

Fletcher, R.J. (1995). *The limits of settlement growth: A theoretical outline.* Cambridge: Cambridge University Press.

Fombrun, C.J. (1996). *Reputation: Realizing value from the corporate image.* Boston: Harvard Business School Press.

Foote, D. (2004, September). Recipe for offshore outsourcing failure: Ignore organization, people issues. *ABA Banking Journal.*

Foster, I., & Kesselman, C. (1999). *The grid: Blueprint for a new infrastructure.* New York: Morgan Kaufmann.

Friedman, B. (2000). Trust online. *Communications of the ACM, 43*(12), 34-40.

Friedman, R. A., & Podolny, J. (1982). Differentiation of boundary spanning roles: Labor negotiations and implications for role conflict. *Administrative Science Quarterly, 1*(37), 28-47.

Fryxell, G.E., Dooley, R.S., & Vryza, M. (2002). After the ink dries: The interaction of trust and control in US-based international joint ventures. *Journal of Management Studies, 39,* 865-886.

Fukuyama, F. (1995). *Trust: The social virtues & the creation of prosperity.* New York: The Free Press.

Gabarro, J.J. (1978). The development of trust, influence, and expectations. In A.G. Athos & J.J. Gabarro (Eds.), *Interpersonal behavior: Communication and understanding in relationships* (pp. 290-303). Englewood Cliffs, NJ: Prentice Hall.

Gambetta, D. (1988). Can we trust trust? In D. Gambetta (Ed.), *Trust: Making and breaking cooperative relations* (pp. 213-237). Oxford: Basil Blackford.

276

Gambetta, D. (Ed.). (1988). *Trust: Making and breaking cooperative relations*. Oxford: Basic Blackwell.

Ganesan, S. (1994). Determinants of long-term orientation in buyer-seller relationships. *Journal of Marketing, 58*, 1-19.

Garfinkel, H. (1963). A conception of, and experiments with, "trust" as a condition of stable concerted action. In O.J. Harvey (Ed.), *Motivation and social interaction* (pp. 187-238). New York: Ronald Press.

Gefen, D. (1997). *Building users' trust in freeware providers and the effects of this trust on users' perception of usefulness, easy of use and intended use of freeware*. Unpublished doctoral dissertation, Georgia State University.

Gefen, D. (2000). E-commerce: The role of familiarity and trust. *Omega: The International Journal of Management Science, 28*(5), 725-737.

Gefen, D. (2002). Reflections on the dimensions of trust and trustworthiness among online consumers. *ACM SIGMIS Database, 33*(3), 38-53.

Gefen, D., & Heart, T. (2006). On the need to include national culture as a central issue in e-commerce trust beliefs. *Journal of Global Information Management, 14*(4), 1-30.

Gefen, D., Karahanna, E., & Straub, D. W. (2003). Inexperience and experience with online stores: The importance of TAM and trust. *IEEE Transactions on Engineering Management, 50*(3), 307-321.

George, J. M., & Jones, G. R. (1996). *Understanding and managing organizational behavior*. Reading, MA: Addison-Wesley Publishing Company.

Gerck, E. (1998). *Towards real-world models of trust: Reliance on received information.*

Giddens, A. (1984). *The constitution of society*. Cambridge, UK: Polity.

Giddens, A. (1990). *The consequences of modernity*. Cambridge, UK: Polity.

Giddens, A. (1991). *Modernity and self-identity*. Cambridge, UK: Polity.

Giddens, A. (1994). Living in a post-traditional society. In U. Beck, A. Giddens, & S. Lash (Eds.), *Reflexive modernisation* (pp. 56-109). Cambridge, UK: Polity.

Giddens, A. (1994). Risk, trust, reflexivity. In *Reflexive Modernization*. U. Beck, A. Giddens, & S. Lash (Eds.), Cambridge, UK: Polity Press, 184-197.

Giffin, K. (1967). The contribution of studies of source credibility to a theory of interpersonal trust in the communication process. *Psychological Bulletin, 68*(2), 104-120.

Gill, M. J. (2003). Biased against "them" more than "him": Stereotype use in group-directed and individual-directed judgments. *Social Cognition, 21*(5), 321-348.

Golembiewski, R.T., & McConkie, M. (1975). The centrality of interpersonal trust in group processes. In G.L. Cooper (Ed.), *Theories of group processes* (pp. 131-85). London: John Wiley.

Gomes-Casseres, B. (1997). *The alliance revolution. The new shape of business rivalry*. Cambridge, MA: Harvard University Press.

Good, D. (1988). Individuals, interpersonal relations, and trust. In D. Gambetta (Ed.), *Trust: making and breaking cooperative relations* (pp. 31-48). New York: Basil Blackwell Ltd.

Grabosky, P. (2001, April 23). The nature of trust online [Electronic Version]. *The Age*, pp. 1-12. Retrieved from http://www.aic.gov.au/publications/other/online_trust.html

Granovetter, M. (1985). Economic action and social structure: A theory of embeddedness. *American Journal of Sociology, 91*(3), 481-510.

Granovetter, M. S. (1973). The strength of weak ties. *American Journal of Sociology, 81*, 1287-1303.

Grant, R.M. (1996, Winter). Toward a knowledge-based theory of the firm. *Strategic Management Journal, 17*, 109-122.

Griffith, D. A., Hu, M. Y., & Ryans, J. K. (2000). Process standardization across intra- and inter-cultural relationships. *Journal of International Business Studies, 31*(2), 303-325.

Griffith, T.L., Mannix, E.A., & Neale, M.A. (2003). Conflict and virtual teams. In C.B. Gibson & S.G. Cohen (Eds.), *Virtual teams that work* (pp. 335-352). San Francisco: Jossey-Bass.

Grossman, M. (2004, September). The role of trust and collaboration in the Internet-enabled supply chain. *The Journal of American Academy of Business*, pp. 391-396.

Gual, J., & Ricart, J. E. (2001). *Enterprise strategies in telecommunications and Internet.* Barcelona: Retevisión Foundation.

Guernsey, L. (2000). Suddenly, everybody's an expert on everything. Retrieved March 9, 2007, from http://www.nytimes.com/library/tech/00/02/circuits/articles/03info.html

Guinan, P. J., & Faraj, S. (1998). Reducing work related uncertainty: The role of communication and control in software development. In *Proceedings of the 31st Annual Hawaii International Conference on Systems Sciences* (pp. 73-82). Retrieved February 21, 2007, from http://ieeexplore.ieee.org/iel5/5217/14260/00654761.pdf

Gulati, R. (1995). Does familiarity breed trust? The implications of repeated ties for contractual choice in alliances. *Academy of Management Journal, 38*(1), 85-112.

Gulati, R. (1995). Does familiarity breed trust? The implications of repeated ties for contractual choice in alliances. *Academy of Management Journal, 38*(1), 85-112.

Gulta, R. (1995). Does Familiarity Breed Trust? The Implications of Repeated Ties for Contractual Choice in Alliances. *Academy of Management Journal 38*, 85-112.

Hagel, J., & Armstrong, A. (1997). *Net gain: Expanding markets through virtual communities.* Boston: Harvard Business School Press.

Hall, E.T. (1976). *Beyond culture.* Garden City, CA: Anchor.

Hall, E.T., & Hall, M. (1990). *Understanding cultural differences.* Yarmouth, Maine: Intercultural Press.

Hampden-Turner, C., & Trompenaars, F. (2000). *Building cross-cultural competence: How to create wealth from conflicting values.* Chichester, UK: John Wiley & Sons.

Hancock, B. (1999). Security Views. *Computers and Security 19*(7), 553-64.

Handy, C. (1995, May-June). Trust and the virtual organization, *Harvard Business Review, 73*(3), 40-50.

Hanneman, R. (2001). *Introduction to social network methods.* Department of Sociology, University of California, Riverside.

Hansen, M. (1999). The search-transfer problem: The role of weak ties in sharing knowledge across organization subunits. *Administrative Science Quarterly, 44,* 82-111.

Haragadon, A. B. (1998). Firms as knowledge brokers: Lessons in pursuing continuous innovation. *Management Review, 40*(3), 209-227.

Hardin, R. (1993). The street-level epistemology of trust. *Politics and Society, 21*(4), 505-529.

Hardin, R. (2006). *Trust.* United Kingdom: Polity.

Hardin, R. (Ed.). (2004). *Distrust.* New York: Russell Sage Foundation.

Hart, R.K., & McLeod, P.L. (2003). Rethinking team building in geographically dispersed teams: One message at a time. *Organizational Dynamics, 31*(4), 352-361.

Hartley, C., Brecht, M., et al. (1977). Subjective time estimates of work tasks by office workers. *Journal of Occupational Psychology, 50,* 23-36.

Hatch, P. J. (2005). Offshore 2005: Research, findings and conclusions (White Paper). Retrieved March 15, 2007, from http://www.ventoro.com/Offshore2005ResearchFindings.pdf

Hayes, I. S. (2003, August). *Ready or not: Offshoring: Is it a win-win game?* McKinsey Global Institute Research Papers.

Heeter, C. (1989). Implications of new interactive technologies for conceptualizing communication. In J.L. Salvaggio & J. Bryant (Eds.), *Media use in the information age: Emerging patterns of adoption and consumer use* (pp. 221-225). Hillsdale, NJ: Lawrence Erlbaum Associates.

Hel, D., van Niekerk, R., Berthon, J.P., & Davies, T. (1999). Going with the flow: Web sites and customer involvement. *Internet Research, 9*(2), 109-116.

Henttonen, K., & Blomqvist, K. (2005). Managing distance in a global virtual team: The evolution of trust through technology-mediated relational communication. *Strategic Change, 14*(2), 107-119.

Hoffman, D. L., Novak, T. P., & Peralta, M. (1999). Building consumer trust online. *Communications of the ACM, 42*(4), 80-85. Association for Computing Machinery.

Hoffman, D., Novak, T.P., & Peralta (1999). Building consumer trust online. *Communications of the ACM, 42*(4), 80-85.

Hoffman, D.L., & Novak, T.P. (1996). Marketing in hypermedia computer-mediated environments: conceptual foundations. *Journal of Marketing 60*, 50-68.

Hofstede, G. (1980). *Culture's consequences: International differences in work related values.* Beverly Hills, CA: Sage.

Hofstede, G. (1980). Motivation, leadership, and organization: Do American theories apply abroad? *Organizational Dynamics, 9*(1), 42-63.

Hofstede, G. (1983). The cultural relativity of organizational practices and theories. *Journal of International Business Studies, 14*(2), 75.

Hofstede, G. (1991). *Cultures and organizations: Software of the mind.* London: McGraw-Hill.

Hofstede, G. (1994). *Cultures and organizations: Software of the mind: Intercultural.* London: HarperCollins.

Hofstede, G. (2001) *Culture's consequences: Comparing values, behaviours, institutions and organizations across nations* (2nd ed.). Beverly Hills, CA: Sage.

Hofstede, G. H. (1980). *Culture's consequences: International differences in work related values.* Beverly Hills, CA: Sage Publications.

Hogg, M.A. (1996). Social identity, self-categorization, and the small group. In E. Witte & J.H. Davis (Eds.), *Understanding group behavior: Small group processes and interpersonal relations* (Vol. 2, pp. 227-254).

Hogg, M.A., Terry, D.J., & White, K.M. (1995). A tale of two theories: A critical comparison of identity theory with social identity theory. *Social Psychology Bulletin, 58*(4), 255-269.

Homburg, C., & Furst, A. (2005). How organizational complaint handling drives customer loyalty: An analysis of the mechanistic and the organic approach. *Journal of Marketing, 69*(3), 95-114.

Hosmer, L. T. (1995). Trust: The connecting link between organizational theory and philosophical ethics. *Academy of Management Review, 20*(2), 379-403.

Hossain, L., & Wigland, R.T. (2004). ICT enabled virtual collaboration through trust. *Journal of Computer-Mediated Communications, 10*(1), 1-22.

Hotopp, U. (2002). Teleworking in the UK: The trends and characteristics of teleworking in the UK and comparisons with other Western countries. *Labor Market Trends, 110*(6).

Huang, W.W., Wei, K.-K., Watson, R.T., & Tan, B.C. (2002). Supporting virtual team-building with a GSS: An empirical investigation. *Decision Support Systems, 34*, 359-367.

Hubbell, A. P., & Chory-Assad, R. M. (2005). Motivating factors: Perceptions of justice and their relationship with managerial and organizational trust. *Communication Studies, 56*(1), 47-70.

Huemer, L. (1998). *Trust in business relations: Economic logic or social interaction?* Umea: Borea bokvorlag.

Huff, L., & Kelley, L. (2005). Is collectivism a liability? The impact of culture on organizational trust and customer orientation. A seven nation study. *Journal of Business Research, 58*(1), 96-102.

Hummels, H., & Roosendaal, H.E. (2001). Trust in scientific publishing. *Journal of Business Ethics, 34*(2), 87-100.

Huntley, H. (2006, March). *Offshore deal failure: What providers need to know.* Gartner Research.

Huws, U., Korte, W. B., & Robinson, S. (1990). *Telework: Towards the elusive office.* New York: John Wiley.

Iacono, C.S., & Weisband, S. (1997). Developing trust in virtual teams. In *Proceedings of the 30th Annual Hawaii International Conference on System Sciences.*

IDC (2005). Poland IT services 2005-2009: Forecast and 2004 vendor shares. Retrieved March 14, 2007, from http://idc.com/getdoc.jsp?containerId=ES05M

Itani, W., & Kayssi, A. (2004). J23ME application-layer end-to-end security for m-commerce. *Journal of Network and Computer Applications, 1,* 13-33.

Jackson, P. J., & van der Wielen, J. M. (1998). *Teleworking: International Perspectives: From Telecommuting to the Virtual Organisation.* Routeledge, London.

Järvenpaa, S. L., & Leidner, D. E. (1998). Communication and trust in global virtual teams. *Journal of Computer-Mediated Communication, 3*(4). Retrieved February 21, 2007, from http://www.ascusc.org/jcmc/vol3/issue4/jarvenpaa.html

Jarvenpaa, S. L., & Leidner, D. E. (1999). Communication and trust in global virtual teams. *Organization Science: A Journal of the Institute of Management Sciences, 10*(6), 791-815.

Jarvenpaa, S. L., Knoll, K., & Leidner, D. E. (1998). Is anybody out there? Antecedents of trust in global virtual teams. *Journal of Management Information Systems, 14*(4), 29-64.

Jarvenpaa, S. L., Tractinsky, N., Saarinen, L., & Vitale, M. (1999). Consumer trust in an Internet store: A cross-cultural validation [Electronic Version]. *Journal of Computer Mediated Communications, 5*(2). Retrieved from http://jcmc.indiana.edu/vol5/issue2/jarvenpaa.html

Jarvenpaa, S., Shaw, T.R., & Staples, D.S. (2004). Toward contextualized theories of trust: The role of trust in global virtual teams. *Information Systems Research, 15*(3), 250-267.

Jarvenpaa, S.L., & Leidner, Dorothy E. (1999). Communication and trust in global virtual teams. *Organization Science, 10*(6), 791-815.

Jarvenpaa, S.L., & Todd, P. (1997). Consumer reactions to electronic shopping. *International Journal of Electronic Commerce, 1*(2), 59-88.

Jarvenpaa, S.L., Knoll, K., & Leidner, D.E. (1998). Is there any body out there? Antecedents of trust in global virtual teams. *Journal of Management Information Systems, 14*(4), 29-64.

Jarvenpaa, S.L., Shaw, T.R., & Staples, D.S. (2004). Toward contextualized theories of trust: The role of trust in global virtual teams. *Information Systems Research, 15*(3), 250.

Javidan, M., House, R.J., Dorfman, P.W., Hanges, P.J., & Luque, M.S.d. (2006). Conceptualizing and measuring cultures and their consequences: A comparative review of GLOBE's and Hofstede's approaches. *Journal of International Business Studies, 37*(6), 897.

Jee, J., & Lee, W-N. (2002). Antecedents and consequences of perceived interactivity: An exploratory study. *Journal of Interactive Advertising, 3*(1). Retrieved March 11, 2007, from http://jiad.org/vol3/no1/jee/index.htm

Jettmar, E. M., & Rapp, M. W. (1996). *Computer mediated communication: A relational perspective.* Paper presented at the Annual Convention of the Western States Communication Association, Pasadena, CA.

Johnson, S.D., Suriya, C., Yoon, S.W., Berrett, J.V., & La Fleur, J. (2002). Team development and group processes of virtual learning teams. *Computers & Education, 39,* 379-393.

Johnson-George, C., & Swap, W. C. (1982). Measurement of specific interpersonal trust: Construction and validation of a scale to access trust in a specific other. *Journal of Personality and Social Psychology, 43*(6), 1306-1317.

Jones, G., & George, J. (1998). The experience and evolution of trust: Implications for cooperation and teamwork. *Academy of Management Review, 23*(3), 531-546.

Jones, J., & Vijayasarathy, L.R. (1998). Internet consumer catalog shopping: Findings from an exploratory study and directions for future research. *Internet Research, 8*(4), 322-333.

Jones, Q. (1997). Virtual-communities, virtual settlements & cyber-archaeology: A theoretical outline. *Journal of Computer-Mediated Communication, 3*(12). Retrieved March 11, 2007, from http://www.ascusc.org/jcmc/vol3/issue3/jones.html

Jones, Q. (2000). Time to split virtually: Expanding virtual publics into vibrant virtual metropolis. In *Proceedings of the 33rd Hawaii International Conference on System Sciences.* Retrieved March 11, 2007, from http://www.computer.org/proceedings/hicss/0493/04936/ 04936003.pdf

Jones, Q. (2001). *The boundaries of virtual communities: From virtual settlements to the discourse of dynamics of virtual publics.* Unpublished PhD thesis, Graduate School of Business, University of Haifa, Israel.

Jones, Q., Ravid, G., & Rafaeli, S. (2001). *Empirical evidence for information overload in mass interaction.* Paper presented at the Conference on Human Factors and in Computing Systems, CHI 2001. Retrieved March 11, 2007, from http://gsb.haifa.ac.il/~sheizaf/publications/chi2001.pdf

Jones, S. (1998). *Cybersociety 2.0: Revisiting computer mediated communication and community.* London: Thousand Oaks.

Jones, S.G. (1995). Understanding community in the information age. In S.G. Jones (Ed.), *Cybersociety: Computer-mediated communication and community.* Thousand Oaks; London: Sage.

Jones, W. (2003). *Offshore outsourcing: Trends, pitfalls, and practices* (Executive Report No. 4(4)). Cutter Business Technology Council.

Julsrud, T. E. & Schiefloe, P. M. (2007). Trust and stability in distributed work groups: A social network perspective. *International Journal of Networking and Virtual Organisations.* Forthcoming.

Julsrud, T., Bakke, J. W., et al. (2004). *Status and strategies for the knowledge intensive Nordic workplace.* Oslo, Norway: Nordic Innovation Center.

Julsrud, T., Schiefloe, P. M., et al. (2006). *Networks and trust in distributed groups.* University of Trondheim: NTNU.

Kahai, S.S., & Cooper, R.B. (2003). Exploring the core concepts of media richness theory: The impact of cue multiplicity and feedback immediacy on decision quality. *Journal of Management Information Systems, 20*(1), 263.

Kahneman, D. (2003). Maps of Bounded Rationality, Psychology for Behavioral Economics. *American Economic Review, 93*(5), 1449-1475.

Kanawattanachai, P., & Yoo, Y. (2002). Dynamic nature of trust in virtual teams. *Strategic Information Systems, 11,* 187-213.

Kanawattanachi, P., & Yoo, Y. (2002). Dynamic nature of trust in virtual teams. *Sprouts: Working Papers on Information, Environments, Systems, and Organizations, 2*(2), 42-58. Retrieved February 25, 2007, from http://sprouts.case.edu/2002/020204.pdf

Kannan, P., Chang, A., & Whinston, A. (2001, January 3-6). *Wireless commerce: Marketing issues and possibilities.* Paper presented at the the 34th Annual Hawaii International Conference on System Sciences (HICSS-34).

Kaplan R.S., & Norton, D.P. (1992, January-February). The balanced scorecard: Measures that drive performance. *Harvard Business Review, 70*(1), 71-79.

Kasper-Fuehrer, E.C., & Ashkanasy, N. (2001). Communicating trustworthiness and building trust in interorganizational virtual organizations. *Journal of Management, 27*(3), 235.

Katsh, E., & Rifkin, J. (2001). *Online dispute resolution.* San Francisco: Jossey Bass.

Katzenback, J.R., & Smith, D.K. (1993). The discipline of teams. *Harvard Business Review,* 111-120.

Kayworth, T., & Leidner, D. (2001, 2002). Leadership effectiveness in global virtual teams. *Journal of Management Information Systems, 18*(3), 7-40.

Keen, P.G.W. (1999) Transforming intellectual property into intellectual capital: Competing in the trust economy. In N. Imparato (Ed.), *Capital for our time: The economic, legal, and management challenges of intellectual capital.* Stanford, CA: Hoover Institution Press.

Keen, P.G.W. (Ed.). (1999). *Electronic commerce relationships: Trust by design.* Englewood Cliffs, NJ: Prentice-Hall.

Kelly, K. (1998). *New rules for the new economy: 10 radical strategies for a connected world.* New York: Viking/Penguin Putnam.

Kern, T., & Willcocks, L. (2003). Exploring relationships in information technology outsourcing: The interaction approach. *European Journal of Information System, 11*(1), 3-23.

Kerridge, E. (1988). *Trade and banking in early modern England.* Manchester, UK: Manchester University Press.

Kielser, S. (1987). The hidden messages of the computer science networks. *Harvard Deusto Business Review, Trim. 1,* 69-78.

Kiesler, S. (1986). The hidden messages in computer networks. *Harvard Business Review, 64*(1), 46.

Kiesler, S., & Sproull, L. (1992). Group decision making and communication technology. *Organizational Behavior and Human Decision Processes, 52*(1), 96-123.

Kiesler, S., Siegel, J., & McGuire, T. (1984). Social psychological aspects of computer-mediated communication. *American Psychologist, 39,* 1123-1134.

Kilduff, M., & Corley, K. G. (2000). Organizational culture from a network perspective. In N. Ashkanasy, C. P. M. Wilderom, & M. F. Peterson (Eds.), *Handbook of organizational culture and climate* (pp. 211-221). New York: Sage.

Kilduff, M., & Tsai, W. (2003). *Social networks and organizations.* London: Sage.

Kim, D. J., Ferrin, D. L., & Rao, H. R. (2003). A study of the effect of consumer trust on consumer expectations and satisfaction: The Korean experience. In *ACM International Conference Proceedings, 50,* 310-315.

Kim, D. J., Ferrin, D. L., & Rao, H. R. (Forthcoming). A trust-based consumer decision making model in electronic commerce: The role of trust, risk, and their antecedents. *Decision Support Systems.*

Kim, D. J., Song, Y. I., Braynov, S. B., & Rao, H. R. (2005). A multi-dimensional trust formation model in B-to-C e-commerce: A conceptual framework and content analyses of academia/practitioner perspective. *Decision Support Systems, 40*(2), 143-165.

Kim, K. K., & Prabhakar, B. (2004). Initial trust and the adoption of B2C e-commerce: The marries of Internet Banking. *Databases advances in information systems, 35*(2), 50-65.

Kipnis, D.(1996). Trust and technology. In R.M. Kramer & T.R. Tyler (Eds.), *Trust in organizations: Frontiers of theory and research* (pp. 39-50). Thousand Oaks, CA: Sage Publications.

Kirchmeyer, C., & Cohen, A. (1992). Multicultural groups, their performance and reactions with constructive conflict. *Group and Organization Management, 17,* 153-170.

Kirkman, B.L., Rosen, B., Gibson, C.B., Tesluk, P.E., & McPherson, S.O. (2002). Five challenges to virtual team success: Lessons from Sabre. *Academy of Management Executive, 16*(3), 67-79.

Kishore, R., Rao, H. R., Nam, K., Rajagopalan S., & Chaudhury, A. A. (2003, December). Relationship perspective on IT outsourcing. *Communications of the ACM, 46*(12), 87-92.

Kluckhohn, F., & Strodtbeck, F.L. (1961). *Variations in value orientations.* Evanston, IL: Peterson.

Ko, J., & Kim, Y. G. (2003). Sense of virtual community: Determinants and moderating role of the virtual community origin. *International Journal of Electronic Commerce, 8*(2), 75-88.

Koehn, D. (2003).The nature of and conditions for online trust. *Journal of Business Ethics, 43*(1), 3-19. Retrieved August 23, 2006, from http://proquest.umi.com

Koeszegi, S.T., Srnka, K.J., & Pesendorfer, E.-M. (in press). Electronic negotiations: A comparison of different support systems. *Die Betriebswirtschaft.*

Kogut, B., & Zander, U. (1992, August). Knowledge of the firm, combinative capabilities, and the replication of technology. *Organization Science, 3*(3), 383-397.

Kollok, P., & Smith, M. (1999). *Communities in cyber-place.* London: Routledge.

Konradt, U., & Hertel, G. (2002). *Management virtueller teams.* Weinheim & Basel: Beltz Verlag.

Koreto, R. (1997). In CPAs we trust. *Journal of Accountancy, 184*(6), 62-64.

Korte, W. B., Steinle, W. J., & Robinson S. (1988). *Telework: Present situation and further development of a new form of work.* Bonn: Elsevier Science.

Kostera, M. (1996). *Postmodernizm w zarządzaniu* [Postmodernism in management]. Warszawa: PWE.

Kostera, M. (2003). *Antropologia organizacji* [Organizational anthropology]. Warszawa: PWN.

Kouzes, J.M., & Posner, B.Z. (2002). *The leadership challenge* (3rd ed.). San Francisco: Jossey-Bass.

Koźmiński, A. K., & Sztompka P. (2004). *O wielkiej przemianie* [On the great change]. Warszawa: Wydawnictwo WSPiZ.

KPMG (1999). *The new mass medium.* USA: Ziff-Davis/Dell/Intel.

Krackhardt, D. (1992). The strength of strong ties: The importance of philos in organizations. In N. Nohria & R. Eccles (Eds.), *Network and organizations: Structure, form and action* (pp. 216-239). Boston: Harvard University Press.

Krackhardt, D. (1999). The ties that torture: Simmelian tie analysis in organizations. *Research in the Sociology of Organizations, 16*, 183-210.

Krackhardt, D., & Brass, D. (1994). Intraorganizational networks. In S. Wasserman & J. Galaskiewicz (Eds.), *The micro side. Advances in social network analysis* (pp. 207-229). Thousand Oaks, CA: Sage.

Krackhardt, D., & Kilduff, M. (2002). Structure, culture and simmelian ties in entrepreneurial firms. *Social Networks, 3*(24), 279-290.

Kramer, R. (1994). The sinister attribution error: Paranoid cognition and collective distrust in organizations. *Motivation and Emotion, 18*(2), 199-230.

Kramer, R. M. (1999). Trust and distrust in organizations: Emerging perspectives, enduring questions. *Annual Review of Psychology, 50*, 569-598.

Kramer, R. M., & Cook, K. S. (2004). Trust and distrust in organizations: Dilemmas and approaches. In R. M. Kramer & K. S. Cook (Eds.), *Trust and distrust in organizations. Dilemmas and approaches* (pp. 1-17). New York: Russel Sage Foundation.

Kramer, R. M., & Tyler, T. R. (1996). *Trust in organizations. Frontiers of theory and research.* Thousand Oaks, CA: Sage Publications.

Kramer, R. M., & Tyler, T. R. (1996). Whither trust. In R. M. Kramer & T. R. Tyler (Eds.), *Trust in organizations: Frontiers of theory and research.* Thousand Oaks, CA: Sage Publications.

Kramer, R.M. (1995). Power, paranoia and distrust in organizations: The distorted view from the top. In R.J. Bies, R.J Lewicki, & B.H. Sheppard (Eds.), *Research on negotiation in organizations* (pp. 119-154). Greenwich, CT: JAI Press.

Kramer, R.M., & Tyler, T.R. (Eds.). (1996). *Trust in organisations: Frontiers of theory and research.* Thousand Oaks, CA: Sage.

Kranawattanachai, P., & Yoo, Y. (2002). Dynamic nature of trust in virtual teams. *Journal of Strategic Information Systems, 11*, 187-213.

Kraut, R.E., Galegher, J., Fish, R., & Chalfonte, B. (1992). Task requirements and media choice in collaborative writing. *Human Computer Interaction, 7*, 375-407.

Kurland, N., & Bailyn, L. (1999). Telework: the advantages and challenges of working here, there, anywhere, and anytime. *Organizational Dynamics, Autumn*, 53-68.

Kyas, O. (1997). *Internet security: Risk analysis, strategies and firewalls*. New York: International Thomson Publishing.

Lacity, M. C., & Willcocks, L.P. (2001). *Global information technology outsourcing: In search of business advantage*. Chichester, West Sussex, England: John Wiley & Sons.

Lam, K.Y., Chung, S.L., Gu, M., & Sun, J.G. (2003). Lightweight security for mobile commerce transactions. *Computer Communications, 26*, 2052-2061.

Lane, C. (1998). Introduction: Theories and issues in the study of trust. In C. Lane & R. Bachmann (Eds.), *Trust within and between organizations. Conceptual issues and empirical applications* (pp. 1-30). Oxford: Oxford University Press.

Lane, C., & Bachman, R. (1996). The social constitution of trust: Supplier relations in Britain and Germany. *Organisation Studies, 17*(3), 365-395.

Lane, C., & Bachman, R. (1997). Co-operation in Inter-Firm Relations in Britain and Germany: The Role of Social Institutions. *British Journal of Sociology 48*(2), 226-54.

Lane, C., & Bachman, R. (Eds.). (1998). *Trust within and between organisations: Conceptual issues and empirical applications*. Oxford, NY: Oxford University Press.

Lane, C., & Bachmann, R. (Eds.). (1998). *Trust within and between organizations: Conceptual issues and empirical applications*. Oxford: Oxford University Press.

Larsen, K.R.T., & McInerney, C.R. (2002). Preparing to work in virtual organization. *Information & Management, 39*(6), 445-456.

Latane, B., Liu, J.H., Nowak, A., Bonevento, M., & Zheng, L. (1995). Distance matters: Physical space and social impact. *Personality and Social Psychology Bulletin, 21*(8), 795-805.

Latour, B. (1986). The powers of association. In: J. Law (Ed.), *Power, action and belief: A new sociology of knowledge?* (pp. 264-280). London: Routledge & Kegan Paul.

Lave, J., & Wenger, E. (1991). *Situated learning. Legitimate peripheral participation*. Cambridge, UK: Cambridge University Press.

Lawler, E.E. (2003). Pay systems for virtual teams. In C.B. Gibson & S.G. Cohen (Eds.), *Virtual teams that work* (pp. 121-144). San Francisco: Jossey-Bass.

Lax, D.A., & Sebenius, J.K. (1986). *The manager as negotiator*. New York: Free Press.

Leach, W. D., & Sabatier, P. A. (2005). To trust an adversary: integrating rational and psychological models of collaborative policymaking. *The American Political Science Review, 99*(4), 491-503.

Lechner, U., & Schmid, B.F. (2000). Communities and media: Towards a reconstruction of communities on media. In *Proceedings of the 33rd Hawai International Conference on System Sciences*. Retrieved March 11, 2007, from http://ieeexplore.ieee.org/xpl/freeabs-alljsp?arnumber=926817

Lee, A. (1994). Electronic mail as a medium for rich communication: An empirical investigation using hermenuetic interpretation. *MIS Quarterly, 18*(2), 143-157.

Lee, A., & Ngwenyama, O. (1997). Communication richness in electronic mail: Critical social theory and the contextuality of meaning. *MIS Quarterly, 21*(2), 145-167.

Lee, M. K. O., & Turban, E. (2001). A trust model for consumer Internet shopping. *International Journal of Electronic Commerce, 6*(1), 75-91.

Lee-Kelly, L. (2006). Locus of control and attitudes to working in virtual teams. *International Journal of Project Management*.

Leidner, D.E., & Kayworth, T. (2006). Review: A review of culture in information systems research: Toward a theory of information technology culture conflict. *MIS Quarterly, 30*(2), 357.

284

Leimeister, J.M., Ebner, W., & Krcmar, H. (2005). Design, implementation, and evaluation of trust-supporting components in virtual communities for patients. *Journal of Management Information System, 21*(4), 101-131.

Lengel, R.H., & Daft, R.L. (1988). The selection of communication media as an executive skill. *The Academy of Management Executive, 2*(3), 225.

Letki, N., & Evans, G. (2005). Endogenizing social trust: Democratization in east-central Europe. *British Journal of Political Science, 35*, 515-529.

Levenson, A.R., & Cohen, S.C. (2003). Meeting the performance challenge: Calculating return in investment for virtual teams. In C.B. Gibson & S.G. Cohen (Eds.), *Virtual teams that work* (pp. 145-174). San Francisco: Jossey-Bass.

Levin, D. Z., Whitener, E. M., & Cross, R. (2006). Perceived trustworthiness of knowledge sources: The moderating impact of relationship length. *Journal of Applied Psychology, 91*(5), 1163-1171.

Lewicki, R. J., & Bunker, B. B. (1995). Developing and maintaining trust in work relationships. In R. M. Kramer & T. R. Tyler (Eds.), *Trust in organizations: Frontiers of theory and research* (pp. 114-139). Thousand Oaks, CA: Sage.

Lewicki, R. J., & Bunker, B. B. (1996). Developing and maintaining trust in work relationships. In R. Kramer & T. R. Tyler (Eds.), *Trust in organizations: Frontiers of theory and research* (pp. 114-139). Thousand Oaks, CA: Sage Publications.

Lewicki, R., McAlliser, D., & Bies, R. (1998). Trust and distrust: New relationships and realities. *Academy of Management Review, 23*(3), 438-458.

Lewicki, R.J., & Bunker, B. (1995). Trust in relationships: A model of trust development and decline. In B.B. Bunker & J.Z. Rubin (Eds.), *Conflict, cooperation and justice* (pp. 133-73). San Francisco: Jossey-Bass.

Lewicki, R.J., & Bunker, B. (1996). Developing and maintaining trust in work relationship. In R. Kramer & T. Tyler (Eds.), *Trust in organisations: Frontiers of theory and research* (pp. 114-139). Thousand Oaks, CA: Sage.

Lewicki, R.J., & Litterer, J. (1985). *Negotiation: Readings, exercises and cases.* Boston: Irwin.

Lewicki, R.J., McAllister, D.J., & Bies, R.J (1998). Trust and distrust: New relationships and realities. *Academy of Management Review, 23*(3), 438-458.

Lewicki, R.J., Saunders, D.M., Minton, J.W., & Barry, B. (2002). *Negotiation: Readings, exercises and cases* (4th ed.). Boston: McGraw-Hill/Irwin.

Lewis, J. D., & Weigert, A. (1985). Trust as a social reality. *Social Forces, 6*(4), 967-985.

Lewis, J. D., & Weigert, A. (1985). Trust as a social reality. *Social Forces, 63*, 967-985.

Lewis, J.D., & Weigert, A. (1985). Social atomism, holism and trust. *Sociological Quarterly, 26*(4), 455-471.

Li, D., & Lin, Z. (2004, December 5-8). Negative reputation rate as the signal of risk in online consumer-to-consumer transactions. In *Proceedings of ICEB 2004.*

Light, D. A. (2001). Sure, you can trust us. *MIT Sloan Management Review, 43*(1), 17.

Lim, K. H., Leung, K., Sia, C. L., & Lee, M. K. (2004). Is eCommerce boundary-less? Effects of individualism-collectivism and uncertainty avoidance on internet shopping. *Journal of International Business Studies, 35*, 545-559.

Linde, C. (1988). The quantitative study of communicative success: Politeness and accidents in aviation discourse. *Language and Society, 17,* 375-399.

Lipnack, J., & Stamps, J. (1997) *Virtual teams: Reaching across space, time, and organizations with technology.* New York: John Wiley & Sons.

Lipnack, J., & Stamps, J. (2000). *Virtual teams: Reaching across space, time and organizations with technology.* New York, Chichester: John Wiley & Sons.

Lippert, S. K. & Swiercz, P. M. (2005). Human resource information systems (HRIS) and technology trust. *Journal of Information Science, 31*(5), 340-353.

Lippert, S. K. (2001). *An exploratory study into the relevance of trust in the context of information systems technology.* Doctoral dissertation, The George Washington University, Washington, DC.

Lippert, S. K. (2007). Investigating post-adoption utilization: An examination into the role of inter-organizational and technology trust. *IEEE Transactions on Engineering Management.*

Lippert, S. K., & Davis, M. (2006). Synthesizing trust and planned change initiatives to enhance information technology adoption behavior. *Journal of Information Science, 32*(5), 434-448.

Lippert, S. K., & Forman, H. (2006). A supply chain study of technology trust and antecedents to technology internalization consequences. *International Journal of Physical Distribution and Logistics Management, 36*(4), 271-288.

Liu, Z.G. (1999). Virtual community presence in the Internet relay chatting. *Journal of Computer-Mediated Communication, 5*(1). Retrieved March 11, 2007, from http://www.ascusc.org/jcmc/ vol5/ issue1/liu.html

Livers, A.B., & Caver, K.A. (2003). *Leading in black and white: Working across the racial divide in corporate America.* San Francisco: Jossey-Bass.

Livingston, J. (2005). How valuable is a good reputation? A sample selection model of Internet auctions. *The Review of Economics and Statistics, 87*(3), 453-465.

Lohse, G. L., & Spiller, P. (1998). Electronic shopping. *Communications of the ACM, 41*(7), 81-87.

Lombard, M., & Snyder-Duch, J. (2001). Interactive advertising and presence: A framework. *Journal of Interactive Advertising, 1*(2).

Luhmann, N. (1979). *Trust and power.* Chichester, UK: Wiley.

Luhmann, N. (1988). Familiarity, confidence, trust: Problems and alternatives. In D. Gambetta (Ed.), *Trust: Making and breaking co-operative relations.* Oxford, UK: Basil Blackwell.

Luhmann, N. (2000). *Vertrauen* (4th ed.). Stuttgart: Lucius & Lucius.

Lui, S.S., & Ngo, H. (2004). The role of trust and contractual safeguards on cooperation in non-equity alliances. *Journal of Management, 30*(4), 471-485.

Lull, J. (1990). *Inside family viewing: Ethnographic research on television's audience.* London: Routledge.

Lunn, R. J., & Suman, M. W. (2002). Experience and trust in online shopping In B. Wellman & C. A. Haythornthwaite (Eds.), *The Internet in everyday life* (pp. 549-577). Blackwell Publishing.

Lurey, J.S., & Raisinghani, M.S. (2001). An empirical study of best practices in virtual teams. *Information & Management,* 523-544.

Maamar, Z. (2003). Commerce, e-commerce, and m-commerce: What comes next? *Communications of the ACM, 46*(12), 251-257.

Magretta, J. (1998, September-October). Fast, global, and entrepreneurial: Supply chain management Hong Kong style. *Harvard Business Review, 76*(5), 102-114.

Malhotra, A., & Segars, A. H. (2005). Investigating wireless Web adoption patterns in the U.S. *Communications of the ACM, 48*(10), 105-110.

Marchington, M., Grimshaw, D. (Eds.). (2005). *Fragmenting work. Blurring organizational boundaries and disordering hierarchies.* Oxford: Oxford University Press.

Márquez-García, A. M., Fuentes-Lombardo, G., & Bruque-Cámara, S. (2006). The role of IT in family firm internationalization through strategic alliances. In S. Martínez-Fierro, J. A. Medina-Garrido, & J. Ruiz-Navarro (Eds.), *Utilizing Information Technology in developing strategic alliances among organizations* (pp. 170-202). Hershey, PA: Idea Group Publishing.

Massey, A.P., Montoya-Weiss, M.M., & Hung, Y.T. (2003). Because time matters: Temporal coordination in global virtual project teams. *Journal of Management Information Systems, 19*(4), 129-155

Matson, E., Patiath, P., & Shavers, T. (2003). Strengthening your organization's internal knowledge market. *Organizational Dynamics, 32*(3), 275-285.

Mayer, M. D., & Tan, F. B. (2002). Beyond models of national culture in information systems research. *Journal of Global Information Management, 10*(1), 24-32.

Mayer, R. C., Davis, J. H., & Schoorman, F. D. (1995). An integrative model of organizational trust. *Academy of Management Review, 20*(3), 709-734.

Mayer, R.C., & Davis, J.H. (1995). An integrative model of organizational trust. *Academy of Management Journal, 20*(3), 709-734.

Mayer, R.C., & Davis, J.H. (1999). The effect of the performance appraisal system on trust for management: A field quasi-experiment. *Journal of Applied Psychology, 84*(1), 123.

Mayer, R.C., & Gavin, M.B. (2005). Trust in management and performance: Who minds the shop while the employees watch the boss? *Academy of Management Journal, 48*(5), 874.

Mayer, R.C., Davis, J.H., & Schoorman, F.D. (1995). An integrative model of organizational trust. *Academy of Management Review, 20*(3), 709-734.

Mayer, R.C., Davis, J.H., & Shoorman, F.D. (1995). An Integrative Model of Organizational Trust. *Academy of Management Review, 20*(3), 709-34.

Maznevski, M. L., & Chudoba, K. M. (2000). Bridging space over time: Global virtual team dynamics and effectiveness. *Organizational Science, 11*(5), 473-492.

Maznevski, M., Davison, S.C., & Barmeyer, C. (2005). Management virtueller teams. In G. Stahl, W. Mayrhofer & T. Kühlmann (Eds.), *Internationales personalmanagement* (pp. 91-114). Mering: Rainer Hampp Verlag.

Maznevski, M.L., & Chudoba, K.M. (2000). Bridging space over time: Global virtual team dynamics and effectiveness. *Organization Science, 11*(5), 473-492.

McAfee, A. P. (2006). Enterprise 2.0: The dawn of emergent collaboration. *Sloan Management Review, 47*(3), 21-31.

McAllister, D.J. (1995). Affect- and cognition-based trust as foundations for interpersonal cooperation in organizations. *Academy of Management Journal, 38*(1), 24-59.

McCroskey, J. C., & Richmond, V. P. (1996). *Fundamentals of human communication: An interpersonal perspective.* Prospect Heights, IL: Waveland Press, Inc.

McEvily, B., & Zaheer, A. (2004). Architects of trust: The role of network facilitators in geographical clusters. In R. M. Kramer & K. S. Cook (Eds.), *Trust and distrust in organizations. Dilemmas and approaches* (pp. 189-213). New York: Russel Sage Foundation.

McEvily, B., Perrone, V., & Zaheer, A. (2003). Trust as an organizing principle. *Organization Science, 14*(1), 91-103.

McGuire, T.W., Kiesler, S., & Siegel, J. (1987). Group and computer-mediated discussion effects in risk decision-making. *Journal of Personality and Social Psychology, 52*, 917-930.

McKnight, D. H., & Chervany, N. L. (1996). *The meanings of trust* (MISRC Working Paper Series No. 96-04). University of Minnesota. Retrieved February 21, 2007 from http://www.misrc.umn.edu/wpaper/wp96-04.htm

McKnight, D. H., & Chervany, N. L. (2002). What trust means in e-commerce customer relationships: An interdisciplinary conceptual typology. *International Journal of Electronic Commerce, 6*(2), 35-60.

McKnight, D. H., Choudhury, V., & Kacmar, C. (2002). Developing and validating trust measures for e-commerce: An integrative typology. *Information Systems Research, 13*(4), 334-359.

McKnight, D. H., Choudhury, V., & Kacmar, C. (2002). The impact of initial consumer trust on intentions to transact with a Web site: A trust building model. *Journal of Strategic Information Systems, 11*(3-4), 297-323.

McKnight, D. H., Cummings, L. L., & Chervany, N. L. (1998). Initial trust formation in new organizational relationship. *Academy of Management Review, 23*(3), 473-490.

McKnight, D. H., Cummings, L. L., et al. (1995). *Trust formations in new organizational relationships* (Information and Decision Sciences Workshop). University of Minnesota.

McKnight, D., Cummings, L., & Chervany, N. (1998). Initial trust formation in new organizational relationships. *Academy of Management Review, 23*(3) 473-491.

McKnight, D.H., & Chervany, N.L. (2002). What trust means in e-commerce customer relationships: An interdisciplinary conceptual typology. *International Journal of Electronic Commerce, 6*(2), 35-59.

McMillan, D.W., & Chavis, D.M. (1986). Sense of community. A definition and theory. *Journal of Community Psychology, 14*(1), 6-23.

McMillan, S.J., Hwang, J., & Lee, G. (2003). Effects of structural and perceptual factors toward the website. *Journal of Advertising Research, 43*(4), 400-409.

Melnik, M.I., & Alm, J. (2002). Does a seller's e-commerce reputation matter? Evidence from eBay auctions. *Journal of Industrial Economics, 50*(3), 337-349.

Menichella, E. (2000). Chat r u ready. *Internet News.* Retrieved March 11, 2007, from http://inews.tecnet.it/Articoli/2000/2000-12/dossier0012.html

Meredith, J.R., & Shafer, S.M. (2007). *Operations management for MBAs* (3rd ed.). Hoboken, NJ: John Wiley & Sons.

Meyerson, D., Weick, K., & Kramer, R. (1996). Swift trust and temporary groups. In R.M. Kramer & T.R. Tyler (Eds.), *Trust in organizations: Frontiers of theory and research* (pp. 166-195). Thousand Oaks, CA: Sage.

Mishira, A. K. (1996). Organizational response to crisis: The centrality of trust. In R. M. Kramer & T. R. Tyler (Eds.), *Trust in organizations: Frontiers of theory and research* (pp. 261-287). Thousand Oaks, CA: Sage.

Mlynarczyk, D. (2004). *Analiza stanu obecnego i perspektywy rozwoju rynku IT* [Analysis of the current condition and the perspectives of the development of the IT market]. Warszawa.

Mnookin, R.H., Peppet, S.R., & Tulumello, A.S. (2000). *Beyond winning: Negotiating to create value in deals and disputes.* Cambridge, MA: The Belknap Press of Harvard University Press.

Mohr, J., & Spekman, R. (1994). Characteristics of partnership success: Partnership attributes, communication behavior, and conflict resolution techniques. *Strategic Management Journal, 15*(2), 135-152.

Möllering, G. (2006). *Trust: Reason, routine, reflexivity.* Oxford: Elsevier.

Montoya-Weiss, M.M., Massey, A.P., & Song, M. (2001). Getting it together: Temporal coordination and conflict resolution in global virtual teams. *Academy of Management Journal, 44*(6), 1251-1262.

Moore, C. (2003). *The mediation process* (3rd ed.). San Francisco: Jossey Bass.

Moore, D., Kurtzberg, T., Thompson, L., & Morris, M. (1999). Long and short routes to success in electronically mediated negotiations: Group affiliations and good vibrations. *Organizational Behavior and Human Decision Processes, 77*(1), 22–43. Retrieved July 3, 2007, from http://www.cbdr.cmu.edu/mpapers/emn.pdf

Moores, S. (1993). *Interpreting audiences: The ethnography of media consumption.* Thousand Oaks, CA: Sage.

Moorhead, G., & Griffin, R. W. (1995). *Organizational behavior: Managing people and organizations* (4th ed.). Boston: Houghton Mifflin Company.

Morgan, N.A., Anderson, E.W., & Mittal, V. (2005). Understanding firms' customer satisfaction information usage. *Journal of Marketing, 69*(3), 131-151.

Morgan, R.M., & Hunt, S.D. (1994). The commitment-trust theory of relationship marketing. *Journal of Marketing, 58,* 20-38.

Morris, M., Nadler, J., Kurtzberg, T., & Thompson, L. (2002). Schmooze or lose: Social friction and lubrication in e-mail negotiations. *Group Dynamics, 6*(1), 89-100. Retrieved August 23, 2006, from http://gateway.uk.ovid.com/gwl/ovidweb.cgi

288

Muir, B. M. (1994). Trust in automation: Part I. Theoretical issues in the study of trust and human intervention in automated systems. *Ergonomics, 37*(11), 1905-1922.

Nadler, J., & Shestowsky, D. (2006). Negotiation, information technology and the problem of the faceless other. In A.W. Kruglanski & J.P. Forgas (Series Eds.) & L. Thompson (Ed.), *Negotiation theory and research. Frontiers of Social Psychology.* New York: Psychology Press.

Naisbitt, J. (1982). *Megatrends: Ten new directions transforming our lives.* New York: Time Warner.

Naquin, C.E., & Paulson, G.D. (2003). Online bargaining and interpersonal trust. *Journal of Applied Psychology, 88*(1), 113-120. Retrieved July 3, 2007, from http://gateway.uk.ovid.com/gwl/ovidweb.cgi

Nass, C., & Moon, Y. (2000). Machines and mindlessness: Social responses to computers. *Journal of Social Issues, 56*(1), 81-103.

Ngwenyama, O.K., & Lee, A.S. (1997). Communication richness in electronic mail: Critical social theory and the contextuality of meaning. *MIS Quarterly, 21*(2), 145.

Nielsen, B. B. (2004). The role of trust in collaborative relationships: A multi-dimensional approach. *M@n@gement, 7*(3), 239-256.

Nilles, M., Carlson, F., Gray, P., & Hanneman, G. (1976). *The telecommunications-transportation tradeoff.* New York: John Wiley & Sons.

Nissenbaum, H. (2004). Will security enhance trust online, or supplant it? In R. M. Kramer & K. S. Cook (Eds.), *Trust and distrust in organizations. Dilemmas and approaches* (pp. 155-185). New York: Russel Sage Foundation.

Nohria, N., & Eccles, R.G. (1992). Face-to-face: Making network organisations work. In N. Nohria & R.G. Eccles (Eds.), *Networks and organisations* (pp. 288-308). Boston: Harvard Business School Press.

Nonaka, I., & Takeuchi, H. (1995). *The knowledge-creating company.* New York: Oxford University Press.

Nooteboom, B. (2002). *Trust: Forms, foundations, functions, failures and figures.* Cheltenham: Edward Elgar.

Novak, T.P., & Hoffman, D.L. (1997). A new marketing paradigm for electronic commerce. *The Information Society: An International Journal 13*(1), 43-54.

Novak, T.P., & Hoffman, D.L. (1997). *Measuring the flow experience among Web users* (Working Paper). Vanderbilt University. Retrieved March 7, 2007, from http://www.2000.osgm.vanderbilt.edu/novak/flow.julv.1997/flow.htm

Novak, T.P., Hoffman, D.L., & Yung, Y.F. (2000). Measuring the customer experience in online environments: A structural modeling approach. *Marketing Science, 19*(1), 22-42.

O'Hara Devereaux, M., & Johansen, R. (1994). *Global work: Bridging distance, culture and time.* San Francisco: Jossey-Bass Publishers.

O'Leary, M., Orlikowski, W., et al. (2002). Distributed work over the centuries. Trust and control in the Hudson Bay Company. In P. Hinds & S. Kiesler (Eds.), *Distributed work.* Cambridge, MA: MIT Press.

Oakes, G. (1990). The sales process and the paradox of trust. *Journal of Business Ethics, 9,* 671-679.

Obłój, K. (2002). *Tworzywo skutecznych strategii* [Material of effective strategies]. Warszawa: Polskie Wydawnictwo Ekonomiczne.

Olson, M. H. (1988). Organizational barriers to telework in telework. In W. B. Korte, W. J. Steinle & S. Robinson (Eds.), *Present situation and further development of a new form of work* (pp. 135-151). Bonn: Elsevier Science.

Ouchi, W. (1981). Theory Z: *How American business can meet the Japanese challenge.* Reading, MA: Addison-Wesley.

Overby, S. (2003). The hidden costs of offshore outsourcing. *CIO Magazine.* Retrieved March 15, 2007, from http://www.cio.com/archive/090103/money.html

Overwalle, F. V., & Heylighen, F. (2006). Talking nets: A multiagent connectionist approach to communication and trust between individuals. *Psychological Review, 113*(3), 606-627.

Oza, N. V., Hall, T., Rainer, A., & Grey, S. (2006). Trust in software outsourcing relationships: An empirical investigation of Indian software companies. *Information and Software Technology, 48*(5), 345-354.

Pack, T. (1999). Can you trust Internet information? *Link - up, 16*(6), 24.

Panteli, N. (2005). Trust in global virtual teams. *Ariadne 43*. Retrieved April 2007 from http://www.ariadne.ac.uk/issue43/panteli/

Park, C., & Jun, J.-K. (2003). A cross-cultural comparison of Internet buying behavior. *International Marketing Review, 20*(5), 534-553.

Parkhe, A. (1998). Building trust in international alliances. *Journal of World Business, 33*(4), 417-437.

Parsons, T. (1951). *The social system.* London: Routledge Keegan & Paul.

Parsons, T. (1969). Research with human subjects and the professional complex. In P. A. Freund (Ed.), *Experimentation with human subjects* (pp. 116-151). New York: George Braziller.

Pastore, A., & Vernuccio M. (2004). *Marketing, innovazione e tecnologie digitali.* Padova: Cedam.

Paul, D.L., & McDaniel, R.R. (2004). A field study of the effect of interpersonal trust on virtual collaborative relationship performance. *MIS Quarterly, 28*(2), 183.

Paul, S., Sheetharaman, P., Samarah, I., & Mykytyn, P.P. (2004). Impact of heterogeneity and collaborative conflict management style on the performance of synchronous global virtual teams. *Information & Management, 41*, 303-321.

Pauleen, D.J., & Yoong, P. (2001). Relationship building and the use of ICT in boundary-crossing virtual teams: A facilitator's perspective. *Journal of Information Technology, 16*, 205-220.

Paulson, G.D., & Naquin, C.E. (2004). Establishing trust via technology: Long distance practices and pitfalls. *International Negotiation, 9*(2), 229-244.

Pavlou, P. A., & Chai, L. (2002). What drives electronic commerce across cultures? A cross-cultural investigation of the theory of planned behavior. *Journal of Electronic Commerce Research, 3*(4), 240-253.

Pepper, G. L., & Larson, G. S. (2006). Overcoming information communication technology problems in a post-acquisition organization. *Organizational Dynamics, 35*(2), 160-169.

Pfannenstein, L., & Tsai, R. (2004). Offshore outsourcing: Current and future effects on American IT industry. *Information Systems Management, 2*(4), 72-80.

Piccoli, G., & Blake, I. (2003). Trust and the unintended effects of behaviour control in virtual teams. *MIS Quarterly, 27*(3), 365-395.

Plank, R. E., Reid, D. A., & Pullins, E. B. (1999). Perceived trust in business-to-business sales: A new measure. *The Journal of Personal Selling & Sales Management, 19*(3), 61-71.

Png, I. P. L., Tan, B. C. Y., & Wee, K.-L. (2001). Dimensions of national culture and corporate adoption of IT infrastructure. *IEEE Transactions on Engineering Management, 48*(1), 36-45.

Podolny, J., & Baron, J. (1997). Resources and relationships, social networks and mobility in the workplace. *American Sociological Review, 62*, 673-693.

Polanyi, M. (1967). *The tacit dimension.* New York: Anchor Day Books.

Poltrock, S.E., & Engelbeck, G. (1999). Requirement for virtual collocation environment. *Information and Software Technology, 41*, 331-339.

Poole, M. S., & Jackson, M. H. (1993). Communication theory and group support systems. In L. M. Jessup & J. S. Valacich (Eds.), *Group support systems: New perspectives.* Old Tappan, NJ: Macmillan.

Poole, M.S. (1999). Organizational challenges for the new forms. In G. DeSanctis & J. Fulk (Eds.), *Shaping organizational form: Communication, connection, and community.* Thousand Oaks, CA: Sage Publications.

Poole, S. (2000). *Trigger happy: The inner life of video games*. London: Fourth Estate.

Porter, C.E. (2004). A typology of virtual communities: A multidisciplinary foundation for future research. *Journal of Computer-Mediated Communication, 10*(1). Retrieved March 11, 2007, from http://jcmc.indiana.edu/vol10/issue1/porter.html

Porter, M. (1985). *Competitive advantage*. New York: Free Press.

Porter, M. (2001). Strategy and the Internet. *Harvard Business Review (March)*, 63-78.

Porter, M.E. (1980). *Competitive strategy: Techniques for analyzing industries and competitors*. New York: Free Press.

Potter, R.E., & Balthazard, P. (2004). The role of individual memory and attention processes during electronic brainstorming. *MIS Quarterly, 28*(4), 621.

Potter, R.E., & Balthazard, P.A. (2002). Understanding human interaction and performance in the virtual teams. *Journal of Information Technology Theory and Application, 4*(1), 1-23.

Pottler, R.E., & Balthazard, P.A. (2002). Virtual team interaction styles: Assessment and effects. *International Journal of Human-Computer Studies, 56*, 423-443.

Powell, W.W. (1990). Neither market nor hierarchy: Network forms of organisation. *Research in Organisational Behavior, 12*, 295-336.

Premkumar, G., Ramamurthy, K., & Crum, M. (1997). Implementation of electronic dates interchange. *Journal of Management Information Systems, 11*(2), 157-186.

Pruitt, D., Rubin, J., & Kim, S. (1994). *Social conflict: Escalation, stalemate, and settlement* (3rd ed.). New York: McGraw Hill.

Pruitt, D.G., & Kimmel, M. (1977). Twenty years of experimental gaming: Critique, synthesis, and suggestions for the future. *Annual Review of Psychology, 28*, 363-392.

Qiu, L. & Benbasat, I. (2005). Online consumer trust and live help interfaces: The effects of text-to-speech voice and three-dimensional avatars. *International Journal of Human-Computer Interaction, 19*(1), 75-94.

Rafaeli, S. (1986). The electronic bulletin board: A computer driven mass medium. *Computers and the Social Sciences, 2*(3), 123-136.

Rafaeli, S., & Sudweeks, F. (1997). Networked interactivity. *Journal of Computer-Mediated Communications, 2*(4). Retrieved March 11, 2007, from http://www.ascusc.org/jmc/vol2/issue4/rafaeli.sudweeks.html

Ragin, C. A. (1994). *Constructing social research*. Thousand Oaks, CA: Sage.

Raiffa, H. (1982). *The art and science of negotiation*. Cambridge, MA: Harvard University Press.

Ratnasingam, P. (2005). E-commerce relationships: The impact of trust on relationship continuity. *International Journal of Commerce and Management, 15*, 1-16.

Ratnasingham, P. (1998). The importance of trust in electronic commerce. *Internet Research, 8*(4), 313-321.

Rea, T. (2001). Engendering trust in electronic environments: Roles for a trusted third party. In C. Castelfranchi & Y.H. Tan (Eds.), *Deception, fraud and trust in virtual societies* (pp. 221-234). Dodrecht, The Netherlands: Kluwer.

Reber, G., & Berry, M. (1999). A role of social and intercultural communication competence in international human resource development. In S. Lähteenmäki, L. Holden, & I. Roberts (Eds.), *HRM and the learning organisation* (pp. 313-343). Publications of the Turku School of Economics and Business Administration, Series A-2.

Reichheld, F., & Shefter, P. (2000). E-loyalty: Your secret weapon on the Web. *Harvard Business Review, 78*(4), 105-114.

Rheingold, H. (1993). *The virtual community: Homesteading on the electronic frontier*. New York: Addison-Wesley.

Rice, R. E. (1980). Computer conferencing. In B. Dervin & M. J. Voight (Eds.), *Progress in communication science*. Norwood, NJ: Ablex.

Richard, O.C. (2000). Racial diversity, business strategy, and firm performance: A resource-based view. *Academy of Management Journal, 43*(2), 164-177.

Ridings, C.M., Gefen, D., & Arinze, D. (2002). Some antecedents and effects of trust in virtual communities. *Journal of Strategic Information Systems, 11*(3-4), 271-295.

Rietjens, B. (2006). Trust and reputation on eBay: Towards a legal framework for feedback intermediaries. *Information & Communications Technology Law, 15*, 55-78.

Rindova, V.P., Williamson, I.O., Petkova, A.P., & Sever, J.M. (2005). Being good or being known: An empirical examination of the dimensions, antecedents, and consequences of organizational reputation. *Academy of Management Journal, 48*, 1033-1049.

Ring, P.S., & Van de Ven, A.H. (1992). Structuring cooperative relationships between organizations. *Strategic Management Journal, 13*(7), 483-498.

Ring, P.S., & Van de Ven, A.H. (1994). Developmental processes of cooperative interorganizational relationships. *The Academy of Management Review, 19*(1), 90-118.

Riopelle, K., Gluesing, J.C., Alcordo, T.C., Baba, M.L., Britt, D., McKether, W., et al. (2003). Context, task and the evolution of technology use in global virtual teams. In C.B. Gibson & S.G. Cohen (Eds.), *Virtual teams that work* (pp. 239-264). San Francisco: Jossey-Bass.

Ripperger, T. (1998). *Ökonomik des Vertrauens: Analyse eines Organisationsprinzips*. Tübingen: Moor Siebeck.

Robinson, M., & Kalakota, R. (2004). *Offshore outsourcing: Business model, ROI, and best practices*. Alpharetta, GA: Mivar Press.

Rocco, E. (1998). Trust breaks down in electronic contexts but can be repaired by some initial face-to-face contact. In *Proceedings of the SIGCHI Conference on Human Factors in Computing Systems* (pp. 496-502).

Los Angeles: ACM Press. Retrieved August 23, 2006, from http://portal.acm.org

Rodriguez, F. (2003). Influence of the new technologies on the organizational behavior. In F. Gil, & C. Alcover (Eds.), *Introduction to the psychology of the organizations* (pp. 179-228). Madrid: Publishing alliance.

Roehm, H.A., & Haugtvedt, C.P. (1999). Understanding interactivity of cyberspace advertising. In D.W. Schumann & E. Thorson (Eds.), *Advertising and the World Wide Web* (pp. 27-39). New Jersey: Lawrence Erlbaum Associates.

Rose, N., & Miller, P. (1992). Political power beyond the state: Problematics of government. *British Journal of Sociology, 43*(2), 173-205.

Ross, W., & LaCroix, J. (1996). Multiple meanings of trust in negotiation theory and research: A literature review and integrative model. *International Journal of Conflict Management, 7*(4), 314-360.

Ross, W.H., & Wieland, C. (1996). Effects of interpersonal trust and time pressure on managerial mediation strategy in a simulated organizational dispute. *Journal of Applied Psychology, 81*(3), 228-248.

Rothberg, H.N., & Erickson, G.S. (2002). Competitive capital: A fourth pillar of intellectual capital? In N. Bontis (Ed.), *World congress on intellectual capital readings* (pp. 94-103). Woburn, MA: Butterworth-Heinemann.

Rothberg, H.N., & Erickson, G.S. (2005). *From knowledge to intelligence: Creating competitive advantage in the next economy*. Woburn, MA: Elsevier Butterworth-Heinemann.

Rotter, J. B. (1967). A new scale for the measurement of interpersonal trust. *Journal of Personality, 35*(4), 651-665.

Rotter, J. B. (1971). Generalized expectancies for interpersonal trust. *American Psychologist, 26*(5), 443-452.

Rotter, J. B. (1980). Interpersonal trust, trustworthiness, and gullibility. *American Psychologist, 35*, 1-7.

Rotter, J.B. (1967). A new scale for the measurement of interpersonal trust. *Journal of Personality, 35*, 651-665.

Rotter, J.B. (1971). Generalized expectancies for interpersonal trust. *American Psychologist, 35*, 1-7.

Rousseau, D. M., Sitkin, S. B., Burt, R. S., & Camerer, C. (1998). Not so different after all: A cross-discipline view of trust. *Academy of Management Review, 23*(3), 393-404.

Roy, M. H., & Dugal, S. S. (1998). Developing trust: The importance of cognitive flexibility and co-operative contexts. *Management Decision, 36*(9), 561-567.

Rule, C. (2002). *Online dispute resolution for business.* San Francisco: Jossey Bass.

Russ, G.S., Daft, R.L., & Lengel, R.H. (1990). Media selection and managerial characteristics in organizational communications. *Management Communication Quarterly: McQ (1986-1998), 4*(2), 151.

Rust, R.T., & Varki, S. (1996). Rising from the ashes of advertising. *Journal of Business Research, 37*(3), 173-181.

Sabel, C. (1993). Studied trust: Building new forms of cooperation in a volatile economy. *Human Relations, 46*(9), 1133-1170.

Sadeh, N. (2002). *M-Commerce: Technologies, services, and business models.* Boston: Wiley.

Sako, M. (1992). *Prices, quality and trust, inter-firm relations in Britain & Japan.* Cambridge: Cambridge University Press.

Sako, M. (1994). Price, quality and trust: Inter-firm relations in Britain and Japan. Cambridge, UK: Cambridge.

Sako, M. (1998). Does trust improve business performance? In C. Lane & R. Bachmann (Eds.), *Trust within and between organizations: Conceptual issues and empirical applications* (pp. 88-117). Oxford: Oxford University Press.

Sako, M., & Helper, S. (1998). Determinants of trust in supplier relations: Evidence from the automotive industry in Japan and the United States. *Journal of Economic Behavior & Organization, 34*(3), 387-417.

Salanova, M., & Schaufeli, W. B. (2000). Exposure to information technologies and its relation to burnout. *Behavior and Information Technology, 19*(5), 385-392.

Saparito, P. A., & Lippert, S. K. (2006). *A typology for building trust in interpersonal relationships within an organizational setting* (Working Paper Series). Drexel University.

Sarker, S., & Sahay, S. (2004). Implication of space and time for distributed work: An interpretive study of US-Norwegian systems development teams. *European Journal of Information Systems, 12*, 3-20.

Schelling, T.C. (1980). *The strategy of conflict* (2nd ed.). Cambridge, MA: Harvard University Press.

Schermerhorn, J. R. (1975). Determinants of interorganizational cooperation. *Academy of Management Journal, 18*(4), 846-856.

Schollmeier, R. (2002). A definition of peer-to-peer networking for the classification of peer-to-peer architecture and applications. In *Proceedings of the First International Conference on Peer-to-Peer Computing.* IEEE Press.

Schoorman, D.F., Mayer, R.C., & Davis, J.H. (1996). Including versus excluding ability from the definition of trust. *Academy of Management Review, 21*(2), 339-340.

Schulz, M., & Jobe, L.A. (2001). Codification and tacitness as knowledge management strategies: An empirical exploration. *Journal of High Technology Management Research, 12*, 139-165.

Seal, W. B. (1998). Relationship banking and the management of organizational trust. *International Journal of Bank Marketing, 16*(3), 102-108.

Seligman, A. (1997). *The problem of trust.* Princeton, USA: Princeton University Press.

Senge, P.M., (1990). *The fifth discipline: The art and practice of the learning organization.* New York: Doubleday Currancy.

Serva, M.A., Fuller, M.A., & Mayer, R.C. (2005). The reciprocal nature of trust: A longitudinal study of interacting teams. *Journal of Organizational Behavior, 26*(6), 625.

Shankar, V., Sultan, F., & Urban, G.L. (2002). Online trust: A stakeholder perspective, concepts, implications, and future directions. *Journal of Strategic Information Systems, 11*(4), 325-344.

Shapiro, C., & Varian, H. (1998). *Information rules: A strategic guide to the network economy.* Cambridge, MA: Harvard Business School Press.

Shapiro, D., Sheppard, B., & Cheraskin, L. (1992). Business on a handshake. *Negotiation Journal, 8*(4), 365-377.

Shapiro, R.M., & Jankowski, M.A. (1998). *The power of nice.* New York: John Wiley.

Shapiro, S. P. (1987). The social control of impersonal trust. *American Journal of Sociology, 93*(3), 623-658.

Sheppard, B. H., Hartwick, J., & Warshaw, P. R. (1988). The theory of reasoned action: A meta analysis of past research with recommendations for modifications in future research. *Journal of Consumer Research, 15*(3), 325-343.

Sheppard, B., & Sherman, D. (1998). The grammar of trust: A model and general implications. *The Academy of Management Review, 23*(3), 422-437.

Shin, Y. (2005). Conflict resolution in virtual teams. *Organizational Dynamics, 34*(4), 331-345.

Shmatikov, V., & Talcott, C. (2005). Reputation-based trust management. *Journal of Computer Security, 13*, 167-190.

Siau, K., Lim, E., & Shen, Z. (2001). Mobil commerce: Promises, challenges, and research agenda. *Journal of Database Management, 12*(3), 4-13.

Siau, K., Sheng, H., & Nah, F. (2003). *Development of a framework for trust in mobile commerce.* Paper presented at the Workshop on HCI Research in MIS.

Siau, S., & Shen, Z. (2003). Building customer trust in mobile commerce. *Communication of ACM, 46*(4), 91-94.

Siegel, J., Dubrovsky, V., Kiesler, S., & McGuire, T.W. (1986). Group processes in computer-mediated communication. *Organizational Behavior and Human Decision Processes, 37*(2), 157.

Silver, A. (1985). Trust in social and political theory. In G.D. Suttles & M.N. Zald (Eds.), *The challenge of social control* (pp. 52-70). Greenwich, CT, USA: Ablex.

Silver, A. (1989). Trust as a moral ideal: An historical approach. *Archives Europeenes de Sociologie, 30*(2), 69-87.

Silver, A. (1998). Two different sorts of commerce: Friendship and strangership in civil society. In J.Weintraub & K. Kumar (Eds.), *Private and public in thought and practice* (pp. 43-74). Chicago: University of Chicago Press.

Silverman, D. (2001). *Interpreting qualitative data.* Thousand Oaks, CA: Sage.

Simpson, G.R., & Bridis, T. (2000, June 19). Oracle hired firm to probe Microsoft allies. *The Wall Street Journal*, p. A48.

Sitkin, S. B., & George, E. (2005). Managerial trust-building through the use of legitimating formal and informal control mechanisms. *International Sociology, 20*(3), 307-338.

Sitkin, S.B., & Roth, N.L. (1993). Explaining the limited effectiveness of legalistic "remedies" for trust/distrust. *Organisation Science, 4*(3), 367-392.

Six, F. (2005). *The trouble with trust: The dynamics of interpersonal trust building.* Cheltenham: Edward Elgar.

Smith, M.A., & Kollock, P. (Ed.) (1999). *Communities in cyberspace.* London: Routledge.

Smith, M.A.. (1992), *Voices from the WELL: The logic of the virtual commons.* Unpublished master's thesis, University of California, Los Angeles.

294

Soh, C., Kien, S. S., & Tay-Yap, J. (2000). Cultural fits and misfits: Is ERP a universal solution? *Communication of ACM, 43*(4), 47-51.

Sohn, D., & Lee, B. (2005). Dimensions of interactivity: Differential effects of social and psychological factors. *Journal of Computer-Mediated Communication, 10*(3). Retrieved March 11, 2007, from http://jcmc.indiana.edu/vol10/issue3/sohn.html

Soliman, K. S., & Janz, B. D. (2004). An exploratory study to identify the critical factors affecting the decision to establish Internet-based interorganizational information systems. *Information & Management, 41*(6), 697-706.

Sproull, L., & Kiesler, S. (1986). Reducing social context cues: Electronic mail in organizational communication. *Management Science, 32,* 1492-1512.

Srite, M., & Karahanna, E. (2006). The role of espoused national cultural values in technology acceptance. *MIS Quarterly, 30*(3), 679.

Stack, L.C. (1988). Trust. In H. London & J.E. Exner, Jr. (Eds.), *Dimensionality of personality* (pp. 561-599). New York: Wiley.

Staniszkis, J. (2001). *Postkomunizm: Próba Opisu* [Postcommunism: Attempt of description]. Gdańsk: Słowo/Obraz Terytoria.

Star, S. L., & Griesmer, J. (1989). Institutional ecology, "translation" and boundary objects: Amateurs and professionals in Berkley's Museum of Vertebrate Zoology. *Social Studies of Science, 19,* 387-420.

Stiffler, D. (2006). Outsourcing transition case study: Customized communication speeds savings.

Strauss, S.G. (1997). Technology, group processes, and group outcomes: Testing the connections in computer-mediated and face-to-face groups. *Human-Computer Interaction, 12,* 227-266.

Strong, K., & Weber, J. (1998). The myth of the trusting culture. *Business & Society, 37*(2), 157-183.

Stumpf, S., & Alexander, T. (1999) Management von Heterogenität und Homogenität in Gruppen. *Personalführung, 5*(99), 36-44.

Sultan, F., Urban, G. L., Shankar, V., & Bart, I. Y. (2002). *Determinants and role of trust in e-business: A large scale empirical study.* MIT Sloan Working Paper, 4282.

Swan, J. E., Bowers, M. R., & Richardson, L. D. (1999). Customer trust in the salesperson: An integrative review and meta-analysis of the empirical literature. *Journal of Business Research, 44*(2), 93-107.

Sydow, J. (1998). Understanding the constitution of interorganizational trust. In C. Lane & R. Bachman (Eds.), *Trust within and between organizations.* (pp. 31-63). New York: Oxford University Press.

Sztompka, P. (1996). Trust and emerging democracy: Lessons from Poland. *International Sociology, 11*(1), 37-62.

Sztompka, P. (1999). *Trust: A sociological theory.* Cambridge, UK: Cambridge University Press.

Sztompka, P. (2005). *Socjologia: analiza społeczeństwa* [Sociology: Analysis of society]. Kraków: Znak.

Tadelis, S. (1999). What's in a name? Reputation as a tradeable asset. *American Economic Review, 89*(3), 548-563.

Tam, P-W. (2002, September 25). High-technology giant duels with nimble knock-off artists. *The Wall Street Journal,* pp. A1, A10.

Tan, B. C. Y., Wei, K.-K., Watson, R. T., & Walczuch, R. M. (1998). Reducing status effects with computer-mediated communication: Evidence from two distinct national cultures. *Journal of Management Information Systems, 15*(1), 119-141.

Tan, B. C. Y., Wei, K.-K., Watson, R. T., Clapper, D. L., & McLean, E. R. (1998). Computer-mediated communication and majority influence: Assessing the impact in an individualistic and a collectivistic culture. *Management Science, 44*(9), 1263-1278.

Tan, Y.H., & Thoen, W. (2000). A generic model of trust in electronic commerce. *International Journal of Electronic Commerce, 5*(2), 61-74.

Tan, Y.H., & Thoen, W. (2002). Formal aspects of a generic model of trust for electronic commerce. *Decision Support Systems, 33*(3), 233-246.

Tarasewich, P., Nickerson, R. C., & Warkentin, M. (2002). Issues in mobile e-commerce. *Communications of the Association for Information Systems, 8*, 41-64.

Team building. Theoretical overview of trust building and teamwork in online environments, pp. 1-8. Handout.

Thomas, A. (1999). Cultural diversity and work group effectiveness. *Journal of Cross-Cultural Psychology, 30*(2), 242-263.

Thomas, A., & Ravlin, E.C. (1996). Effect of cultural diversity in work groups. In P.A. Bamberger, M. Erez, & S.B. Bacharach (Eds.), *Research in the sociology of organizations* (pp. 1-33). Greenwich: JAI Press.

Thompson, L. (2001). *The mind and heart of the negotiator* (2nd ed.). Upper Saddle River, NJ: Prentice Hall.

Thompson, L., & Nadler, J. (2002). Negotiating via information technology: Theory and application. *Journal of Social Issues, 58*(1), 109-124.

Tomer, A., & Ken, F. (2005). *Improvement of mapping by odometry.* Retrieved February 21, 2007, from Computer Science Department, Intelligent Systems Laboratory, Israel Institute of Technology: http://www.cs.technion.ac.il/Labs/Isl/MARS/FProjects/previous/ Mapping/rep/odo_report.pdf

Tompkins, P. (2003). *Truth and trust in cyberspace.* Paper presented at the Conference on Communication Ethics and Virtual Reality. Co-sponsored by Brigham Young University, University of Illinois, and WACC.

Townsend, A.M., DeMarie, S.M., & Hendrickson, A.R. (1998). Virtual teams: Technology and the workplace of the future. *Academy of Management Executive, 12*(3), 17-29.

Travis, L., & Durocher, C. (2003, November). Strategic use of offshore resources can double labor savings (AMR Research Report).

Treaty of Maastricht (1992). Retrieved June 29, 2007, from http://www.eurotreaties/com/masstrichtext.html

Tremayne, M. (2005). Lessons learned from experiments with interactivity on the web. *Journal of Interactive Advertising, 5*(2). Retrieved November 5, 2005 from http://www.jiad.org/vol5/no2/tremayne/

Trevino, L.K., Lengel, R.H., & Daft, R.L. (1987). Media symbolism, media richness, and media choice in organizations. *Communication Research*, 553-574.

Trevino, L.K., Lengel, R.H., Bodensteiner, W., Gerloff, E.A., & Muir, N.K. (1990). The richness imperative and cognitives style: The role of individual differences in media choice behavior. *Management Communication Quarterly: McQ (1986-1998), 4*(2), 176.

Trompenaars, F. (1993) *Riding the waves of culture.* London: N. Brealey.

Turner, J.C. (1987). *Rediscovering the social group: A self categorization theory.* New York: Basil Blackwell Inc.

Turner, J.C., Oakes, P.J., Haslam, S.A., & McGarty, C. (1994). Self and collective: Cognition and social context. *Personality and Social Psychology Bulletin, 20*(5), 454.

Tyler, T.R., & Kramer, R.M. (1996). Whither trust? In R.M. Kramer & T.R. Tyler (Eds.), *Trust in organizations: Frontiers of theory and research* (pp. 1-15). Thousand Oaks, CA: Sage Publications.

Tyran, C.K., Dennis, A.R., Vogel, D.R., & Nunamaker, J.F. (1992). The application of electronic meeting technology to support strategic management. *MIS Quarterly, 16*(3), 313-334.

Ullman-Margalit, E. (2004). Trust, distrust, and in between. In R. Hardin (Ed.), *Distrust* (pp. 60-83). New York: Russell Sage Foundation.

Urban, G. L., Sultan, F., & Qualls, W. J. (2000). Placing trust at the center of your Internet strategy. *Sloan Management Review, 42*(1), 39-48.

Urban, G.L. (2003). *The trust imperative* (MIT CeB Working Paper, 175).

Urban, G.L., Sultan, F., & Qualls, W.J. (2000). Placing trust at the center of your Internet strategy. *Sloan Management Review, 42*(1), 38-48.

296

Urban, G.L., Sultan, F., & Qualls, W.J. (2000). Placing trust at the center of your Internet strategy. *Sloan Management Review, 42*(1), 39-48.

Urry, J. (2000). *Sociology beyond societies.* London: Routledge.

Ury, W. (1991). *Getting past no.* New York: Bantam Books.

Valacich, J.S., Dennis, A.R., & Nunamaker, J.F. (1992). Group size and anonymity effects on computer-mediated idea generation. *Small Group Research, 23*(1), 49-73.

Valley, K. L., Moag, J., & Bazerman, M. H. (1998). A matter of trust: Effects of communication on the efficiency and distribution of outcomes. *Journal of Economic Behavior in Organizations, 34*(2), 211-238.

Venkatesh, V., & Davis, F. D. (2000). A theoretical extension of the technology acceptance model: Four longitudinal field studies. *Management Science, 46*(2), 186-204.

Violino, B. (2002, June 17). Building B2B trust. *Computerworld.* Retrieved March 12, 2007, from http://www.computerworld.com/managementtopics/ebusiness/story/0,10801,71988,00.html

Violino, B. (2002, June 17). Gated communities yield B2B trust. *Computerworld.* Retrieved March 12, 2007, from http://www.computerworld.com/managementtopics/ebusiness/story/0,10801,71929,00.html

Walczuch, R., & Lundgren, H. (2004). Psychological antecedents of institution-based consumer trust in e-retailing. *Information & Management, 42,* 159-177.

Walczuch, R., Seelen, J., & Lundgren, H. (2001). Psychological determinants for consumer trust in e-retailing. *Eighth Research Symposium on Emerging Electronic Markets (RSEEM 01).*

Wall, J. A., Stark, J. B., & Standifer, R. L. (2001). A current review and theory development. *Journal of Conflict Resolution, 45*(3), 370-391.

Wallace, P. (1999). *The psychology of the Internet.* New York: Cambridge University Press.

Walther, J. B., & Bunz, U. (2005). The rules of virtual groups: Trust, liking, and performance in computer-mediated communication. *Journal of Communication, 55*(4), 828-846.

Walther, J.B. (1992). Interpersonal effects in computer-mediated interaction: A relational perpective. *Communication Research, 19*(1), 52.

Walther, J.B. (1993). Impression development in computer-mediated interaction. *Western Journal of Communication, 57,* 381-398.

Walther, J.B. (1995). Relational aspects of computer-mediated communication: Experimental observations over time. *Organization Science, 6,* 186-203.

Walther, J.B., & Bunz, U. (2005). The rules of virtual groups: Trust, liking and performance in computer-mediated communication. *Journal of Communication, 55*(4), 828-846. Retrieved July 3, 2007, from http://bunz.comm.fsu.edu/JoC2005_55_4_virtual.pdf

Walther, J.B., & Burgoon, J.K. (1992). Relational communication in computer-mediated interaction. *Human Communication Research, 19,* 50-88.

Wang, H., Lee, M.K.O., & Wang, C. (1998). Consumer privacy concerns about Internet marketing. *Communications of the ACM, 41*(3), 63-70.

Wang, Y.D., & Emurian, H.H. (2005). An overview of online trust: Concepts, elements, and implications [Electronic version]. *Computers in Human Behavior, 21*(1), 105-125. Retrieved July 3, 2007, from http://nasa1.ifsm.umbc.edu/cv/TrustCHB.pdf

Wasko, M., & Faraj, S. (2000). It is what one does; Why people participate and help others in electronic communities of practices. *Journal of Strategic Information Systems, 9*(2/3), 155-173.

Wasserman, S., & Faust, K. (1994). *Social network analysis. Methods and applications.* New York: Cambridge University Press.

Watson, W.E., Johnson, L., & Zgourides, G.D. (2002). The influence of ethnic diversity on leadership. Group process, and performance: An examination of learning teams. *International Journal of Intercultural Relations, 26*, 1-16.

Watson, W.E., Kumar, K., & Michaelsen, L.K. (1993). Cultural diversity's impact on interaction process and performance: Comparing homogeneous and diverse task groups. *Academy of Management Journal, 36*(3), 590-602.

Weber, M. (1978). *Economy and society*. Berkeley: University of California Press.

Weick, K.E. (1990). P. Goodman, L. Sproull and Associates (Eds.) Technology as equivoque: Sense making in new technologies. In *Technology and organizations* (pp. 1-44). San Francisco: Jossey-Bass.

Welch, M. R., Rivera, R. E. N., Conway, B. P., Yonkoski, J., Lupton, P. M., & Giancola, R. (2005). Determinants and consequences of social trust. *Sociological Inquiry, 75*(4), 453-473.

Wellman, B. (2002). Little boxes, globalization, and networked individualism? In M. Tanabe, P. v. d. Besselaar, & T. Ishida (Eds.), *Digital cities II: Computational and sociological approaches* (pp. 10-25). Berlin: Springer.

Wellman, B., & Gulia, M. (1999). Net surfers don't ride alone: Virtual communities as communities. In P. Kollock & M. Smith (Eds.), *Communities in cyberspace*. Berkeley: University of California Press.

Wellman, B., Quan-Haase, A., Boase, J., Chen, W., Hampton, K., Diaz, I., et al. (2003) The social affordances of the internet for networked individualism. *Journal of Computer Mediated Communication, 8*(3). Retrieved April 2007 from http://www.ascusc.org/jcmc/vol8/issue3/wellman.html

Wellman, B., Salaff, J., Dimitrova, D., Garton, L., Gulia, M., & Haythornthwaite, C. (1996). Computer networks as social networks: Collaborative work, telework, and virtual community. *Annual Review of Sociology, 22*(1), 213-238.

Wenger, E. (1998). *Communities of praxis. Learning, meaning and identity*. Cambridge: Cambridge University Press.

Whittaker, S., Terveen L., Hill, W., & Cherny, L. (1998). *The dynamics of mass interaction*. In Proceedings of CSCW, Seattle, Washington. Retrieved March 11, 2007, from http://citeseer.nj.nec.com/cache/papers/cs/14002/http: zSzzSzwww.research.att.comzSz~stevewzSzcscw98-published.pdf/the-dynamics-of-mass.pdf

Wicks, A. C., Berman, S. L., & Jones, T. M. (1999). The structure of optimal trust: Moral and strategic implications. *Academy of Management Review, 24*(1), 99-116.

Williams, R. (1974). Television, technology and cultural form. London, UK: Fontana.

Williamson, O. (1993). Calculativeness, trust and economic organisation. *Journal of Law and Economics, 30*.

Williamson, O. E. (1985). *The economic institution of capitalism*. New York: Free Press.

Wilson, J. M., Straus, S. G., et al. (2006). All in due time: The development of trust in computer-mediated and face-to-face teams. *Organizational Behavior and Human Decision Processes, 99*, 16-33.

Windley, P.J., Tew, K., & Daley, D (2006). *A framework for building reputation systems*. Retrieved March 9, 2007, from http://www.windley.com/essays/2006/dim2006/framework_for_building_reputation_systems

Witt, L.A., Andrews, M.C., & Kacmar, K.M. (2000). The role of participation in decision-making in the organizational politics-job satisfaction relationship. *Human Relations, 53*, 341-358.

Wolff, K. H. (Ed.). (1950). *The sociology of Georg Simmel*. New York: The Free Press.

Wong, S.-S., & Burton, R.M. (2000). Virtual teams: What are their characteristics and impact on team performance? *Computational & Mathematical Organization Theory, 6*, 339-260.

Workman, M., Kahnweiler, W., & Bommer, W. (2003). The effects of cognitive style and media richness in commitment to telework and virtual teams. *Journal of Vocational Behaviour, 63*, 199-219.

298

Yamagishi, T., & Yamagishi, M. (1994). Trust and commitment in the United States and Japan. *Motivation and Emotion, 18*(2), 129-166.

Yeo, J., & Huang, W. (2003). Mobile E-commerce outlook. *International Journal of Information Technology & Decision Making, 2*(2), 313-332.

Yoo, Y., & Alavi, M. (2004). Emergent leadership in virtual teams: What do emergent leaders do? *Information and Organization, 14*, 27-58.

Young, M. (1988). *The metronomic society: Natural rhythms and human time keeping.* London: Thames and Hudson.

Yu, B., & Singh, M.P. (2000). A social mechanism of reputation management in electronic communities. In M. Klusch & L. Kerschberg (Eds.), *Proceedings of the 4th International Workshop on Cooperative Information Agents.*

Yuan, S.T., & Sung, H. (2004). A learning-enabled integrative trust model for e-markets. *Applied Artificial Intelligence, 18*, 69-95.

Zaccaro, S.J. & Brader, P. (2003). E-leadership and the challenges of leading e-teams: Minimizing the bad and maximizing the good. *Organizational Dynamics, 31*(4), 377-387.

Zacharia, G., & Maes, P. (2000). Trust management through reputation mechanisms. *Applied Artificial Intelligence, 14*(9), 881-907.

Zack, M.A. (1999, Spring). Developing a knowledge strategy. *California Management Review, 41*(3), 125-145.

Zack, M.A. (1999, Summer). Managing codified knowledge. *Sloan Management Review, 40*(4), 45-58.

Zaheer, A., & Venkatraman, N. (1995). Relational governance as an inter-organizational strategy: An empirical test of the role of trust in economic exchange. *Strategic Management Journal, 16*(5), 373-392.

Zaheer, A., McEvily, B., & Perrone, V. (1998). Does trust matter? Exploring the effects of interorganizational and interpersonal trust on performance. *Organization Science, 9*(2), 141-159.

Zahra, S. (2003). International expansion of US manufacturing family business: The effect of ownership and involvement. *Journal of Business Venturing, 18*(4), 495-511.

Zahra, S. (2005). Entrepreneurial risk taking in family firms. *Family Business Review, 18*(1), 23-40.

Zahra, S. A., Yavuz, R. I., & Ucbasaran, D. (2006). How much do you trust me? The dark side of relational trust in new business creation in established companies. *Entrepreneurship Theory and Practice, 30*(4), 541-555.

Zand, D.E. (1972). Trust and managerial problem solving. *Administrative Science Quarterly, 17*(2), 229-239.

Zander, U., & Kogut, B. (1995, January-February). Knowledge and the speed of transfer and imitation of organizational capabilities: An empirical test. *Organization Science, 6*(1), 76-92.

Zheng, J., Veinott, E., Bos, N., Olson, J.S., & Olson, G.M. (2002). Trust without touch: Jumpstarting long-distance trust with initial social activities [Electronic version]. In *Proceedings of the SIGCHI 2002 Conference on Human Factors in Computing Systems* (pp. 141-146).

Zigurs, I. (2003). Leadership in virtual teams: Oxymoron or opportunity. *Organizational Dynamics, 31*(4), 339-351.

Zolin, R., & Hinds, P. (2002). Trust in context: The development of interpesonal trust in geographically distributed work teams (CIFE Working Paper). S. University. Stanford, Center for Integrated Facility Engineering.

Zolin, R., Hinds, P.J., Fruchter, R., & Levit, R.E. (2004). Interpersonal trust in cross-functional geographically distributed work: A longitudinal study. *Information & Organization, 14*(1), 1-26.

Zuboff, S. (1988). *In the age of the smart machine. The future of work and power.* New York: Basic Books.

Zucker, L. (1986). Production of trust: Institutional sources of economic structure (1840-1920). *Research in Organizational Behavior, 8*, 53-111.

About the Contributors

Linda L. Brennan is an associate professor of management in the School of Business at Mercer University in Macon. She conducts research and consults in the areas of project management and control, with a particular emphasis on project definition, team processes, and change management. Dr. Brennan's prior work experience includes management positions at The Quaker Oats Company and marketing and systems engineering experience with the IBM Corporation. She received her PhD from Northwestern University, her MBA from the University of Chicago, and her BIE from the Georgia Institute of Technology. She is a licensed professional engineer and a certified project management professional.

Victoria Johnson is dean of the School of Business and Economics at Georgia Gwinnett College. Dr. Johnson is dean of the School of Business and professor of management at Georgia Gwinnett College. She received her PhD in public administration from the University of Georgia and holds a Master of Public Administration and a BA in English from Georgia State University. Her research interests and publications are in the areas of corporate strategy and policy analysis, international business ethics, and corporate social responsibility. She is co-editor of the book, *Social, Ethical, and Political Implications of Information Technology* and serves on the editorial board of *Global Business and Economic Review*.

* * *

Terry R. Adler earned a PhD from the University of Cincinnati. Dr. Adler is a well-known researcher in long-term contractual agreements. His interests have led to many publications in trust, distrust, and transaction cost economics issues. He currently is an associate professor at New Mexico State University and a consultant for many S&P 500 companies.

Werner Auer-Rizzi is an associate professor of business administration at the Department of International Management at the Johannes Kepler University in Linz. After his studies in business administration and informatics, he worked as a software developer and as a system analyst at IBM. He spent one year as a visiting scholar at the Rotman School of Management of the University of Toronto and teaches regularly at the Turku School of Economics. His research and teaching are grounded in all areas of the organizational behavior field. His current research interests include decision making, trust in organizations, and cross cultural aspects of management and leadership.

John Willy Bakke is a research scientist at Telenor Research and Innovation, where his primary areas of research are flexible work arrangements and user interpretations and user acceptance of technologies. Recent projects include studies of teleworking, the role of ICTs and workplace design for work task execution and collaboration, and identity and belonging in distributed work groups. He is co-author and co-editor of the books *Handbook on Teleworking, New Ways of Working*; and *The Electronic Nomad* (all in Norwegian).

Sebastián Bruque-Cámara is a professor of management in the Department of Business Administration, Accounting, and Sociology at the University of Jaén (Spain). He currently conducts research on organizational factors affecting IT adoption and social networks effects on workers' acceptance of technological changes. His research has appeared in the *European Journal of Information Systems, Internet Research, Journal of High Technology Management Research,* and *Strategic Management and Technology Analysis Journal.*

Rosa D'Ettorre has been assistant researcher in marketing at the University of Lecce, Italy, where she graduated in economics and business. She has also been manager in customer relationship management at the Polytechnic of Bari, and in Web marketing at Webberia Srl, a private company. She has been consultant for several national firms and she is now financial advisor and accountant in Lecce, where she is attending the Superior School for Secondary Learning.

Mohamed Daassi holds a PhD in management information systems and is a temporary teaching and research assistant at the University of Grenoble (France). He received his MS in MIS from the University of Savoie, France (2002), and his BS in operation research from El Manar University, Tunisia (2001). Dr. Daassi has published his research work in *International Journal of Business Information Systems, International Journal of Networking and Virtual Organizations,* and some book chapters. His current research and teaching interests include information management, collaborative technologies, virtual teams, and social aspects of human-computer interaction.

Robert Easter is the founder of the Qwest Group consultancy. He has been identifying and developing creative technology applications to help organizations improve their operating performance for over 25 years. His focus is on wireless applications for untethered communications as well as on data mining and modeling techniques to help his customers make critical business decisions. Mr. Easter has a BS in finance from the University of Illinois and an MBA in marketing and finance from Northwestern University.

Noam Ebner, an attorney-mediator and negotiation consultant, is co-director of Tachlit Mediation and Negotiation Training in Jerusalem, Israel. He is a visiting professor at Sabanci University in Istanbul, Turkey, teaching in the graduate program on conflict analysis and resolution as well as in the MBA and EMBA programs. He is also a senior fellow at the UN's University for Peace in Costa Rica. In addition, he conducts training in face-to-face and e-negotiation in many corporate, academic, and organizational settings.

G. Scott Erickson is associate professor and chair of marketing/law in the School of Business, Ithaca College. He holds a PhD from Lehigh University. Research interests include knowledge management, intellectual property, and trade secrets. His book with Helen Rothberg, *From Knowledge to Intelligence: Creating Competitive Advantage in the Next Economy,* was published by Butterworth-Heinemann/Elsevier in 2005.

Marc Favier is professor of information systems at the University of Grenoble (France). Dr. Favier is a graduate in computer science studies. He has published 28 articles in scientific journals and eight books or chapters of books. He has supervised IS PhD students and directed various graduate MBA or PhD programs. His current research and teaching interests include information management, strategy and IS, e-business, DSS, collaborative technologies, and virtual teams.

Iris Fischlmayr currently holds the position of an assistant professor of business administration at the Department of International Management at the Johannes Kepler University in Linz. She is engaged in teaching and research in the field of cross-cultural management and organizational behaviour. Her focus is in expatriation, women in international management, virtual multicultural teams, and cultural sensitivity. She is involved in various international projects with partner universities (e.g., Turku School of Business and Economics, Finland; Richard Ivey School of Business in London, Ontario or ESADE in Barcelona, Spain) and multinational companies. Currently, she is writing her habilitation (postdoctoral lecture qualification) on different behavioural influences on virtual multicultural team work.

Mark A. Fuller is the chair of the Department of Information Systems at Washington State University, and holds the Philip L. Kays distinguished professorship in information systems. His research focuses on virtual teamwork, technology supported learning, and trust in technology-mediated environments, and has appeared in outlets such as *Information Systems Research, Management Information Systems Quarterly, The Journal of Management Information Systems,* and *Decision Support Systems.*

Michael D. Glissmeyer is a doctoral student at New Mexico State University. His current research interests include organizational trust, simultaneous trust and distrust, and transaction cost economics.

Gianluigi Guido (PhD, management sciences, University of Cambridge, UK) is a professor of marketing at the Faculty of Economics of the University of Lecce, and at the LUISS University of Rome, Italy. He has been professor at the University of Padua and at the University of Rome "La Sapienza," Italy, visiting researcher at the University of Florida at Gainesville, and Fulbright Scholar at the University of Stanford. He has published more than 80 articles on national and international journals and 10 books and as many chapters in edited books. He has been listed in the last ten editions of *Marquis Who's Who in the World* and in the last six editions of *Marquis Who's Who in Finance & Industry.*

Nabila Jawadi is a PhD candidate in information systems at the University of Paris Dauphine. Her current research interests include information management, virtual teams management, trust, and leadership exercise in virtual teams. She also has past researches in e-learning adoption among workers.

Tom Erik Julsrud works as a research scientist at Telenor Research and Innovation and at Studio Apertura at the Norwegian University of Science and Technology (NTNU). He has graduated in psychology and sociology and holds a master's degree in media and communication studies from the University of Oslo. He has bee co-authoring books on telework and distributed work and has published articles and papers on collaboration in distributed and virtual teams.

Michel Kalika is a full professor at Paris Dauphine University (France). He is the director of the Center for Research in Management & Organization and the PhD program in information systems (e-management). Dr. Kalika's research focus is the use of e-mail by managers and the impact on meetings, IT strategic alignment, and the link between IT and performance. He is the author or co-author of books on organizational structure, management, e-management, and numerous articles in academic journals.

Dan J. Kim is an associate professor of computer information systems at University of Houston Clear Lake. He earned his PhD in MIS from SUNY at Buffalo. Recently he has focused on trust in electronic commerce, wireless and mobile commerce, and information security and assurance. His research work has been published or is forthcoming and includes more than 50 papers in refereed journals and conferences including *Communications of ACM, Decision Support Systems, Electronic Market, Journal of Global Information Management, International Journal of Mobile Communications, IEEE IT Professional, ICIS, and HICSS.* He received the best-paper runner-up award at the ICIS 2003 and the best research paper award at AMCIS 2005.

Rachna Kumar is a professor of information systems and technology at Alliant International University in San Diego, California. Her research, publishing, and consulting interests include software offshoring processes, transfer of information systems skills in binational contexts, computer-mediated learning environments, software productivity measures, and online education. She is the co-founder of GlobalMind: Center for Strategic Consulting at the Marshall Goldsmith School of Management, providing training and consultation on topics related to technology and strategy in global contexts. Dr. Kumar has consulted for several high tech firms in the U.S., Singapore, Mexico, and India. She has a PhD in information systems from Stern School of Business at New York University, New York.

Dominika Latusek is a faculty member at the Leon Kozminski Academy of Entrepreneurship and Management in Warsaw, Poland. Before joining LKAEM, she was studying and conducting research in Poland, Germany, and the U.S. Her current research interests focuses on the dynamics of trust and distrust, and practices of cooperation in high-tech business. Her previous research includes field studies of culture and management practices in international corporations on the emerging markets.

Beverly Leeds, BA (Hons.), MBA MA, MCIM, MILT, is a principal lecturer at Lancashire Business School, University of Central Lancashire, Preston, UK. Her teaching responsibilities include employability, professional development, and marketing, as well as continuing professional development in e-marketing and contact center management. Previously she was employed by ICL Fujitsu as a design consultant and project manager where she worked with a large number of both private and public sector organizations. She runs a number of projects that partner students with local small businesses. Her research interests are concerned with the management of teleworkers and particularly the issues of trust and time. She has acted as a remote working consultant to both public and private organizations and has produced a number of handbooks on telework as well as conference papers and journal articles. She is a chartered marketer, a member of the Institute for Learning and Teaching and a board member of the International Telework Academy (ITA).

Susan K. Lippert was educated at The George Washington University and received a PhD and MBA. Her work experience includes 10 years of IT consulting with public, private, and nonprofit organizations within the Washington, DC area. She is currently serving as vice chair for membership for the Special Interest Group on Human Computer Interaction (SIGHCI) and secretary/treasurer for the Special Interest Group on Adoption of Information Technology (SIGADIT). She is engaged in research and teaching at Drexel University in Philadelphia. Her current research interests include understanding post-adoption behavior of supply chain affiliates and the role of technology trust on the use and management of information technology.

Alfonso Miguel Márquez-García is associate professor of management in the Department of Business Administration, Accounting, and Sociology at the University of Jaén (Spain). He also serves as virtual tutor at the Catalonia Open University (UOC). He has participated in international research projects granted by the EU. His research interests are about cooperation, strategic alliances, and trust. Some results have been published in *Revesco, ICADE*, book chapters at IGI Global and Kluwer Academic Publishers, the Workshop on Trust Within and Between Organisations (EIASM), and Conference Proceedings of the European Academy of Management (EURAM).

Roger C. Mayer is a professor of management at the University of Akron. He previously served on the faculties of Notre Dame, Purdue, and Baylor. Dr. Mayer received a PhD in organizational behavior and human resource management from the Krannert Graduate School of Management at Purdue University. Mayer's research is focused on trust, employee decision making, attitudes, and productivity, and has been published in such leading journals as *The Academy of Management Journal, The Academy of Management Review, The Journal of Applied Psychology, The Journal of Organizational Behavior, and The Strategic Management Journal*. He is an editorial review board member for *The Academy of Management Journal*.

Ronald E. Pike is a doctoral student in the Department of Information Systems at Washington State University. He holds an MBA from California State University, Chico, and has 17 years experience in the design and implementation of telecommunications and data communications systems. Mr. Pike's research is focused on trust and information privacy. His background in telecommunications systems has resulted in a particular interest in the development of trust and privacy within a computer mediated environment.

M. Irene Prete is a doctoral candidate in economic and quantitative methods for market analysis at the University of Lecce, Italy, and assistant researcher in marketing and territorial marketing at the same university. She holds a research MS de Gestion, at the University of Paris XII (France), and a Master in Banking and Finance, at the University of Tor Vergata, Rome (Italy). She has been lecturer in several national and international conferences on consumer behavior, political marketing, customer satisfaction, and human resources. She is also a marketing and management consultant for primary companies.

Helen N. Rothberg is associate professor of strategy in the School of Management, Marist College. She holds a PhD from City University Graduate Center. She is also the principal consultant for HNR Associates, a network of knowledge focusing on strategic change, competitive intelligence, and knowledge management challenges. Her research interests include competitive intelligence and knowledge management.

Loong Wong has worked at senior levels within industry and has also lectured at various universities in the Asia-Pacific region. He is currently at the University of Canberra and has taught e-commerce, international business, and strategic management. He has published widely and some of his research can be found in *Prometheus, Information Society, Journal of Contemporary Asia,* and *Asian Business and Management.* He maintains an active interest in the political economy of the Asian-Pacific region.

Marty Yopp is a professor in adult and organizational learning at the University of Idaho Boise Center. She has been teaching and advising graduate students in adult education and human resource development for more than a decade. She also offers online classes to graduate students using a virtual format. Dr. Yopp earned her EdD from The George Washington University in Washington, DC, in higher education with an emphasis in human resource development and adult education. She is the past-president of the Idaho Lifelong Learning Association and a member of the Treasure Valley Chapter of ASTD.

Index